WITHDRAWN

PAINTS & COATINGS HANDBOOK
Second Edition

PAINTS & COATINGS Handbook

Second Edition

For Contractors, Architects Builders and Engineers

by Abel Banov

Structures Publishing Company, Farmington, Michigan 1978

Copyright © 1978   Structures Publishing Co.
Farmington, Mich.

All rights reserved, including those of translation. This book, or parts thereof, may not be reproduced in any form without permission of the copyright owner. Neither the author, nor the publisher, by publication of data in this book, ensure to anyone the use of such data against liability of any kind, including infringement of any patent. Publication of data in this book does not constitute a recommendation of any patent or proprietary right that may be involved.

Manufactured in the United States of America.

Book designed by Richard Kinney.

Current Printing (last digit)
10 9 8 7 6 5 4 3 2 1

International Standard Book Number: 0-912336-66-8
Library of Congress Number: 78-54102

Structures Publishing Co.
Box 1002, Farmington, Mich. 48024

To Joan H. Banov
Wife, and unbelievably patient helpmate

Contents

Foreword ix

Acknowledgements xi

Introduction—What Makes a Coating Work 1

1. **What Goes in a Coating** 15
   Kinds of coatings; resin types: solvent and water-thinned; pigments; solvents; extenders; and additives.

2. **Testing** 83
   Drying time; gloss; flexibility; hardness; light fastness; adhesion; water and chemical resistance; abrasion resistance; hide; scrubs; washability; stain removal; weathering; and film thickness.

3. **Surface Preparation** 108
   General guidelines; hand and power cleaning; blasting; chemical treatment; preparing cementitious, wood, and metal surfaces.

4. **Calks and Sealants** 143
   Low-priced and high grade calks: butyl and acrylic; elastomeric sealants: neoprene, acrylic, Hypalon; chemical curing: polysulfides, polyurethanes, silicones.

5. **Application of Coatings** 160
   Tools: brushes, rollers, flat pads; spray guns; rules for application; materials; notes on selection, use, and care of applicators; types of spray: air, airless, electrostatic; comparisons.

6. **Deterioration of Coatings** 185
   General; water sorption and release; permeability; absorption; defects due to water: on wood and plywood, cementitious materials, metal (emphasis on corrosion); defects due to faulty structural design.

7. **Selection and Specification—for Exterior Surfaces (Sections 700-708.8)** 212

Introduction—701; *cementitious* surfaces: general and problem conditions—702; *metal* surfaces: general and problem conditions—703; *wood* surfaces—704.
Performance Comparative Charts, Exterior 258-273

8. **Selection and Specification—for Interior Surfaces (Sections 800-808.3)** 274

Introduction—800-801; *cementitious, metal, wood* surfaces: general purpose—803; high-performance—804; *cementitious* only—805; *metal* only—806; *wood* only—807; *floor* finishes—808.
Performance Comparative Charts, Interior 312-321

9. **Fire Retardant Coatings** 322

General; fire resistant coatings; intumescence; temporary benefit; examples of benefits; rating methods; types; performance as paints; specifying; intumescent mastics.

10. **Economics of Structural Coatings** 330

General; high-quality coatings; systems against corrosion: material costs, application, surface preparation, shop work vs. field work; material cost vs. preparation and application; high cost of low-cost systems; return on investment of high-quality paint; non-corrosive atmospheres; wood flooring.

Appendix "A" 345

Guide to GPC Specifications. GPC numbers are conveniently grouped by type of surface.

Appendix "B" 352

Characteristics for each of 77 Guides to Paints and Coatings. (GPC's).

Appendix "C" 397

Ultra High Performance Coatings.

Appendix "D" 405

Coatings particularly resistant to chemical corrosion or immersion.

Appendix "E" 409

Urethane Foam Insulation Elastomer Roof Coating.

Appendix "F" 414

Comparison Corrosion Resistance Chart.

Index 431

Foreword

"Frequently used, but seldom read" is the rule that Structures has formulated for handbooks in the construction field. Abel Banov, in this excellent work on Paints and Coatings, has done an outstanding job of making your answers to questions in this field readily available, without reading the whole book.

Economics—As a builder and a developer, I found myself fascinated by the Chapter on Economics. The "after tax" effect of the selection of a poor coating as opposed to a good coating proved to be an eye-opener. Above all, this is the first book I have ever seen that made the decision of what paint to specify simple and direct for builders, architects, contractors, plant maintenance engineers and others.

Now, by simply stating the surface to be coated and its exposure, we can state for our painting contractor the 'GPC' numbers of several coatings which will do the job. No longer must we be dependent on claims of paint manufacturers or subcontractors. Mr. Banov aptly subtitles his second chapter "The art of making sure."

At the same time, those of us who really want to know what makes a coating work, chemical formulations, testing, etc., will find this information presented concisely in a readily readable manner.

I would suggest that the 'GPC' numbers which Mr. Banov innovates here will become standard designations in the trade. They are simple and a direct way to specify the performance characteristics.

R. J. Lytle
Publisher

December, 1972

Acknowledgements

The mountain of technical material on which this book is based was furnished by knowledgeable persons from all levels of the paint and coatings industry. To thank them all here would be impossible.

Special gratitude is due Sidney B. Levinson, president of David Litter Laboratories, whose pioneer work in graphic presentations of coatings performance characteristics suggested many of the approaches used here. He and the technical director of his laboratory, Saul Spindel, made their extensive library of specifications available and were always able to find time for counsel.

Grateful acknowledgement is also extended to Joseph Lattimer, architectural consultant, Glidden-Durkee Div., SCM Corp., and Edward E. Gleason, vice president, MCP Facilities, for their assistance in providing facts and figures for the section on economics of coatings; and to Don E. Neese, Chemical Div., Goodyear Tire and Rubber Co., for his enthusiastic help at a crucial period.

Thanks are in order to Jack Connors, Lehman Bros. Corp.; Alan Gemberling, The Sherwin-Williams Co.; Steven Murphy, Pratt & Lambert; Burton L. Savitt, Savitt Bros., Inc., Chicago, for Elliot Paint Co.; Herbert R. Hillman, Eaglo Paints, Inc.; John Ballard, Kurfee's Coatings, Inc.; Abel Schwartz, Carboline Co.; Mel Lheven, Master Mechanics; William A. Wright, Gibson-Homans Co.; Dr. Leroy Shuger, Baltimore Paint Corp.; Mervin Reed, Toch Bros.; Nymer A. Christiansen, Hughson Chemical Co.

Norman Gaynes, of Hoboken Paint Co., Inc., was particularly helpful in the use of Federal specifications.

Material on outstanding commercial specifications came from the following, in addition to Mr. Neese: Morton Gruber and David Sutherland, Ciba-Geigy Corp.; Gerould Allyn and Nelson E. Stefany, Rohm & Haas Co.; John C. Becker, Jr., Borden Chemical Co.; Ray Smith, Dow Chemical Co.; Edward Raswyck, Reichold Chemical Co.; Everett Sklarz, Ashland Chemical Co.; Greg Bruxelles, Hercules, Inc.; Fred Gartenlaub, Northeastern Laboratories Co., Inc.; Al Boardman, Cargill Corp.; Moe Bauman and Benjamin Farber, Farnow, Inc.; Meyer Budman, Superior Materials, Inc.; Irwin Young, Jesse S. Young, Co.; Art Nortman, Burgess Pigments; Alan J. Dankwerth, Engelhard Minerals and Chemicals Corp.

The architect's problems were clarified through the kind collaboration of Bill Duffy, chief of specifications, Frost Associates, Architects, and his assistant, Tim Kirby. Help in this direction, and encouragement, were received from Walter F. Geisinger, of the technical directorate, The Construction Specifications Institute; and Robert L. Petterson, secretary, Production Systems for Architects & Engineers, Inc., publishers of MasterSpecs.

Problems of the interpretation of architects' material were explained by Gerard F. Jansen, of L&L Painting Co., Inc.; Alan Stein and William Chayefsky, of United Painting and Decorating Co.; and Martin Geraghty, of the New York State Div. of Housing and Community Renewal.

Thanks are also due Royal A. Brown, technical director, National Paint and Coatings Association, and Mrs. Beth A. Mathes, assistant director of public relations of NPCA; and Frank Borrelle, assistant executive secretary, and Rosemary Falvey, of the executive staff, Federation of Societies for Paint Technology; and Mrs. Stella Miller, of NPCA.

For the helpful material provided from the voluminous and authoritative files of the *American Paint Journal* the *American Painting Contractor,* and the *American Paint* and *Wallcoverings Dealer,* thanks are due A. F. Voss, Jr., and W. Clark Voss, publishers, and the editors of the magazines. Miss Alice Marens, and Edward Salas, my co-workers in the New York office, are accorded warm thanks for enduring the year in which the book was written.

Introduction

Coatings technology, which is both an art and a science, enables today's architect or builder to select finishes with performance characteristics and potential durability that can be matched to almost any set of simple or challenging circumstances.

And yet even when shielded by coatings, metal is humbled by corrosion; concrete and cement crumble before the bite of acid or the onslaught of driven rain; and wood still rots. The reason, in many instances, is selection of the wrong coating; but other reasons are just as important, namely improper surface preparation or improper application.

The selection of coatings for long-term protection is far more complex than most people realize. Some architectural firms, aware of the huge technological strides that vaulted the coatings industry from its primitive pre-World War II stage to its present sophistication, have assigned specialists to keep abreast of coating developments. Others have done the best they can; and, depending on the extent of their design-reach, this may not have been enough.

It's assumed here that most architects, and others in the construction industry or at least many such individuals, are far too occupied with the intricacies of their own demanding professions to have mastered all that is to be known about surface protection.

It seems, then, that simple methods for specifying maximum surface protection in even the most demanding circumstances should be welcome. Also welcome, we believe, will be a method for specifying protective materials, application procedures, and just as important, if not more so, techniques for preparing surfaces. These methods are aimed at making the once-onerous specification chore so easy that a "special effort" will be required to do it wrong.

Accordingly, this book describes the various types of surfaces and the problems usually involved with each. For each surface and problem references will be made to the proper materials described in Chapters 7 and 8, which are devoted to materials and performance specifications, so that the specifier can readily understand the problem and select his specifications with minimum effort. Wherever possible, basic specifications will be based on rigorously-tested and highly successful U.S. government standards. References are made in Chapters 7 and 8 to sections in Chapter 3 covering proper *surface preparation*. Specifications for *application methods* will be found in Chapter 5.

Since a professional usually likes to know why requirements are set up, and since only a few fundamentals are needed to explain them, the remainder of this introduction will tell in a general way what goes into a coating and how these constituents are selected and combined to produce protection for specific surfaces and prevailing circumstances.

From this background, the specifier will have reinforcement for his belief that care must be exercised in specifying coatings and in stipulating performance characteristics.

A coating can be defined as a mixture, which, when applied in a thin film, provides protection or decoration, or both. The term "thin film" seems self-evident, but this can describe a film ranging from 1/1000 of an inch to 40/1000 inch; and too thin or too heavy a film may have an important bearing on adhesion, durability or surface protection.

Of greatest significance, though, is the word "mixture." Only rarely will a specific type of coating—such as an exterior house paint—have the same components when made by different manufacturers. Each formulator has a broad choice of mixtures for a particular kind of coating. One formulator may put less pigment and more resin in his house paint mixture than another formulator and still have an effective paint; or he can use a different resin or combination of pigments; or he can

Introduction

use less pigment and more "supplemental" pigments, which are also known as extenders, or fillers, or inerts. He may end up with just as good a product as his competitor, or even better; but if he selects his components unwisely, he may end up with a product that is not quite as good, or even almost as good.

Thus it is important to understand why each formulator comes up with his own mixture and what can happen. First, and easiest to explain, is why formulations differ, and why any one company is often obliged to change its formulations. The simplest explanation involves production costs. Raw materials in coatings, like most other chemicals, are bought by the pound, which is a unit of weight, but are sold by the gallon, which is a unit of volume. Each chemical varies from others in the number of pounds that make a gallon. For this reason, the formulator tries to make the best possible paint with the *quality ingredients* that can help him fill up his gallon cans with the bulkiest, or lightest weight materials. Each formulator has his own combination of components for producing a high quality paint; and similarly, he has combinations for making low- and medium-quality paint—all of which are usually identified by price structure, and usually, brand name.

If titanium dioxide, the most important hiding pigment, costs $0.50 per pound, for example, the formulator can make a lower quality paint, and identify it as such, by cutting back on the amount of titanium dioxide and increasing the amount of clay, calcium carbonate, or talc, which are known as extender pigments, and which, in proper amounts, serve very worthwhile purposes.

Other formulators for their high quality, or medium, or low quality paints may get equally good paints in each price range, with more or less titanium dioxide. Similarly, they may increase the percentage of binder resins to get a better paint, or may cut back on binder and increase the amount of extender pigment—thus making a lower quality paint.

For the same reason, a formulator may maintain the same amount of binder resin but may use a less desirable resin in place of the one he uses for the top-of-the-line product. For an alkyd-resin enamel based on soybean oil, he may substitute an alkyd based on linseed oil and menhaden oil, which at the time may be considerably less expensive than soybean oil; but he will very likely get an enamel that will not have the color and tint retention of the soybean oil alkyd. (An alkyd, incidentally, is really a vegetable oil that has had its faults cooked out of it by modification with some fatty alcohol, like glycerine or pentaerythritol, and with an organic acid of a type known as dibasic, of which phthalic anhydride is typical.)

One formulator may pick a soybean oil alkyd because of its color fidelity, and another may prefer to use a safflower oil alkyd for the same reason. One may choose a combination of oils and fatty alcohols that give extra flexibility on wood substrates, and another may want to add components that provide scrubbability or impact resistance.

Quite often, the chemist will have to combine components to yield a slate of performance characteristics that are really compromises. He may prefer only a desirable level of scrubbability, let's say, rather than the ultimate, so that he doesn't cut too deeply into stain removal ease. He must, thus, compromise in one performance area in order to achieve superior characteristics in another area.

From the specifier's standpoint, it is important that the product he selects has been thoroughly tested in use under circumstances that approximate the conditions under which the material is expected to perform. The integrity of the manufacturing firm providing the coating is important in that its reputation is being accepted as an endorsement of the pre-testing of the product.

From the foregoing it may appear that the mixture making up a surface coating consists of nothing more than a pigment and a binder, but actually organic solvents are used in many paints to provide a liquid carrier—or water may serve this function. Constituents are also added to get the various components to combine more readily; others are used to make the mixture flow better on the substrate or to help prevent brush marks in the dried film, and still other materials are added to prevent "skinning" in the can, or sagging after the coating has been applied and is still wet.

Additives are used in water-thinnable paints to get the tiny emulsion particles in the binder to coalesce into a coherent film. Fungicides are needed to fight mildew on exterior surfaces, and bactericides are added to prevent spoilage in the closed can. Still other materials may be stirred in to make the coating dull or glossy.

Introduction

Then, we have supplemental pigments—clay, calcium carbonate, talc, mica, silica, nepheline syenite—all serving a proper and beneficial purpose—when used correctly. Often they are used importantly in the manipulation of mixture; hence, they will come in for close attention.

When coating ingredients are manipulated it is usually to alter an important relationship between the quantity of pigment and the quantity of binder-resin, a relationship expressed as Pigment Volume Concentration, or PVC, a term encountered in coatings industry jargon as often as horsepower is heard in automotive comparisons.

Extenders or supplemental pigments, it should be said here, are counted as pigments for purposes of computing PVC.

The PVC is determined by dividing the volume of pigments alone into the combined volume of resin-plus-pigments. Hence, if 12 gallons of pigment (let's say titanium dioxide, which is a real hiding pigment, and clay or talc, which are extender pigments or fillers) are used in a formulation with 36 gallons of alkyd resin, then the PVC (Pigment Volume Concentration) is obtained by dividing 12 (the volume of pigment and extender) by 48 (the combination of 36 gallons of resin and the 12 gallons of pigment and extender). This comes out to 25 percent and is expressed as a PVC of 25.

Since Pigment Volume Concentration refers to the amount of pigment in proportion to resin, it follows that a low PVC means considerable resin is being used in relation to pigment. A high PVC in a mixture means less resin in relation to pigment. In other words, as we increase the proportion of pigment, we increase the PVC; as we decrease the pigment and increase the proportion of resin, we lower the Pigment Volume Concentration, or PVC.

With increased proportions of resin, you would expect the coating to be more flexibile, able to take more washing and scrubbing, and to be more durable than one with a decreased amount of resin (assuming the pigment volume stays the same). That's why, in general, a paint with a low PVC (more resin in relation to pigment) is more flexible, washable, and durable. As we will see, it's also glossier, and low PVC mixtures are usually known as enamels.

A paint with a high PVC (more pigment in relation to resin) is usually more buttery and brushes on more easily, has better hiding power (if the added pigments are hiding pigments such as titanium dioxide, instead of supplemental pigments such as clay, calcium carbonate, mica, or talc), and usually costs less to produce because of the lower raw material costs. Also, because the mixture has relatively less resin it is thicker and does not penetrate as well into porous surfaces as do the more fluid mixtures that are richer in binder. High PVC paints, thus, are used where penetration should be limited, as in primer-sealers, which are used to prevent waste of paint in coating porous substrates such as wall board or plaster.

A conscientious formulator will stick close to desirable PVC rates when he is obliged to tinker with his mixture. But, since a cost-conscious formulator can achieve his desired PVC by substituting low-priced supplemental pigments in place of more expensive hiding pigments, he may end up by using too much substitute and cause undesirable developments. One of several possibilities is excessive water absorption (particularly if he uses too much hydrous clay or calcium carbonate, both of them helpful materials in limited quantities). Recent findings point to this as a cause of early failure, particularly if the primer coat and the topcoat absorb and then release water at different rates.

Two outstanding paint scientists, T. Kirk Hay and Garmond G. Schur, carried out a series of classic experiments showing that harm results when water absorption and desorption rates of the primer differ from those of the topcoat. Above certain percents of humidity, water enters the coating system (by a process resembling capillarity) and then cannot leave the primer fast enough because the topcoat releases its moisture too slowly. A primer with a PVC that is too high has a lot of soft, permeable pigment and holds the water longer than a coating with a low PVC (and, thus, less pigment) in the topcoat; and all sorts of troubles can result, such as corrosion of metal, fungus, peeling, and stripping.

The results of tinkering with PVC are offered here to hint at what can happen if you select paint too casually.

Many reliable firms—just about all the major national manufacturers and most of the well-established regional firms—jealously guard the sanctity of their good names and are zealous in their quality control and in their assurance that their production departments have selections of proven formulas to use for their product mixes. But the price-manufac-

Introduction

turers, who live by cutting corners, are another matter.

An observation is in order here: the few dollars saved on a job by buying questionable coatings is inconsequential when you consider that the big cost of painting is surface preparation and application. (This will be covered in detail in Chapter 10, Economics of Structural Coatings.)

If by this time, we've understood what is meant by Pigment Volume Concentration (the ratio of pigments—both hiding and supplemental—to the combined volume of these same pigments and the vehicle), then we're ready to understand how they are controlled to get desirable properties in a coating.

First, we must understand that pigment particles have very tiny voids or air spaces separating them, much of it caused by the irregular surfaces of these ground and milled solids. For each quantity of a pigment or combination of pigments, a given amount of a binder is required to fill up all the tiny voids and air spaces.

If we start with dry pigment, as we add resin, or binder, we gradually reach the point where all the voids are filled. This stage is known as the Critical Pigment Volume Concentration. If we add more binder, it has no pigment to bind, so we are said to have "free binder," which could be said to be an excess of binder, since it has no binding duty to perform. As we shall see, it has other roles to play.

At the stage denoted as Critical Pigment Volume Concentration, we have no voids among the pigment particles and no free binder in the mixture. This is known as CPVC, and is important because it becomes a reference point when a formulator seeks to develop properties of a coating.

Coatings with Pigment Volume Concentrations above the *critical* point, in effect, have excess pigment. This means that between the pigment particles voids exist. Let's say we have materials for a paint, which when mixed would have a Critical Pigment Volume Concentration of 60. If we start out with dry powder and begin to add resin, we have a PVC well in excess of 60. In fact, we may not get any feeling of fluidity until we have about 10 percent binder present. Now, remember all of the pigment particles will not be surrounded by binder until we move through PVC's of 80, 75, 70, 65 and come to a PVC of 60, the Critical Pigment Volume Concentration. That means that even when we have added enough binder to attain a PVC of 62, we still have some pigment particles that have voids around them.

At all those PVC's above 60, the coatings are somewhat soft because of the excess pigment; so these PVC's lend themselves to coatings used for easy sanding undercoats, where you want a material that isn't very hard and which can be easily abraded to form a good tie-coat for the top finish.

House paints, on the other hand, need a small excess of binder so they will level and form a smooth coat and impart a small amount of gloss, so they are formulated just below the Critical Pigment Volume Concentration. Just below CPVC, some of the binder, then, is not needed to fill voids between the pigment particles; and this is available to add fluidity. Fluidity, in this case, in addition to aiding leveling and imparting slight gloss also helps house paints to penetrate and absorb small particles of chalky substances that may be present on the surface from deteriorating coats that had been on the house previously.

Metal protective paints provide maximum durability at approximately CPVC, because at that point pigment and binder are well-balanced, with essentially no space for harmful moisture to ease its way to the susceptible metal beneath. In fact, one of the tests for determining the CPVC of a mixture of a particular binder and a variety of pigments and extenders is to make up a range of combinations of these materials, each with a different PVC. These are formulated into paints and are applied to bare steel plates. All formulations that will keep the steel plates from rusting after prolonged exposure in salt-fog are regarded as being below the Critical Pigment Volume Concentration, since none of them has enough voids between the pigments to permit the salt fog to penetrate.

However, only one is the CPVC, the highest Pigment Volume Concentration among them, because anything above that PVC has voids between the pigment particles which is shown if rust occurs, meaning that moisture entered; and anything below has free binder, which means it is below the CPVC.

Another simple method for determining CPVC is to take sample formulations like those above, put coatings of equal thickness on the same substrate, and let them dry thoroughly. A high gloss enamel is then painted over the dried films. Soon it will be noticed that some of the surfaces of the high gloss topcoats will be altered. They will become less glossy. That's

Introduction

because some of the test formulations, serving now as primers for the purpose of the test, have high Pigment Volume Concentrations; and some of the pigment in the test materials are drawing solvent and some resin from the high gloss topcoats. As we remove resin and solvent, we pull out the very ingredients whose relatively high proportions result in such a high gloss. Thus we know that resin and solvent are being absorbed from some of the test formulations.

The Critical Pigment Volume Concentration of this mixture of binder and pigment in combination is somewhere in a very narrow range of samples. The range is between the formulation with the highest PVC that didn't affect the gloss; and the formulation with the lowest PVC affecting it. In other words, CPVC is just below the PVC of the sample that had just a little bit too much pigment; and it's just above the PVC of the formulation that had too little pigment to draw out the resin and solvent. The one with Critical Pigment Volume Concentration should have just enough binder to hold the pigment in its grip so that it won't be wooed by the siren call of fresh, free binder in the superimposed high gloss enamel topcoat.

This test illustrates another important point in paint specification: the primer selected for use with an enamel topcoat should be a mixture with a Pigment Volume Concentration somewhat below the Critical Pigment Volume Concentration so that no excess pigment or extender pigment will be available in the primer to absorb solvent or vehicle from the enamel, thus causing loss of gloss.

You will, later, see frequent reference to *enamel holdout* as a characteristic of primers used with enamels. It is important, then, to make certain that primers specified for use with enamels have the proper PVC so that loss of gloss will be avoided.

So much attention has been given to Pigment Volume Concentration and Critical Pigment Volume Concentration because it is essential to an understanding of what makes a coating work, or not work. In a sense, knowing about PVC and CPVC starts a specifier on the road to understanding at least some of the problems of paint formulation and also some of the variables in paint manufacture, which if abused can lead to trouble.

Little would be gained by going into the fine points of

formulation, except as they aid in specifying. Later, helpful details will be presented with that in mind and in the appropriate section.

Before getting to the real heart of the book, which is to relate surfaces to be protected to the circumstances in which they will function and the materials that will protect them—plus the methods of surface preparation and application that will make these materials work—it will be helpful to become acquainted with the constituents of the coatings that will be considered later.

This book has been designed primarily to aid architects, builders, prime contractors, industrial maintenance engineers, and others in specifying proper coatings. The book contains information to aid in understanding surface preparation, application, and type of coating to use for a given job. Ideally the specifier should read the text from cover to cover to get an idea of the interrelationship of type of materials, testing methods, expected wear characteristics, and type of coating required. This careful, detailed reading would prepare an architect to specify the original construction surfaces—wood, concrete, or metal—while at the same time being aware of the limitations, advantages, and disadvantages of each type of material so far as coatings are concerned. This would help in avoiding the all too common practice of designing and constructing a building and then—as an afterthought—making a decision on painting. The point is merely that an integrated knowledge of painting and coatings and construction surfaces is desirable if a coating system is to be an integral system of a building rather than simply something added on at the end. The result of thinking of coating last is apt to be either an inadequate or costly coating system.

We recognize, however, that: (1) such a careful reading may not be possible and (2) when the construction surface is already in existence, quick specifying data is desirable in order to pick out the best coating system.

The quickest way to develop a set of specifications for a particular job is to turn directly to the marked section at the end of the book. The following procedure is suggested:

1) Check the Guide to Detailed Specification Charts. This guide is broken down according to type of surface, for

Introduction

example, "Cementitious Surfaces, Exterior." Here you will find listed all the GPC specification numbers suitable for the surface you are interested in (GPC stands for Guide to Paints and Coatings). You can then make a list of the GPCs generally applicable to the surface.

2) Turn to the Detailed Specification Charts immediately following the Guide. Each GPC is described in detail in these Charts. Study the Charts carefully, keeping in mind the serviceability characteristics and cost requirements for your job. Based on these, select a GPC suitable for the job.

At this stage you may also wish to check the Performance Comparative Charts at the end of Chapter 7 (exterior surfaces) and Chapter 8 (interior surfaces). These Charts make comparisons of certain of the GPCs to enable you to see more clearly exactly how the specifications differ from one another.

Also at this stage, you may want to glance at the relevant sections in Chapters 7 and 8 for further information on the particular type of surface for which you are specifying. If you are specifying for an exterior cementitious surface, for example, it would be a good idea to reread Section 702.

3) Once the specific GPC is selected, choose a proprietary product which meets the specification. Normally there will be several such products which either meet or exceed the specification. We have *not* listed proprietary products in this book. You will need to contact the architectural representative or sales representative for the paint manufacturer that you deal with. Provide this representative with the required specification and let him indicate the coatings available from his company which meet the specification. Most large manufacturers will have standard formulations meeting most of the GPC specifications. The average architect or paint specifier, dealing with anywhere from two to a dozen reputable paint manufacturers, should have little difficulty in finding a standard, off-the-shelf product to match the required specification.

Readers may have occasion to set specifications more rigid than any particular GPC, and can, in effect, write their own specifications. You may, for instance, desire more rigorous scrubbability characteristics. Then you could specify a particular GPC plus the additional scrubbability characteristics desired. It may be that a standard proprietary product already meets your more strenuous requirement. In other instances, if your particular job is large enough, the manufacturer may be willing to reformulate his product to meet the more demanding specification.

While the quick method for using this book may be suitable for many situations, the following three examples illustrate a more thorough method, which in many instances will justify the extra time required.

In the first example, cedar shingles are to be protected in an environment with wide daytime-nighttime and seasonal temperature swings and considerable rain. Our first step is to turn to the Table of Contents where we see that Chapter 7 covers selection of coatings for exterior surfaces. We note that Paragraph 704 is on wood surfaces. Turning to that paragraph we continue to sub-paragraph 704.1 and Table 704.1a, which are about primers for woods such as cedar, which have dyes that can stain. In this paragraph we are referred to a Performance Comparative Chart at the end of the chapter, which compares several stain-stopping primers, including GPC 65, which is based on a specification developed by the U.S. Housing and Urban Development Department to meet health bans on lead-in-paint, and also GPC 66, a commercial acrylic primer. As with all specifications identified for ready-reference by GPC numbers, these are found in greater detail in the edge-marked pages in the back of the book.

For a topcoat, Paragraph 704.2 cites several specifications, including a linseed oil flat (GPC 41). Details of this specification, of course, are found, along with other related coatings, in the edge-marked pages. In making a choice, special attention, naturally, will be given to weather-resistance, color retention, and mildew resistance.

To designate proper surface preparation, the specifier may wish to read Paragraph 308, of Chapter 3, on preparation of wood surfaces; and Chapter 5 to assure himself about proper application methods; and, finally, he may want to review

Introduction

Paragraphs 604 and 605, of Chapter 6, which deal with prevention of deterioration of wood surfaces and their protective coatings. All these subjects are identified in the Table of Contents and the Index at the back of the book.

Our second example is a concrete wall in an elementary school corridor. Chapter 8, the Table of Contents informs us, covers interior surfaces. For the battering encountered in a school, we need a tough, impact-resistant coating, able to resist numerous scrubbings and the abrasion of kicking feet. A selection may have to balance cost against performance. An alkyd, such as Federal Specification TTP 30 (GPC 2), and TTP 29 (GPC 1) a latex paint, are among the least expensive coatings for the purpose; but high performance coatings, such as Federal Specification TTC 542 (GPC 20), a urethane; or TTP 95, which calls for either a chlorinated rubber, or styrene paint, or GPC 16, a commercial epoxy, will last far longer and be less costly in the long run since repainting of surfaces with cheaper material usually means repeating application costs and surface preparation expenses, which can be avoided, along with coatings costs, when long-term materials are used. (See Chapter 10, Economics of Structural Coatings.)

If high performance coatings are used, surface preparation is important, so it will pay to read Paragraph 307, of Chapter 3, on preparation of cementitious surfaces. To understand how to prevent deterioration of coatings on cementitious surfaces, Paragraph 606, of Chapter 6, will be helpful.

For our third example, a steel surface to be exposed near the seashore is to be protected against wind-driven rain and atmospheric chemicals.

The Table of Contents lists Chapter 7 for "Selection and Specification of Coatings—Exterior Surfaces." Paragraph 703, we find, covers exterior metal surfaces, and sub-paragraph 703.1 refers to various pigments in metal primers and their advantages. Performance Comparative Charts at the end of the chapter are cited to enable the specifier to choose among several specifications, including TTP 645 (GPC 55), which has an anti-corrosive pigment, and a commercial version of an epoxy polyamide. After deciding which of the specifications look most promising for our purpose, we note the GPC numbers and turn to the edge-marked pages in the appendix where the GPC's are found.

With the detailed specifications furnished in each GPC a specifier can select an existing set of performance characteristics or can write his own specifications on the basis of the furnished information.

The Table of Contents serves as a guide. Cross-references are frequently made in the text to guide the specifier to pertinent information on surface preparation, application techniques, prevention of deterioration of coatings and substrates, fire retardant coatings, and calks and sealants.

chapter 1

What Goes in a Coating

DEFINITIONS AND CLASSIFICATIONS

A wide variety of raw materials may be combined to produce three basic classes of coatings. These three basic classes are *paints, varnishes,* and *lacquers.*

101 Paints. A paint is, basically, a mixture of a hiding pigment and a material to bind the pigments together, the purpose of which is the protection of a surface from its environment. The *binder* may be a vegetable oil or a synthetic resin, and it may be thinned in an organic solvent or dispersed in water. Those dissolved in solvents are called *solvent-reduced paints.* Those dispersed in water are usually known as *latex paints* or *latex emulsion paints.*

Solvent thinners or water dispersants aid in paint application. These carriers evaporate when the mixture is spread. As a carrier leaves the mixture, after first rising to the spreadout film's surface, the binder dries and hardens. Most solvent-thinned paints dry and harden when their binders are exposed to oxygen in the atmosphere. The binders are oxidized as the solvents evaporate from the dissolved binder and make room for the oxygen.

Water-thinned paints harden when their solid, dispersed resin binders are freed of water by evaporation. These tiny resin particles are drawn together as they dry by the action of coalescing agents and form a continuous film.

A third method of hardening is associated with solvent-reduced coatings. Hardening in this method is accomplished by curing. Paint that hardens this way usually is sold in two containers, one containing a partially-complete basic resin binder plus pigments, and the other a resinous material that chemically completes the basic resin so that it can bind pigment and harden when spread on a surface. If the curing portion had been added and had completed the resin at the factory, the hardening reaction would have occurred in the closed can where it would obviously do no good.

Coatings of this type usually have a limited time period or "pot life" during which they must be used, once they are mixed. Often, these two-component products are high-performance materials and are used mainly by professional painters rather than do-it-yourselfers, although the latter can safely use most of them.

An interesting curing system uses the moisture present in the atmosphere as the necessarily missing curing ingredient. Since moisture is the curing agent, only one package is needed. This product will be described later under moisture-cured urethanes (Section 110.6a).

Paint requires more ingredients than mere pigment and binder, of course. As more sophisticated and durable synthetic materials became available, chemicals were needed to make them work, as in the example of the coalescing agents cited above in connection with the film-forming action of water-thinned latex paints.

As we shall see, a number of additives contribute significantly to coating performance. To name a few, cellulosic thickeners aid brushing and help prevent sag; mica contributes to the hardness and durability of exterior latex paints; anti-microbials help prevent mildew; and anti-foaming agents and antifreeze are also used—all of these additives are part of what makes paint work.

102 Varnish. Unlike paint, varnish is a single-element coating. In its pure form, varnish consists of the element in a paint that we described as the binder. In a varnish, the resin has nothing to bind, its sole mission being the protection of surfaces without hiding their natural beauty.

These varnishes, or transparent vehicles, hold up well on interior or exterior surfaces. The difficulty on exteriors, however, is preventing the harmful components of sunlight—the ultraviolet rays—from fading and deteriorating the varnish. Ultraviolet absorbers have been somewhat successful in this.

Exterior varnishes were once made almost exclusively of tung oil combined with phenolic resins, which are close relatives of Bakelite, one of the first commercial synthetics. Recently, some success has been had with clear, nonyellowing urethane varnishes.

Alkyd, epoxy, and urethane resins are used for interior varnishes, sometimes with colloidal silica to cut the gloss. Floor varnish was once mainly shellac, but newly developed urethanes and epoxies are now taking over.

103 Stains. Stains are used mainly on wood substrates and are a hybrid between paint and varnish. They contain small amounts of coloring material of one kind or another but never enough to hide the substrate. Stains for exterior surfaces contain light-fast pigments, such as iron oxide, titanium dioxide, and carbon black, while interior stains, because interior light is not an important deteriorating factor, may contain organic dyes and less sensitive pigments. No matter what coloring material is used, stains are almost always transparent.

104 Lacquers. Architects and builders will rarely have an occasion to specify the application of a lacquer in the field. However, they may want to specify factory-coated building materials with lacquers having certain properties; so a nodding acquaintance with this third major category is desirable.

A lacquer, to use a simple definition, is a solution consisting of resins in an organic solvent. Their ability to harden on a surface depends on the evaporation of the solvent. No oxidation, curing, or chemical coalescing agent is needed. The film simply forms because the solvents have taken flight. The fugitive solvents are hastened on their way, most often, in baking ovens. Lacquers are fast-drying and are usually very durable, very smooth, and serve effectively to provide various degrees of protection and decoration.

Coil coatings used for metal sidings are often lacquers. Lacquer sealant coatings, as well as topcoats for many kinds of masonry, such as concrete blocks, brick, and tile, are available.

105 Primers. Most coatings are used in systems usually consisting of a primer coat and one or more overcoats. Some coatings are designed to provide adequate coverage with one coat.

Primers are usually designed to link the substrate to the coat or coats that provide the greater part of the protection. Sometime, however, primers themselves play an important part in protection; outstanding examples of this are the primers in anti-corrosive systems. In these systems, pigments serve to check the effects of water and oxygen on metal. One such pigment, zinc dust, as we shall see later, serves as a sacrifice to the more or less satanical gods of corrosion, in the service of steel. Involved is a whole interplay of infinitely small electric charges, which the zinc dust captures in a process known as cathodic protection and, thus, diverts the damaging charge from the metal substrate. The process is illustrated in Chapter 7, Fig. 703.1.

Some primers serve also as sealers. These function on porous substrates, such as fibrous woods or the paper used as facing for gypsum wallboard. Without these nonpenetrating sealers, porous substrates would soak up paint needlessly. Other primers check the natural dyes in wood, keeping them from migrating to the surface and leaving unsightly stains.

106 Topcoats. Topcoats are more than just the frosting on the cake. They add heft to the protective system and contain the components that provide, in varying degrees, resistance to weather, chemicals, dirt, scrubbing, and staining.

Three major types of topcoats are usually available in paints, varnishes, and lacquers. They are flat, semigloss, and gloss. The latter two are known as enamels.

Flats contain higher percentages of pigment and extender pigment than the other two; semigloss coatings have more pigment than gloss enamels. As the quantity of pigment is lowered, the paint is harder as well as glossier; and usually it is more resistant to dirt, stains, and abrasion.

Flats are widely used for wall paints where decoration is a main consideration. Semigloss or gloss enamels are usually selected for trim, molding, and doors.

Where meticulous cleanup is necessary, as in dairies, pharmaceutical plants, and food canning plants, walls may

What Goes in a Coating

be coated with the hardest possible enamel to make the job of cleaning as thorough as possible by utilizing the slipperiness of a hard, glossy surface for easy dirt removal. Surface protection problems usually arise from the environment; thus, rough-house youngsters in a school corridor need a coating able to take a lot of impact and abrasion. Exterior topcoats on a chemical factory, or near one, may have to resist chemical fumes.

In selecting an enamel for industrial or home interior walls or equipment, care must be taken to avoid excessively shiny versions because of the possible visual discomfort and physical weariness that can be induced. Hard, durable enamels are available with the gloss eliminated by the addition of flatting agents.

107 Plastisols, Organosols, Powders, and Electrocoats. Several types of factory-applied coatings that may be found on structural parts desired by an architect or builder merit attention and will be covered in greater detail later in this chapter. (See Section 110.7b.) They do not fit into the three categories of coatings defined earlier. They are *plastisols* and *organosols*, which, like emulsion latex paints, are dispersed in their carriers rather than dissolved. The plastisols are dispersed in liquids that also serve as plasticisers or flexibilizers in the finished dried coating, while organosols are dispersed in organic materials that meet the definition of a solvent, although they are not used as solvents. Here they serve as the liquid carriers for the dispersed components.

Both plastisols and organosols offer high impact resistance. They differ from most conventional industrial coatings in that they must be exposed to considerable heat in an oven to achieve film coalescence.

The third type of coating that differs from paint, varnish, and lacquer consists of a *powder* that is deposited on either a hot or electrically grounded surface where it fuses and yields a tough, high-build film. The powder is applied either in a fluidized bed, which is a container in which the heated object is put so that the fluffed-up powder in the bed can settle on its surface from an updraft of air blown from beneath; or it may be applied by an electrospray gun which draws the powder from a miniature fluidized bed attached to it.

In the spray method, the object must be conductive and must be grounded.

Materials used for *electrocoats,* or *electrodeposition,* are closely related to both emulsion paints and solvent-thinned coatings, because they are primarily solutions, and they are dissolved in a water carrier. They are also related to plating systems, and the process could be called plating with paint, because the metal objects to be coated are electrically grounded and are actually plated with the solids in the paint when an electric charge passes from the tank to the grounded objects, through the paint.

An advantage of both the fluidized-bed powder system and electrodeposition, as compared with conventional spray systems, is that wraparound or complete coverage of front and back and edge surfaces is possible. Thus these methods enhance corrosion-prevention because all the areas, including edges, that are likely to be receptive to moisture or chemical contaminants are equally covered. Edges are often missed by conventional spraying.

Electrodeposition has the disadvantage of limited film thickness, generally about one-half mil, while powder can be deposited up to 40 mils.

Plastisols and Organosols are mainly based on vinyl chloride and modified vinyls. Powders can be based on almost any solid resin with a melting point that can be reached in available ovens and which is not adversely affected by the high heat. Epoxies, polyesters, nylons, vinyls, acrylics, and polyester-polyimide resins are most widely used at this time.

MATERIALS

108 General. Now that we have classified coatings into paint, varnish and lacquer—as the main types—and plastisols, organosols, powder and electrocoats as newer and still less important versions used mainly for factory-applied coatings, we're ready to become acquainted with the materials that are brought together by the formulator to make the final products that provide protection for structural surfaces.

Main emphasis will be given to binders and pigments, since they are the major factors in a coating. Theoretically, these two can protect a surface without help from other auxiliary

What Goes in a Coating

materials. Actually, however, the resulting protective film would most likely be short-lived, ineffective, and disappointing.

To round out the checklist of what goes into a coating, a brief description will be given of the additives that have opened up the use of numerous durable, resinous binders that would never have been serviceable without them.

109 Solvent-Reduced Binders. Solvent-reduced binders are the oldest type, going back into antiquity. Most of these dry by oxidation, which means that some part of the binder's molecule, usually aided by a catalyst or drier, combines with oxygen in the atmosphere and hardens. Some solvent-reduced binders harden by combining with other materials added just before use. These are described as two-component systems and will be covered in this section when we reach epoxies and urethanes. Table 109A shows the performance characteristics of solvent-thinned binders, and use and service characteristics are shown in Table 109B. In addition, Table 109C provides comparisons of the in-use characteristics of a number of the paint systems

Table 109A. *Performance Characteristics of Solvent-Thinned Binders.*

Use For	Alkyd	Vinyl	2-Pkg. Epoxy	Oil	Phenolic	Rubber Base	Moisture- Cure Urethane
Adhesion	VG	F	E	VG	G	G	G
Hardness	G	G	VG	F	VG	VG	E
Flexibility	G	E	E	G	G	G	VG
Resistance to:							
Abrasion	G	VG	VG	F	VG	G	E
Acid	F	E	G	P	VG	VG	E
Alkali	F	E	E	P	G	VG	VG
Detergent	F	E	E	F	VG	VG	VG
Heat	G	P	G	F	G	VG	G
Strong solvents	F	F	E	P	VG	F	E
Water	G	E	G	G	E	VG	VG

Source: David Litter Laboratories.
 Key:
 E   Excellent, outstanding
 VG Very good
 G   Good or average
 F   Fair
 P   Poor
 NR Not recommended

Table 109b. *Use and Service Characteristics of Solvent-Thinned Binders.*

Use On	Alkyd	Vinyl	2-Pkg. Epoxy	Oil	Phenolic	Rubber Base	Moisture-Cure Urethane
Wood	G	NR	G	G	G	NR	G
Fresh, dry concrete	NR	VG	VG	NR	NR	VG	G
Metal	VG	VG	VG	VG	VG	G	G
Interior	G	G	G	G	G	G	G
Exterior							
Rural	G	VG	G	G	G	G	G
Seashore	G	E	VG	G	VG	VG	VG
Industrial areas	F	E	E	VG	VG	VG	VG
Fade resistance	VG	E	G	G	G	G	F
Chalk resistance	G	E	F	G	G	G	F

Source: David Litter Laboratories.
Key:
E   Excellent, outstanding
VG Very good
G   Good or average
F   Fair
P   Poor
NR Not recommended

discussed in this section. It will be useful, in reading the material that follows, to refer back to this table for service characteristics.

109.1 Straight Oil Paints. Straight oil paints, based on vegetable oils, which consist mainly of fatty acids, dry by oxidation with the help of metal salts that act as catalysts or driers. Oil paints have been declining in importance. The only significant one still used in volume is linseed oil house paint. For some years, this has been steadily losing ground to latex house paints, which have the advantages of faster drying and easier clean up. Some signs, since 1969, have been pointing to a revival of linseed oil paints because several large manufacturers have begun offering high-quality versions, with extra-large quantities of heavy-bodied oil. These provide more durable protection than the older types, but at a higher price.

Soybean oil, menhaden oil, safflower oil, tung oil, castor oil, and tall-oil fatty acids, which are derived from a by-product of paper manufacture, are all used in making alkyd resins but are infrequently used as binders by themselves.

What Goes in a Coating

Table 109c. High Quality Coatings Performance Comparatives.

	Two-Package Epoxy Polyester	Two-Package Catalyzed Epoxy	Two-Component Urethane	Two Component Polyester	Air-Dry Alkyd
Resin cost for a dry film 1 mil thick ($ per sq.ft.)	.0050	.0050	.0070	.0030	.0020
Applied cost of 8-mil coating ($ per sq.ft.)	.25-.40	.30-.50	.40-.55	.20-.35	.15-.25
Estimated years of wear	10-25	10-25	10-25	10-25	5-8
Number of coats to obtain 8-mil film	2	3	2-3	1-2	3
Gloss retention	Excellent	Poor	Poor	Poor	Good
Pot life	16-48 Hours	4-24 Hours	2-24 Hours	1/4-2 Hours	Unlimited
Nonyellowing properties	Excellent	Poor	Poor	Fair	Fair

109.2 Alkyd Resins. Alkyd resins have, to a great extent, supplanted oils in paint. Essentially alkyds are treated vegetable oils, which have some of their bad habits knocked out of them in a high-heat reaction between the fatty acids in the oil and a so-called fatty alcohol—such as glycerine or pentaerythritol—and a very efficient and versatile material called a dibasic acid, such as phthalic anhydride, or isophthalic anhydride, or maleic anhydride, among others.

Since the reaction of an alcohol and an acid meets the definition of an ester, alkyds qualify for that term. We can even use a more glamorous term and say alkyds are really polyesters. Another phrase to describe an alkyd is *polymeric ester*.

No matter what you call them, paints based on alkyds have a lot going for them. They're low in cost and have good durability, flexibility, and gloss retention. They have acceptable solvent resistance, toughness, and heat resistance, and fair color retention. Like vegetable oils, they dry by oxidation, meaning they combine with atmospheric oxygen, and harden fairly rapidly. They are used in architectural coatings for flats or enamels and in factory-applied finishes, either for air-drying or oven-baked finishes. They are compatible with lots of other resins and oils and impart desirable alkyd characteristics to them when necessary.

Alkyds are sometimes described as ideal binders for pigments, because they have the ability to wet and disperse them readily.

They can handle pigments in a wide range of volume concentrations. (Now, we're back to our old friend Pigment Volume Concentration again.) It is because of this ability to find and hold varying amounts of pigments that they are used in coatings that can penetrate surfaces when necessary without leaving soft troublesome pigment residues on the surface; or they can serve in flat house paints that do not penetrate much but flow and level with the best of them; or they make first-rate high gloss enamels, with lots of resin used with relatively little pigment.

Alkyds are classified as long-oil, medium-oil, or short-oil alkyds. Long- and medium-oil alkyds are encountered in the architectural field, and short and medium are used in industrial finishes. Long-oil alkyds contain from 55 to 70 percent oil;

and very-long-oil alkyds run from 70 percent oil on up. The more oil the more flexible the film, and also the softer, the less viscous, and the more easily damaged by solvents.

Alkyds used in house paints are very-long-oil, since a fluent, easy spread is desired to help the coating penetrate and adhere to the surface. Medium-length alkyds are used extensively for gloss and semi-gloss enamels and until the recent advent of latex enamels, they practically monopolized the field.

Color retention of alkyds is usually better than that of the vegetable oils included in their structure. However, when yellowing oils, such as linseed, are used, some discoloration may occur in the oxidation process. When soybean oil, safflower, or tall oil is used, color integrity is usually better, because their fatty acids are more stable.

Alkyd-based paints should not be used over freshly made alkaline surfaces such as concrete, masonry, or plaster; and

Table 109.2. *Salient Facts about Alkyds.*

Brushability	G	Gloss retention	E
Odor	Mild	Color, initial	E
Method of cure	Oxygen	Yellowing (clear)	Slight
Speed of cure		Fade resistance	E
50°–90° F	G	Hardness	G
Below 50° F	F	Adhesion	G
Film build	G	Flexibility	G
Safety	G	Resistance to:	
Use on wood	E	Abrasion	G
Use on fresh concrete	NR	Water	F
Use on metal	E	Detergents	F
Min. surface	F	Acid	F
prep'n class for metal		Alkali	F
Use as clear	E	Strong solvents	P
Use in aluminum paint	G	Heat	G
Choice of gloss	All	Moisture permeability	Mod.
Service:			
Interior	E		
Normal exposure	E		
Marine exposure	F		
Corrosive exposure	F		

Key:
Excellent . . . . E
Good . . . . G
Fair . . . . F
Poor . . . . P
Very Poor . . . . VP
Not recommended . . . . NR
Not applicable X

they have only fair durability in mildly corrosive environments. Otherwise, alkyds serve as good workhorse resins, economical and effective. The most important characteristics of alkyds are rated in Table 109.2.

109.3 Epoxies. These have been available commercially in this country since about 1950. About 35 percent of the sizable volume consumed in the U.S. is used for heavy-duty plant maintenance paints; and a growing quantity is used for tilelike coatings in homes, hospitals, dairies, and other places where maximum cleanliness is needed on walls and floors or where chemical resistance is important. They are also used where the surface is likely to be subjected to corrosion, or to heavy impact and abrasion or, in general, where it is likely to take a considerable beating. Epoxy characteristics are evaluated in Table 109.3.

Table 109.3. *Salient Facts about Eposies.*

Brushability	F	Gloss retention	P
Odor	Strong	Color, initial	G
Method of cure	Chemical	Yellowing (clear)	X
Speed of dry		Fade resistance	F
50°-90° F	G	Hardness	E
Below 50° F	F	Adhesion	E
Film build	E	Flexibility	E
Safety	F	Resistance to:	
Use on wood	E	Abrasion	E
Use on fresh concrete	E	Water	G
Use on metal	E	Detergents	E
Min. surface	F	Acid	G
prep'n on metal		Alkali	E
Use as clear	NR	Strong solvents	E
Use in aluminum	G	Heat	G
paint		Moisture permeability	Low
Choice of gloss	All		
Service:			
Interior	E		
Normal exposure	E		
Marine exposure	E		
Corrosive exposure	E		

Key:
Excellent .... E
Good .... G
Fair .... F
Poor .... P
Very poor ....VP
Not recommended ....NR
Not applicable X

What Goes in a Coating

Epoxy-based coatings are tough—and usually expensive, but not so costly that their anticipated long life span won't make them economical in the long run. (See Chapter 10, "Economics of Structural Coatings.")

A sizable amount is used for pipe coating, particularly if the pipe will serve under difficult conditions.

Another favorable detail leveling out the high cost is the high-build films possible with epoxies, which means fewer coats are needed to attain a desirable thickness. Since it is the cost of application that is usually most expensive, the resultant savings may offset all or part of the extra paint cost.

Epoxy resins by themselves have limited use. First, their molecules are fairly low in weight, and they need beefing up. Fortunately, supplementary resins are available, which add to their weight while reacting with the epoxide groups to form complex molecules that have a number of desired properties. Various "curing" agents are available to enable the formulator to enhance the particular characteristics needed for a given set of circumstances.

Epoxies used in architectural finishes are usually "cured" with polyamide resins, a family distantly related to nylon and used extensively in factory-applied finishes. Other epoxy types also used for architectural coatings will be mentioned later. Promising water-reducible versions have appeared.

While almost any kind of surface can be protected by epoxy paints, epoxies may chalk if used on exteriors. This is a slow process and takes anywhere from 6 to 12 months to show. On industrial buildings where gloss is not important, chalking can help keep the structure clean; rain washes he chalk off, taking dirt along as well. New curing agents have led to epoxies that show promise of chalk resistance and longer weatherability than previous products.

The chalking process, though dulling, usually takes a long time to wear away the coating. Some oil storage tanks, for example, show scarcely discernable loss of film thickness after more than eight years' exposure. Where rain or abrasion is significant, though, loss of film is apparent but not necessarily crucial. For instance, on a cone roof storage tank 2 mils (0.002 inch) were lost after six years' exposure; but four additional mils remained.

High-build epoxy coatings are available with as much as 40 mils dry film thickness obtainable in one single coat. By careful attention to Pigment Volume Concentration (our old friend again), these high-build epoxies can range from semigloss to dead flat finishes. High-gloss epoxy enamels for high-build are unavailable because they require "free binder" (that's the resin in a mixture that plays no role in filling the voids between the pigment particles, remember?) and if we had enough excess binder to get a high-gloss epoxy our coating without the plumping effect of a large amount of pigment would lack sufficient body to hug the surface, and it would sag and look awful.

109.3a Coal-Tar Epoxies. Epoxies combine with coal-tar pitch to provide black coatings that withstand both fresh and salt water and chemicals, including a number of inorganic acids. They are used extensively and successfully underground and underwater.

109.3b Epoxy Esters. Epoxy esters, usually made by combining an epoxy resin and a vegetable oil or resin, are used for coatings intended for use as metal primers for corrosion resistance. These materials in one form or another are used also for floor varnishes and factory area coatings because of their durability.

109.3c Epoxy Polyesters. Epoxy polyesters make up high performance systems combining the best features of epoxies and polyesters. By varying the rate of epoxy to polyester, coating systems are created with tailored characteristics. For example, by increasing the relative amount of polyester, solvent and acid resistance can be increased. Gloss retention, flexibility, adhesion, and alkali resistance are enhanced by increasing the epoxy portion. Also, when hardness and toughness are added by sufficiently boosting the polyester portion, we have coatings that resemble tile with the added advantage of being grout-free.

Epoxy polyester coatings can be made to dry quickly or over a 10-hour period of time by adjusting the amount of polyester used. Some systems air-dry in less than an hour.

Brilliant gloss and gloss retention, outstanding adhesion, and unusually high film-build are features of these coatings, which have already seen important service in atomic energy plants, dairies, pharmaceutical houses, and wherever a long

What Goes in a Coating 29

life resistance to stains and many chemicals is needed, and where meticulous washability is important.

110 Inorganic Vehicles. Inorganic binders usually are based on sodium, potassium, lithium, or ethyl silicate. Their main use is in metal primers with zinc dust. The inorganics react with zinc leaving a very hard corrosion-resistant film. Some of these primers also contain lead oxides, which react with the silicates and zinc to add to the corrosion resistance.

These inorganic coatings with heavy zinc content are known as zinc-rich paints, a name also given to organic primers with large quantities of zinc dust.

110.1 Zinc-Containing Silicate Primers. The zinc-containing silicate primers are of three general types:

1) *A three-package system* made up of a water solution of sodium, potassium, or lithium silicate; zinc dust, with a little finely pulverized lead metal; and a low-pH curing material, which is applied when a mixture of the other two has been applied and allowed to dry. One disadvantage of this type of zinc silicate system is that five to seven days are needed for curing and then the catalyst residue must be washed off.
2) *A two-package system,* described as self-curing, with one container having some ethyl silicate, such as tetraethyl silicate, in an alcohol or an ester; and the other containing zinc dust. This mixture requires only one or two days to cure and is free of water-sensitivity after it has dried, usually in one to two hours. Topcoats may be applied when it is cured, and washing is unnecessary.
3) *A one-package dispersion of zinc dust in a zinc silicate binder.* The binder will conduct the minor electric charge that causes corrosion when scratches or voids in a coating on iron or steel surfaces permit water to reach the substrate; the metal in the mixture takes the charge instead of the iron or steel. This process is known as "cathodic" protection. In this way, the electric charge erodes the zinc metal rather than the surface it protects. To render cathodic protection, these mixtures must have at least 80 percent zinc dust (usually they have 90 percent) and

they should have a dry film thickness between 2 to 5 mils, but no more, or else they will crack; and an expensive removal operation is necessary.

Since high humidity or rain will damage these zinc silicates while curing, application specifications should always include a surface test for proper humidity (see Section 211.2) as well as careful requirements for sandblast cleaning of metal so that the primer will be in direct contact with the bare steel surface rather than with its ultrathin coating of oxide or mill scale that affects long-term adhesion.

110.2 Phenolic Resins. Phenolics were among the first synthetic resins; in recent years they have fluctuated in importance, because they were formerly dependent on tung oil, the prices of which have swung very high at times, consequently dis-

Table 110.2. *Salient Facts about Phenolics.*

Brushability	G	Gloss retention	E
Odor	Mild	Color, initial	P
Method of cure	Oxygen	Yellowing (clear)	Considerable
Speed of dry		Fade resistance	G
50°–90° F	G	Hardness	E
Below 50° F	F	Adhesion	G
Film build	G	Flexibility	G
Safety	G	Resistance to:	
Use on wood	E	Abrasion	E
Use on fresh concrete	NR	Water	E
Use on metal	E	Detergents	E
Min. surface	Class F	Acid	E
prep'n on metal		Alkali	G
Use as clear	E	Strong solvents	G
Use in aluminum paint	E	Heat	G
Choice of gloss	All	Moisture permeability	Low
Service:			
Interior	E		
Normal exposure	E		
Marine exposure	G		
Corrosive exposure	G		

Key:
Excellent . . . . E
Good . . . . G
Fair . . . . F
Poor . . . . P
Very poor . . . . VP
Not recommended . . . . NR

couraging their use. Recently, chemists have found a way to treat the less expensive slower-drying oils with tolylene diisocyanate, the key component in urethane resins. The treated oil is then combined with an oil-soluble phenolic. Experts declare this to be *almost as good* as the best phenolic-tung oil for use as an exterior clear varnish on wood.

Phenolic binders are also used as mixing vehicles for making aluminum paints with excellent durability. Pigmented phenolic paints serve as metal topcoats in extremely humid atmospheres and as primers where a surface will be immersed in fresh water. Phenolics offer excellent abrasion and water resistance and serve effectively in mild chemical environments. One disadvantage is a tendency to darken on exposure, so whites and light tints are not used. See Table 110.2.

110.3 Phenolic Alkyds. Phenolic alkyds are a blend of phenolics and alkyds to get the hardness and resistance of phenolics and the color advantages of alkyds.

110.4 Rubber-Based Resins. Rubber-based binders are to be distinguished from rubber-base latex paints, which are water-reducible and more versatile. The solvent-thinnable binders are chlorinated rubber, styrene acrylate, vinyl toluene-butadiene, and styrene-butadiene. The first two are mainly used for masonry and for surfaces exposed to moisture and high humidity or surfaces that are washed frequently. They are often used for swimming pools and cellar walls because they are highly resistant to alkali materials. Chlorinated rubber also offers resistance to acids and oxidation and has low permeability to water and water vapor which has led to use as a corrosion fighter. Its nonflammability and high electrical resistance also offer decided advantages.

Because it has about 35 percent rubber, it is flexible, as are other members of the rubber-based family.

110.4a Styrene Acrylate Copolymer Resins. Styrene acrylate copolymer resins form tough, hard films with good adhesion to a variety of surfaces. One big advantage over chlorinated rubber is their ready solubility in inexpensive mineral spirits. With air pollution laws becoming more and more restrictive on exotic solvents of the aromatic family, chlorinated rubber may be at a distinct disadvantage relative to styrene acrylate.

Texture paints made with styrene acrylate systems and Perlite have been used for exterior or interior use. Government specifications for withstanding wind-driven rain were met by paints based on this combination as were federal specifications requiring the coating to withstand 4 psi hydrostatic pressure. Texture paints and conventional coatings based on this resin have excellent scrubbability and hiding power.

110.4b Vinyl Toluene-Butadiene. Vinyl toluene-butadiene, like styrene acrylate, requires inexpensive low-odor aliphatic solvents. Its primary use is as a waterproofing sealer for masonry. While it is only recommended, thus far, for interior waterproofing, tests approximating a rigid U.S. government specification for wind-driven rain indicate that it has possibilities for outside use as well.

An important advantage for vinyl toluene-butadiene over alkyd and latex dry-powder sealing systems is its suitability for use on wet surfaces.

110.4c Styrene-Butadiene. The other notable member of this family, styrene-butadiene, can be combined with chlorinated plasticisers, silicone resins, and aluminum pigments to produce heat resistant paints capable of withstanding as much as 1200° F. See Table 110.4.

110.5 Silicone Resins. Because carbon, the basic building block of organic chemistry, and silicon occupy the same position on adjoining rings of the periodic table—which is an indication that they should have closely parallel reactions—it was theorized that a whole body of chemistry could be built around the silicon atom, just as one was built around carbon.

An outgrowth of this theory was the development of a family of chemicals known as *silicones*. A silicone is defined as any chemical compound that has two silicon atoms each bound to the same atom of oxygen, but in such a way that the silicon atoms are still free to combine with other atoms or molecules.

Since silicon, like carbon that is similarly bound to oxygen or hydrogen, has bondable tentacles available to form compounds, a host of interesting silicone resins have been made available to the coatings field.

What Goes in a Coating

Table 110.4 *Salient Facts about Rubber-Based Binders.*

Brushability	F	Gloss retention	G
Odor	Mod.	Color, initial	E
Method of cure	Evap.	Yellowing (clear)	X
Speed		Fade resistance	G
50°–90° F	E	Hardness	E
Below 50° F	G	Adhesion	G
Film build	G	Flexibility	G
Safety	G	Resistance to:	
Use on wood	NR	Abrasion	G
Use on fresh concrete	E	Water	E
Use on metal	G	Detergents	E
Min. surface	Class F	Acid	E
prep'n on metal		Alkali	E
Use as clear	NR	Strong solvents	E
Use in aluminum paint	*	Heat	*
Choice of gloss	All	Moisture permeability	Low
Service:			
Interior	E		
Normal exposure	E		
Marine exposure	E		
Corrosive exposure	G		

*Styrene-butadiene, silicone and chlorinated plasticisers, for high-heat aluminum coatings.

Key:
Excellent .... E
Good .... G
Fair .... F
Poor .... P
Very poor .... VP
Not recommended .... NR
Not applicable X

One important characteristic of these resins is their indifference to oxygen. Because of this they maintain excellent color and gloss even when continuously exposed to temperatures in the 500°–600° range. Styrene butadiene, and aluminum pigments combined with silicone, have withstood temperatures of 1200° F. Apparently, the results volatilize and disappear, leaving a porcelain-like shield of aluminum. Silicone's resistance to oxygen, a great spoiler of films, was soon exploited in coatings requiring long-term weatherability; mainly in factory-applied gloss enamels.

Now, about 15 years after this weatherability was discovered, silicones are being combined with organic resins, notably alkyds, epoxies, and acrylics to make air-dried films for field application on architectural surfaces and baking enamels for the coil coatings used for architectural siding. Some of the latter bear long-term guarantees (usually 20 years) against fading and film failure.

Most of the products based on silicone and alkyds, or silicone and epoxies, or silicone and acrylics are proprietary; hence, they are not offered nearly as widely as the older, more common types of coatings. However, where long-term survival of finishes is important, particularly in a corrosive atmosphere, or where an architect's client is farsighted enough to understand the economy of these expensive materials and is willing to bear their initial expense, they merit serious consideration.

Likewise, the high cost of silicones is justified, usually, for protecting architectural metal, such as bronze doors, or natural metal window frames, or silver or gold embellishments.

110.6 Urethanes. Certain molecules, because of their structure, have what are known as reactive sites. This means that they will react with oxygen, nitrogen, or hydrogen, or other atoms existing in another molecule, or they may have ready reactivity with such combinations—to name a few, as oxygen and hydrogen (known as hydroxyls) or combinations of carbon, hydrogen, and oxygen (known as carboxyls), or combinations of nitrogen and hydrogen.

A whole family of these reactive molecules called the isocyanates fit this description because its members react with materials containing an active hydrogen, that is, a hydrogen atom attached to an oxygen or nitrogen atom. These isocyanates when reacted with suitable materials containing active hydrogen end up as urethane resins.

To get urethanes in proper perspective as coating resins, it may help to think of their simpler forms as superalkyds. They do pretty much what alkyds do, except they often do them better. In fact, these simpler urethanes are often called urethane alkyds, or "uralkyds." These basic urethanes are practically made by substituting an isocyanate for phthalic anhydride—a key ingredient in alkyd manufacture.

When this substitution is made, the resulting product cures faster and makes a harder, more flexible film than its alkyd counterpart and has considerably better abrasion, solvent, and chemical resistance. These uralkyds, however, have their disadvantages, too. They don't hold a gloss on exterior exposure

What Goes in a Coating

as well as alkyds; and they cost more. And, whites and pastels tend to lose their color fidelity.

Recent developments have diminished these shortcomings. First, the use of safflower oil as the fatty acid helps curb yellowing; and expanded use, and lower cost, of nonyellowing isocyanates have helped improve economics of the product. Finally, availability of improved ultraviolet absorbers has diminished the harmful effects of sunlight.

110.6a Moisture-Cured Urethanes. A second member of the urethane coatings group is cured by moisture in the atmosphere. Like the uralkyds, these materials are ready-made; no materials have to be added. They are made in such a way that one component, containing combinations of oxygen-hydrogen, or nitrogen-carbon-oxygen, is packaged in the can with a strong thirst for water to complete its reaction.

Table 110.6a. *Salient Facts about Moisture-Cured Urethanes.*

Brushability	F	Gloss retention	F
Odor	Strong	Color, initial	E
Method of cure	Moisture	Yellowing (clear)	Moderate
Speed of dry		Fade resistance	F
50°-90° F	E	Hardness	E
Below 50° F	G	Adhesion	E
Film build	E	Flexibility	E
Safety	F	Resistance to:	
Use on wood	E	Abrasion	E
Use on fresh concrete	G	Water	E
Use on metal	G	Detergents	E
Min. surface	Class P	Acid	E
prep'n for metal		Alkali	E
Use as clear	E	Strong solvents	E
Use in aluminum paint	NR	Heat	G
Choice of gloss	High	Moisture permeability	Low
Service			
Interior	E		
Normal exposure	E		
Marine exposure	E		
Corrosive exposure	G		

Key:
Excellent .... E
Good .... G
Fair .... F
Very poor .... VP
Not recommended .... NR
Not applicable X

These moisture-cured urethanes when spread in thin films are able to cure in moist air. If the humidity is less than 30 percent, these materials take too long to cure; while if the humidity is too high, curing occurs so fast that bubbly surfaces may be produced. This can be a touchy product to use, unless the formulation contains a slowly volatilizing solvent to regulate the speed of cure.

An example of this type of urethane is an exterior varnish often used for redwood. Usually, these materials are harder than the uralkyds, have better chemical resistance but do not weather as well. They also pose a possible problem: excess isocyanate may be released; and this is irritating. See Table 110.6a.

110.6b Heat-Cured Urethane. A third type of one-package urethane is heat-cured; and since it is used primarily for coating magnet wires for coil windings, we will ignore it.

110.6c Two-Package Urethanes. The final two types are described as two-package systems, since some of the components must be separated to prevent premature curing.

The first of these is similar to the one-package, moisture-cured urethane, except that a catalyst—cobalt or lead naphthenate—is added to speed the reaction. The advantage of this moisture-cured version over the one-package is a substantially harder film.

The final form of urethane resin is a two-component system with more versatility than the others. By using a single unchanging isocyanate as one component and by varying the content of the polyester used as the second package, we have a potentially broad selection of products. Because two-package urethanes of the polyester type can be designed for extremely hard, yet flexible coatings, results resembling high quality baked films can be obtained with them. Sometimes they even surpass these baked films. Originally all urethanes tended to discolor to some extent upon exposure to sun and weather. Excellent new two-package versions substitute less-sensitive curing agents—known as aliphatics—for the offending aromatics. Some aliphatic urethanes are highly solvent and stain resistant and are used on public buildings as graffiti-preventive coatings.

What Goes in a Coating

110.6d Descriptions of Urethane Resins. To help understand the various types of urethanes, a committee of the American Society for Testing Materials was formed to define and characterize them. Table 110.6d was developed to provide broad descriptions of the five basic types covering a considerable number of variants.

The possible product characteristics derivable from isocyanates can be seen in a two-package system that has in one package a copolymer consisting of vinyl chloride-vinyl acetate and acetate vinyl alcohol combined with an alcohol made from a derivative of castor oil. An isocyanate material treated with another castor oil derivative is contained in the second package. The copolymer in the first package is loaded with the same oxygen-hydrogen (hydroxyl) groups that we found before which are able to trigger the active hydrogen in the isocyanate. As if these vinyl supplied O-H's aren't enough, the castor oil-derived alcohol has its share of O-H's, so the isocyanate's active hydrogen atoms literally have an orgy.

The result is a resin that serves as the binder for a vinyl urethane coating with unusually good flexibility, toughness, and durability as proven by four years' exposure under grueling conditions without significant change.

Recently, two-package moisture-cured coatings were made by supplying O-H (hydroxyls) to the castor-oil-treated isocyanate by adding the proper nitrocellulose and the proper hydroxyl-containing acrylic resin.

As we will see in Chapters 7 and 8, urethanes, like epoxies, polyesters, and silicones, provide a body of sophisticated chemicals offering a coatings specifier materials for all requirements. Urethanes of the moisture-cure type are found on wood floors or as vital components of durable seamless floors, and they can be found on walls where chemical resistance and impact resistance are important; and they can be found as durable exterior films, either pigmented or as clear or dye-containing varnishes.

110.7 Vinyls (Solvent-Thinned). Some purists would use the chemical classification for vinyls and include under that name such important resins as acrylics, polyethylene, polystyrene, and maleates. In the coatings field, the term *vinyl* usually refers to solvent-thinned resins of polyvinyl chloride and their

Table 110.6d. Urethanes.

Description	Air Cured	Moisture-Cured	Heat Cured	Catalyzed Prepolymer	Two-Can Polyester Polyiscoyanate
Free monomer, with possible hazard %	0	3-10	0	3-10	0-1.5
Hardness (Sward)	20-40	20-60	40-85	25-70	30-60
Chemical resistance	Poor-good	Fair-very good	Excellent	Fair-very good	Good-excellent
Weather durability	Good-very good	Poor-very good	Excellent	Poor-very good	Good-outstanding
Pigmentation	Conventional	Slurry grind	Conventional	Slurry grind	Pigment polyol part

What Goes in a Coating

copolymers and/or their modified versions.

Some confusion is caused by the popularity of water-thinned members of the vinyl family, which are usually polyvinyl acetate copolymers; and the confusion is further compounded by the fact that this same vinyl acetate family provides members that are combined with vinyl chloride to make desirable resins for various purposes. If it has vinyl chloride in it and it requires solvent-thinning, we'll call it a vinyl resin. If it is water thinnable and it has vinyl acetate in the molecule, we'll call it vinyl acetate. The latter will be discussed at length in the section on water-thinnables. (See Section 111.)

Familiarity with vinyl resins is important, not so much because an architect or builder is likely to specify it very often for direct application, but mainly because he may want to specify some prefinished building material, such as siding, where he wants ability to withstand dimensional changes—such as the stresses in wood, or extreme drawing, stamping, or bending of metal. Here he may want to require the building material manufacturer to use a factory-applied vinyl system.

110.7a Vinyl Solutions. Siding made of aluminum, or galvanized or cold-rolled steel is often coated with a plasticized solution of vinyl chloride-vinyl acetate copolymer that has been modified to improve its adhesion to metal. If the metal finishes are to be baked, passivating is necessary to lessen future rusting if surface breaks should occur. This is done with wash primers made of vinyl butyral; or maleic acid-modified vinyl copolymers may be used.

For factory-applied wood siding or decorative interior paneling, vinyl copolymers have been effectively modified by castor oil-derived urethanes (see Section 110.6 on urethanes).

Other modified vinyls are added to alkyds and melamine resins to provide unpigmented films for various decorated wood such as paneling, printed hardboard, and chipboard. These films protect against chemicals and household cleaning compounds.

Vinyl systems, including the urethane combination mentioned previously as a primer, have been successful in intensive tests on exterior plywood, even where Southern yellow pine (the least dimensionally stable of woods) was used. The topcoat in this system is a blend of an unmodified vinyl copolymer and another polymer with hydroxyl-modification, which, in plain language, means that its vinyl acetate component was treated to have available oxygen-hydrogen (hydroxyl) matings ready for the vinyl chloride merger. This improves its adhesion to the primer. See Table 110.7a.

Table 110.7a. *Salient Facts about Vinyls (Solvent-Thinned).*

Brushability	None	Gloss retention	E
Odor	Strong	Color, initial	E
Method of cure	Evap.	Yellowing (clear)	X
Speed of dry		Fade resistance	E
50°–90° F	E	Hardness	G
Below 50° F	G	Adhesion	F
Film build	F	Flexibility	E
Safety	F	Resistance to:	
Use on wood	NR	Abrasion	E
Use on fresh concrete	E	Water	E
Use on metal	E	Detergents	E
Use as clear	NR	Acid	E
Use in aluminum paint	G	Alkali	E
Choice of gloss	Low	Strong solvents	F
Service		Heat	P
Interior	NR	Moisture permeability	Low
Normal exposure	E		
Marine exposure	E		
Corrosive exposure	E		

Key:
Excellent .... E
Good .... G
Fair .... F
Poor .... P
Very poor .... VP
Not recommended .... NR
Not applicable X

110.7b Vinyl Dispersions. The foregoing are described as vinyl solutions. Vinyl resins with their copolymers are also dispersed in their liquid phase rather than being dissolved as in the solution type. The dispersions are further divided into two types: *plastisols,* the liquid portion of which consists entirely of plasticizers; and *organosols* which differ from plastisols by being dispersed in volatile organic solvents.

Plastisols are of interest where films with high impact resistance and tensile strength are needed, particularly if flexibility is also desired.

Organosols offer higher pigmentation possibilities and a greater range of film hardness. Because they can be effective at low viscosity, they are usually easier to apply.

Plastisols and organosols are used entirely for factory-coating, since heat is needed to fuse the resin to the substrate. Fusion is believed to occur when the molecules of the liquid in which the resins are dispersed get so agitated by the heat energy that they change their role from dispersant to solvent. Each combination of solvent and resin has its own fusion temperature. Baking time is determined by the mass of the material and its thermal conductivity.

The tensile strength of the baked film is directly related to the completeness of fusion. Hence, in dealing with factory-finished materials coated with organosols or plastisols, it is important that the finisher keep careful control of his bake schedule. This can be confirmed by testing for tensile strength. (See the earlier discussion of plastisols and organosols in Section 107.)

110.8 Polyvinyl Acetals. Of the two forms of polyvinyl acetals used in coatings—*polyvinyl butyral* and *polyvinyl formal*—only the former is of interest to the construction field because of its use as a metal conditioner, and as a sealer for wood finish systems, including use as a knot sealer.

110.8a Vinyl Butyral. The most important use of vinyl butyral resins is in metal conditioners, or wash primers. Applied as extremely thin films containing a zinc chromate rust inhibitor, they serve the dual purpose of passivating, or preventing future undercutting by rust and making possible the use of coating systems that would not otherwise adhere to metal.

Vinyl butyral wash primers should never be more than 0.0005 inch (one-half mil) thick. They should always be overcoated with primer and topcoat prior to outside exposure, but indoors they can serve as a temporary protection for metal that has been sandblasted or pickled prior to finishing. They provide an excellent anchor for phenolic, oleoresinous, and alkyd paints. Sometimes these resins serve to modify coatings that would not otherwise adhere to vinyl butyral; for example, nitrocellulose lacquers would not adhere to it without having an alkyd modifier.

Most vinyl chloride resins described in the previous subsection do not adhere to these wash primers. The exceptions are the hydroxyl-modified versions (the ones with oxygen-hydrogen pairs provided by special treatment). These versions are used to blend with other vinyls to provide adhesion. Acrylic resins, in general, do not adhere to vinyl butyrals.

111 Water-Thinned Resins. (Latex)

111.1 General. Water-reducible architectural coatings are generally known as *latex* paints, and the resins on which they are based are emulsion resins. The term *latex* was originally used to describe the milky form in which natural rubber is found; this is an emulsion of about 35 percent rubber and 60 percent water. The suspension of natural latex is aided by the other 5 percent, which are foreign substances that apparently provide the kick to keep the rubber particles separated and scattered throughout the liquid.

Emulsion resins are literally born as emulsions, because they are formed in little droplets made up of water combined with an emulsifying agent and the monomers, which are minuscule particles of basic materials that join together to form the resin.

Like natural latex rubber whose tiny particles are separated and stabilized by certain natural materials, the newly born emulsion resin is suspended by emulsifiers, which can take hold of water molecules on one end of their structure and resin on the other, like a mother holding one pugnacious but large child with each of her hands, separating the children as they struggle to punch each other in the nose. For this emulsifying job, surfactants are used; these are very closely related to—and sometimes are—detergents, whose function is to take grease and water by the hands and march them together away from soiled surfaces.

Since the resins formed in the tiny emulsified droplets are longer and more sinuous, and could easily intertwine and overwhelm the almost infinitely small emulsifier, the polymer chemist must use materials known as *protective colloids*, which are water soluble and, thus, slither in and among the water molecules much more readily than the merely dispersed resins. These are added as soon as the batch of emulsion is completed, to keep the tiny tendrils of resin off each other's backs.

What Goes in a Coating

These protective colloids are added in minute quantities to the resin, but later at the paint plant when pigments are added small additional quantities of them enter the picture as the paintmaker tries to keep the pigment particles completely coated so they will not combine and form small lumps, or agglomerations.

Protective colloids such as methylcellulose, hydroxyethylcellulose, sodium carboxymethylcellulose, sodium alginate, vinyl alcohol, among others, even when held down to as little as 0.1 percent of the paint, impart characteristics to the coating that counteract its settling tendency and improve its ability to flow onto the substrate from the application device—brush or roller. Usually, about 1 percent of a latex paint is a protective colloid.

The blessings brought by these protective colloids, as is

Table 111.1 *Salient Facts about Water-Thinned Paints (Latex).*

Brushability	E	Gloss retention	°
Odor	Mild	Color, initial	E
Method of cure	Coalescence	Yellowing (clear)	X
Speed of dry		Fade resistance	E
50° F–90° F	E	Hardness	G
Below 50° F	P	Adhesion	G
Film build	G	Flexibility	E
Safety	E	Resistance to:	
Use on wood	E	Abrasion	G
Use on fresh concrete	E	Water	F
Use on metal	NR°	Detergents	G
Min. surface	°	Acid	G
prep'n on metal		Alkali	G
Use as clear	NR	Strong solvents	G
Use in aluminum paint	NR	Heat	G
Choice of gloss	Flat°°	Moisture permeability	High
Service			
Interior	E		
Normal exposure	E		
Marine exposure	F		
Corrosive exposure	NR		

°Exceptions can be used.
°°Acceptable semigloss are available.

Key:
Excellent .... E
Good .... G
Fair .... F
Poor .... P
Very poor .... VP
Not recommended .... NR
Not applicable X

usually the case with benefits, come at some price. Because they are water-soluble and stay in the dried film, they leave a trail of water-sensitive resin winding throughout the entire film, thus affecting its scrub resistance and washability. See Table 111.1.

111.2 Formulation of Latex Paints. Some idea of the care required to formulate latex paints can be had by reference to what goes in them and what can go wrong.

After the pigment is ground with the protective colloid and a wetting agent is added to drive off the air surrounding the pigment particles and to fill the existing voids, latex is added to it. This could easily cause lots of foaming, because emulsifiers are surface-active agents and notorious bubble-formers; so an anti-foaming agent or bubble-breaker must be added.

If the mixture turns acid, the protective colloid will become insoluble, and the latex particles will strangle each other; so an alkali, such as ammonia, will be added to keep the pH on the alkaline side.

A bactericide is needed to combat the bacteria that produce enzymes which rip apart the cellulose thickeners while the paint is still in the can, leaving a water-thin mess that can't be applied; and a fungicide is needed to prevent mildew, if the latex is to be used outdoors.

If the suspended droplets of resin emulsion are to come together and form a continuous protective film, help is necessary; so a coalescing agent must be added. With so much water present, the canned paint may freeze in winter if an antifreeze is missing; and finally, antirust agents are sometimes needed to keep the can from rusting.

It's a testimonial to the wizardry of the paint chemist that he summons the accumulated knowledge of the physical chemist (colloids), the biochemist (fungicides and bactericides), the rheologists (thickeners), the polymer chemist (latex, coalescing agents, and plasticizers), and the mathematician (computing components to get balanced results)—and he ends up with a coating that works, and works well.

111.3 Using Latex Paints. If an architect or builder decides to utilize latex paints, he has one important responsibility that should not be minimized. He must take steps to prevent application when atmospheric temperatures are below 40° F. Certain coalescing agents—mainly the higher glycols and their ethers—are sometimes used to permit coalescence below 40° F, but care must be exercised to see that the product is designed for these temperatures, which, in any case, won't be less than 36° F.

Another important step in utilizing latex paints is to make certain that the personnel applying the paint spread it on heavily enough. Because emulsion paints, as a general rule, spread more readily than oil paints, painters, either intentionally or otherwise, tend to get more "mileage" per gallon. The result is too thin a film and less protection and durability.

Insistence that prime contractors use only established, highly reputable painting contractors is important on any construction project; but where latex paints are concerned extra care is necessary, even to the point of using inspectors with film-thickness gauges.

(The curious may be interested in the reason for the high spread rates of seemingly thick latex paints. Remember our friends Pigment Volume Concentration and Critical Pigment Volume Concentration encountered earlier? Well, for a number of complex reasons, the CPVC values are usually low for latex paints. As a consequence, they have less pigment and extender in relation to resin. This means, unfortunately, that with low-price water to cut the resin a paint can be made very spreadable at low cost.

(In fact, in a typical latex house paint, water constitutes about the same 14 to 15 percent by volume in the finished product that mineral spirits constitutes in a solvent-thinned house paint, but the temptation to add an ingredient as inexpensive as water is stronger and potentially more profitable than the temptation to use mineral spirits. [See Table 111.3, "Variables Affecting Durability of Interior Flat Latex Paints" and footnotes.]

(So, then, if PVC is relatively low, why do latex paints seem creamy, and sometimes buttery? The answer lies in various additives. First of all, in addition to the nearly 15 percent water, a typical can of latex house paint has about 20 percent fluid constituents that, in addition to their basic role, give it body but add nothing to the film except the negative factor of water sensitivity.

Table III.3. Variables Affecting Durability of Interior Flat Latex Paints.*

Formula	TiO$_2$ (lbs.)	Clay (lbs.)	PVC	Resin (lbs.)	Water (lbs.)	Scrubs
a. High hiding						
1	250	200	52.3	364	204.5	2,300
b. Good scrub resistance						
2	280	120	56.0	279	197.5	1,900
c. Good burnish resistance						
3	230	150	53.4	301.3	221.5	1,800
a. Good scrub resistance						
4	225	125	47.5	364	149	3,800
5	180	130	55.0	254.7	129	1,800
b. Low cost						
6	75	233	58.2	230	110.5	800
a. Top quality						
b. High hiding						
7	275	175	57.0	300	112.5	1,100
This is federal specification TTP 29D.						
8	125	315	56	282	203	1,000

*Based on formulations using Rhoplex AC-22, an acrylic emulsion, supplied by Rohm & Haas Co.

Note: It is possible to indicate the usual cost of the formulas listed above. Formula 6 is the least expensive on a per gallon basis; the costs of the other formulas are listed as the cost of Formula 6 plus an additional increment: Formula 1, $.65; Formula 2, $.58; Formula 3, $.48; Formula 4, $.58; Formula 5, $.32; Formula 7, $.69; Formula 8, $.18.

What Goes in a Coating

Note: Formulas 2 and 7 have the most titanium dioxide among those in the chart above. Since TiO_2 is the premier hiding pigment, these formulas obviously have the highest hiding power; but Formula 7, although costing more to make than Formula 4, has far less scrubbability, the usual indicator of interior durability. The reason is that Formula 4 uses 364 lbs. of resin per hundred gallons of paint, while the more-costly Formula 7 uses only 300 lbs.; and the resin content is related to interior durability. Formula 4 and Formula 2 have the same raw material costs, but Formula 2 can take only one-half as many scrubs. The reason for this is its substantially higher pigment volume content (PVC)—about 20 percentage points more; and it is generally understood that scrub resistance (durability) of an interior paint improves with lower PVC. So how would a specifier select a formulation from among the eight listed above? 1. If he wanted highest hiding in a single coat, he would select Formula 7 because it not only has the most TiO_2, but it also has the lowest water content of the high TiO_2 formulas; 2. If he wanted relatively high hiding while making his paint go further than Formula 7, he would select Formula 2, because it has 197.5 lbs. of water per 100 gallons of paint as against 112.5 lbs. in Formula 7; 3. If he wanted the best all-around interior flat latex, he would probably want Formula 1, because it has excellent scrub resistance (durability), high hiding (since it has 250 lbs. of TiO_2), and as much resin (364 lbs.) as Formula 4; and, in addition, it has 204.5 lbs. of water as against 149 for Formula 4 and 112.5 for Formula 7, so it will cover more surface than either. The specifier has to decide what he needs most: high hiding at all costs—then Formula 7; maximum durability—then Formula 4; best compromise between hiding and durability—then Formula 1. If he wants the cheapest paint for a miserly builder, then Formula 6 is a miserably successful choice, because it will make the builder think he is getting a bargain, and it will serve him right because: (1) He will get poor spreading since Formula 6 only uses 110.5 lbs. of water; (2) he will have to use several coats, which means the labor costs will be multiplied; and (3) he will have to repaint soon. If the specifier is selecting paint for a federal structure, he may be able to get by with Formula 8, since that meets Federal Specification TTP 29 D, governing selection of interior flat emulsion paints. For some reason, federal specifications are minimal for common everyday housepaints, and conscientious specifiers should try to exceed them for their own requirements; but in sophisticated high-performance coatings, such as epoxies, urethanes, vinyls, polyesters, and solvent-thinned rubbers, federal performance and material demands are, in the main, exemplary.

(One of these additives, usually a cellulosic, is what's known as a thixotropic agent, which gives the paint enough body to adhere to the brush or roller without dripping and to the surface being painted without running or sagging. This adhesion occurs because for a short time, as the coating is picked up on the brush and as it is spread on the surface, a remarkable development occurs: the creamy paint thins out for the instant that that pressure is applied.

(It's a neat trick and depends on the pressure of the brush on the paint in the can to thin the paint so that it will adhere to the bristle and then thicken again once the brush pressure stops. Then, when the brush with the thickened paint is momentarily drawn across the surface to be painted, the pressure exerted by the painter provides the energy to thin the paint again for spreading. When the brush pressure is removed, the paint film, by now only a few thousandths of an inch thick, merges the various bristle marks and thickens again, leaving a smooth, level coating.

(That ability to go through thick and thin is known as thixotropy.)

111.4 Polyvinyl Acetate. Most important of the three major resin families used for latex paints are the copolymers of vinyl acetate. These may consist of vinyl acetate monomer combined with maleic anhydride, one of the acids used in alkyd manufacture; or it may be combined with fumaric acid, which is closely related to maleic anhydride; or finally, it may be combined with an acrylate monomer to get some of the characteristics of acrylic latexes, which, under their own name, are more expensive, more widely advertised, and more glamorous in the eyes of the consumer.

The various copolymers are needed because vinyl acetate alone is too hard and requires plasticizing for flexibility. The maleic or fumaric copolymers were widely used as internal plasticisers until the mid-1960s when the acrylate copolymers grew in importance as exterior latex house paints began to replace linseed oil paints. Many paint manufacturers, to simplify production, began using the polyvinyl acetate acrylates for interior as well as exterior paints, a tendency that was strengthened when chemical manufacturers over-extended their produc-

What Goes in a Coating

tion facilities for acrylate components, and their prices dropped as a result.

Some claim the maleates and fumarate versions are as good as the acrylates, but hardly anyone says the acrylates are less good than the others.

In recent years, resin manufacturers have been offering copolymers of ethylene and vinyl acetate, which are reported to have desirable pigment-dispersing properties in addition to greater film strength when applied at low temperatures. Prospects for widespread growth of this latex are limited because few paint companies have the sophisticated heat and pressure reactors needed to achieve the polymerization of vinyl acetate and ethylene; whereas many national and regional companies can make their own polyvinyl acetate copolymers with maleic anhydride or fumaric acid or an acrylate, because these require the same relatively simple reactor used for making alkyds—which many companies have.

Consequently, the makers of polyvinyl acetate/ethylene latex have only been able to sell these materials to a limited number of regional manufacturers who have not been making their own resins.

For many years, polyvinyl acetate latexes were used only on interior walls and in flat paints only. After 1965, successful versions were introduced for exterior use where they have joined the so-called straight acrylic latexes and, together, they account for well over half of the exterior house paint market.

For several years, polyvinyl acetate manufacturers have been trying to crack the market for enamels, but their success has been limited. However, straight acrylics (as distinguished from polyvinyl acetate/acrylates) have garnered a sizable part of the semigloss enamel market.

111.5 Acrylics. The so-called straight acrylics result from the chemical combination, or copolymerization, of two acrylates. The principal acrylate used in paint-grade resins is ethyl acrylate, which provides color stability and resistance to heat, light, weathering, and most of the other troublesome influences contributing to the deterioration of a coating. It also imparts flexibility and strength.

The trouble with ethyl acrylate alone is that it is too soft to provide abrasion and scrub resistance. To overcome this shortcoming, it is combined with methyl methacrylate, a hard, tough, chemical compound whose major claim to fame is that it is the main component in Plexiglass and Lucite.

By adjusting the proportions of these two components, the resin maker can come up with a whole family of outstanding latex paints. One combination makes an excellent house paint; another yields a fine interior paint; and still another is the basis for an interior semigloss enamel that has been making ever-widening inroads into the once-sacred alkyd preserves.

Acrylic latex paints are, in the main, more expensive than polyvinyl acetate because the acrylate monomers are more costly.

Some knowledgeable chemists say that the performance of the acrylic latex paints, in many circumstances, does not justify the extra cost, as compared with polyvinyl acetate/acrylate copolymers. It is generally acknowledged, however, that for cement surfaces containing salts that may emerge later and cause what is called *efflorescence* (a dewy, unattractive look), the acrylics are more effective. Also, tests have shown that they offer better, more trouble-free coatings over southern yellow pine, which is a bothersome, low-cost substrate with irregular hard and soft growth layers that expand and contract.

111.6 Additional Latexes. The probability that harsh air-pollution regulations will eliminate solvents needed for high-performance maintenance paints has led to development of epoxy latex and urethane latex. Although many epoxy and urethane characteristics are claimed for them, detailed test data should be required before use, since performance tests require years of exposure.

112 Pigments.

112.1 General. The function of a pigment in a coating is either to hide to a greater or lesser degree the underlying surface or to serve as a chemical capable of combining with some substance in the coating to form a third function, such as corrosion-inhibition or ultraviolet absorption, or, finally, to help decorate.

Its hiding function may vary: (1) from complete obliteration of the surface; (2) to partial concealment, where the coating is made translucent by certain opalescent materials; or (3) to altered transparency, as in the use of small amounts of

What Goes in a Coating

iron oxide in varnish to stain it red, brown, or yellow.

Some pigments hide and decorate; others hide and decorate and also combine chemically with one or more components of the mixture. To some extent, all white pigments hide and decorate; but some, such as zinc oxide, and all of the white lead pigments, and antimony oxide, also are reactive and may become components of chemicals that serve some other purpose in a coating.

The white pigments can be used for complete hiding (we're not considering extender pigments—clay, calcium carbonate, etc., as white hiding pigments, although under some circumstances they may hide), or they can help play a role in modifying transparency or translucency.

Most colored pigments, on the other hand, will not hide by themselves, and usually require some white hiding pigment—almost always titanium dioxide—to impart hiding.

That's why titanium dioxide is the kingpin of the pigment field. Coatings in this country, in 1972, probably required about 800,000,000 lbs., or about 60 percent of all titanium dioxide produced in the U.S.

112.2 Titanium Dioxide. Starting about 1932, titanium dioxide began to assume dominance in the white hiding-pigment field, replacing zinc oxide, lithopone, and zinc sulfide. Its early acceptance was hindered by the severe chalking that characterized the first form in which it was available—the anatase. Acceptance improved as modified grades reduced chalking in exterior usage, where a limited amount of this surface erosion is helpful. The coating as it chalks becomes self-cleaning inasmuch as the chalk together with surface dirt is usually washed off by the rain.

The benefits of titanium dioxide really became outstandingly advantageous only when the rutile form became available. Now, anatase is used only in limited amounts and in exterior paints only to take advantage of its chalking and self-cleaning. Here, a fault kept under control becomes a virtue.

112.2a Hiding Power. Rutile, besides being nonchalking, hides more effectively; and, hence, less of it is required to achieve a given degree of hiding.

Table 112.2a shows the hiding power in square feet per

pound of various white hiding pigments. These figures are accepted for comparison purposes and are based on the performance in a given paint system. However, the actual hiding power will vary with the concentration in the paint film.

The basis for titanium dioxide's superior hiding efficiency is its refractive index, or its ability in a coating to bounce back all the components of white light that are radiated to it from a natural or synthetic source without absorbing any significant amounts of those light components.

Table 112.2a. *Hiding Power of White Pigments.*

Pigment	Hiding Power (Sq.ft./lb.)
Rutile titanium dioxide (special)	157
Rutile titanium dioxide (normal)	147
Anatase titanium dioxide	115
Zinc sulfide	58
Lithopone	27
Antimony oxide	22
Zinc oxide	20
35% Leaded zinc oxide	20
Basic carbonate white lead	18
Basic sulfate white lead	14
Basic silicate white lead	12

112.2b Light Refraction. To understand this, we must recognize that red surfaces, for example, are red because they absorb all the wave lengths of light, except the red component, which is bounced back to the viewer. The same limited bounce-back factor determines other colors, except white, which, as we said, bounces back the entire visible portion of the spectrum.

But other white pigments also have refractive indexes that let them bounce back the entire visible span of the spectrum. In a coating, however, another factor enters in to improve the efficiency of the bounce—and that's the refractive index of the binder as well. The *wider the gap* between the refractive index of the pigment and that of the binder the higher will be the hiding power of the pigment in a system using that binder.

Most vehicles used in coatings have a refractive index ranging from 148 to 168. Table 112.2b shows the refractive

What Goes in a Coating

indices for various white pigments and some major resins. Since the widest spread is between titanium dioxide and all the binders, and since the spread between the indices of the pigment and the binder has been found to be the determinant of hiding power in a coating, it follows that TiO_2 is the winner.

Table 112.2b. *Refractive Index of Common Coating Materials.*

	R.I.
Hiding Pigments	
Rutile titanium dioxide	2.76
Anatase titanium dioxide	2.55
Zinc sulfide	2.37
Antimony oxide	2.09
Zinc oxide	2.02
Basic lead carbonate	2.00
Basic lead sulfate	1.93
Extenders	
Barytes (barium sulfate)	1.64
Calcium sulfate (gypsum)	1.59
Magnesium silicate (talc)	1.59
Calcium carbonate	1.57
China clay	1.56
Silica	1.55
Binders	
Phenolic resins	1.55–1.68
Melamine resins	1.55–1.68
Urea-formaldehyde resins	1.55–1.60
Alkyd resins	1.50–1.60
Natural resins	1.50–1.55
Tung oil	1.52
Linseed oil	1.48
Soybean oil	1.48

Even at that, TiO_2 must conform to some conditions in order to perform most effectively. If particle size is too large, opacity will be reduced because too few opportunities will be present for the pigment and the binder to combine to use their joint power to bounce back the oncoming white light.

The smaller we have the particle size, naturally, the more pigment surfaces we have and the more chance there is for pigment-binder interfaces to form; and that's where the bouncing occurs—where the two come together. It has been found that if the particle size is too small, the TiO_2 will lose its hiding power; so a theory for an optimum particle size was developed. This called for the size to be about half the wavelength of visible light, or about 0.3 microns, a theory that calls for an assumption—that each randomly separated particle was about 50 wavelengths of light away from each other.

Well, theories are fine, but practical pigment men found from experience that the best results for white opacity are obtained when the pigment particle size of TiO_2 is about 0.15 to 0.20 micron. That size provides the greatest number of opportunities in a square foot, let's say, for the surfaces of the pigments to team up with the surrounding binder to toss back the greatest amount of whole white light to the beholder.

112.2c The Cost of White Pigments. Whether or not the reader has taken the trouble to understand the foregoing discussion of the physics of hiding pigments, suffice it to say that titanium dioxide is far and away the most effective hiding agent available for coatings.

Now we come to the reason for wanting to establish TiO_2's role in hiding. This hiding ability makes it necessary for the architect or builder to make certain that TiO_2 isn't overused at the expense of other white pigments that have some helpful roles to play.

Zinc oxide, for one, is a white hiding pigment that is only about one-eighth as effective as rutile titanium dioxide in hiding; but it is a chemically reactive pigment, unlike the nonreactive TiO_2, and plays a specific role in house paints: for example, it chemically reacts with acids in linseed oil, or other acids, and imparts color fidelity and mildew resistance.

Since zinc oxide is usually significantly less expensive than TiO_2, and performs a chemical role as well as hiding well, it is natural to ask the question: who would want to substitute the more expensive TiO_2 for zinc oxide, which has a definite role to play?

That's because even at 50 cents per pound, TiO_2 hides a square foot of surface for less money than a pound of zinc oxide or other white pigment. Look at Table 112.C. It takes 35 lbs. of rutile TiO_2 to fill a gallon pail, but it takes 47.1 lbs. of zinc

What Goes in a Coating

oxide and 35.8 lbs. of lithopone to fill the same pail. Get the idea? The paint manufacturer buys his raw materials by the pound, through no fault of his own, but he sells it by the gallon. The architect or builder has to buy it by hiding power, or power to protect. When he buys zinc oxide, he buys protection, and he should be certain that when this is needed he does not get a non-reactive, non-protective pigment.

The figures in Table 112.2c show that a gallon pail full of rutile TiO_2 (the kind that doesn't chalk), when used in the proper amount of paint will hide 5,845 sq.ft.; and that this hiding is at a cost of one dollar per 334 sq.ft. of hiding. By comparison, the same gallon pail of zinc oxide will hide only 942 sq.ft., at a cost of one dollar per 55 sq.ft. The white lead filling up a gallon pail will provide enough hiding for 1,000 sq.ft, but its higher per-pound cost means that a dollar will only buy 88 sq.ft. of hiding.

So coverage by rutile TiO_2 is one-sixth as costly as zinc oxide; and one-third that of moly white. It, therefore, behooves a specifier to make certain that the benefits of zinc oxide and moly white are not lost in those paints where they are important because a more efficient, less costly whitener is used. It may be necessary to specify their quantity, as is done in some of the specifications outlined in Chapters 7 and 8 and in the Detailed Specification Charts at the end of the book.

Table 112.2c. *Volume vs. Weight; Cost vs. Hiding*

Pigment	Lbs. per gal.	Cost per lb.	Cost per gal.	Hiding per gal. (sq.ft.)	Hiding* Power per $ (sq.ft.)
Rutile TiO_2	35.0	$0.50	$17.50	5845	334
Anatase TiO_2	32.5	0.48	15.60	3738	239
Lithopone	35.8	0.25	8.95	967	108
Zinc oxide	47.1	0.36 ½	17.19	942	55
Calcium molybdate (aqueous)	24.9	0.58 ¼	14.50	1461	101
Antimony oxide	47.9	1.80	86.22	1054	12
Calcium molybdate (oil-base)	42.1	0.89	37.47	1461	39

*These are based on 1978 prices. The hiding power per $ will vary as prices change, but the principle will remain.

Table 112.2c also shows the relative cost of hiding power. The last four white pigments listed are the reactive pigments.

Their deficiencies as hiding pigments are very clear. Their virtues as reactive chemicals will be described after we leave the nonreactive pigments.

A few years ago the best bargain among non-reactive hiding pigments was a material known as C-pigment, or rutile calcium. Rutile calcium was described as a composite pigment, which means that although it was made of calcium sulfate (gypsum) and TiO_2, it was formed at the same time and was so intimately compounded that it was virtually one substance. The actual product had a core of calcium sulfate, with the hiding TiO_2 deposited around it. The gypsum core, in other words, provided a built-in extender; and for a number of involved reasons based on optical physics, high hiding was achieved at minimum cost.

For certain types of paint, oil-reducible interior paints mainly, the so-called C-pigments were universally favored. Then, came a sad day, and the sole manufacturer of these materials came to a rude awakening and found out that it was costing money to provide these efficient products; so the company did what any self-respecting business firm would—it stopped production.

All of this economic background, and a tale of woe, is recounted to alert specifiers that a lot of tried and true interior formulations based on C-pigments have been changed. The C-pigment producer—and all other rutile titanium dioxide manufacturers—devoted a huge effort to developing substitute formulations. They were aided, also, by manufacturers of clay and other extenders, who were anxious to fill the gap left by the demise of C-pigments—a market encompassing an estimated 100,000,000 lbs. of rutile TiO_2, plus about 200,000,000 lbs. of extender.

All of which gets back to the fact that many formulations have changed. It is highly probable that most established regional and national coatings manufacturers for some time have had product formulations in their files that have replaced those with C-pigment, and that have been thoroughly tested. All of the established companies are extremely careful about maintaining their reputations for dependability; and the changes that are almost certain to have been completed by now will result in equivalent products at the very least.

What Goes in a Coating

112.3 Other Nonreactive White Pigments. These include lithopone and zinc sulfide.

112.3a Lithopone. Two other more or less inert white pigments have had their day. Lithopone, which is zinc sulfide on a barium sulfate or calcium sulfate base, was the first product made by depositing a hiding pigment on an inert core, the same method that made C-pigments so advantageous.

The reason C-pigments were used in preference to lithopone is the latter's *low bulking value*. That means that it has a tighter molecule than the bulky C-pigment. It takes 35.8 lbs. to make a gallon of lithopone, as against 27.1 lbs. of 30 percent C-pigment. And as Table 112.2c shows, the hiding power dollar spent for lithopone buys coverage of 309 sq.ft., as against 625 for 30 percent C-pigment.

Lithopone has not been put in limbo, however. It serves several purposes, virtually as an additive, rather than as a reactive pigment, which it is not. Primarily, it seems to hold other white pigments in an intimate mixture with colored pigments and prevents separation in the can and on the protected surface. This latter separation, called *floating*, is a condition in which colored pigments slip away from the necessary white hiding pigments and form ugly swirls and blotches.

Lithopone serves another helpful function, it aids washability; and it is required in some government specifications that call for prolonged washability.

112.3b Zinc Sulfide. For awhile paint formulators wanted more concentrated whiteness than lithopone could offer so they demanded its parent—zinc sulfide, which was actually responsible for lithopone's hiding. Zinc sulfide's slight yellowish cast virtually eliminates it today from white paints, but it is used occasionally in tinted house paints, where it chalks and eventually causes failure.

Also, a limited amount is used to increase the zinc sulfide content of paints needing lithopone.

112.4 Reactive White Pigments. The reactive white pigments hold a low place on the hiding totem. With one exception, their importance comes from the function served by the end-products they make in combination with acid components in the formulation. The exception is antimony oxide, which serves as a catalyst in certain fire-resistant paints.

112.4a Basic Carbonate White Lead. Old-time painters, the real old hard-handed fellows, still swear by white lead and linseed oil for outside house paint, but their number has dwindled, and no matter, since legislation will soon ban lead.

Today, white lead, or basic carbonate white lead, is out of house paint despite its extensive use by modern formulators in exterior house paint primers. Various public interest groups have succeeded in getting Congress to ban it from interior and exterior surfaces of residences. The reason is to protect unsuspecting youngsters from biting and swallowing scrapings or chips with significant quantities of toxic, or potentially toxic, lead compounds. As a result, the very desirable benefits of white lead are now lost to users.

If so, an ancient product will have had its term, because this material goes all the way back to ancient Greece where early pigment makers developed a process that, with refinements to speed things up, endures to this day.

Simply put, you expose metallic lead to vinegar (acetic acid) and at the right time carbon dioxide is reacted with the resulting lead acetate. The Dutch refined the process by speeding the reaction through the use of manure, to provide natural heat, and lots of carbon dioxide. Today, a more sophisticated system is used involving anodes and cathodes, but it all goes back to acetic acid, lead, and carbon dioxide.

The resulting white lead rarely constituted more than 35 percent of the total pigments used in a house paint, because that seems enough to react with fatty acids in the vegetable oils to form the lead soaps that impart toughness to a film and help make it elastic and able to adhere to new surfaces; moreover, the white lead reacts with acidic breakdown products in the paint and stabilizes the dried film, adding to its life.

112.4b Basic Sulfate White Lead. Performance of basic sulfate white lead is about the same as basic white carbonate, except that it erodes, or chalks, more quickly. It has the advantage

What Goes in a Coating

of filling out a gallon container with only 53 lbs., as against 56.2 lbs. for the carbonate. It is important mainly because it is a component of basic silicate white lead.

112.4c Basic Silicate White Lead. Now, we're back to our composite pigments again: where bulky, low-priced fillers provide the heft and a performing pigment gives the zing. The reactive performer here is lead sulfate, and the filler is a silica.

It was reasoned that since the various white lead pigments use only their surfaces to react with fatty acids, why not put a thin surface on a dead core. The result was the composite, represented here by basic silicate white lead, and also later we will find the principle used for basic silico lead chromate, a colored pigment that has added significant protective scope to anticorrosive coatings.

Basic silicate white lead, because of its silica core, can fill a gallon container with only 33.3 lbs., as against the carbonate's 56.2 lbs.; and since it reacts at pretty much the same rate as the more tightly packed molecules of basic white lead, it is much less expensive for those uses where its poor hiding does not matter. These are usable in industrial coatings.

112.4d Dibasic Lead Phosphite. A bad bargain when it comes to hiding and the number of pounds required to round out a gallon container, dibasic lead phosphite is unique among white pigments: It offers protection against corrosion, particularly in areas where salt water is a problem. This lead salt of common phosphorous acid also stabilizes chlorinated resins, thereby slowing degradation. These, too, can be used in industrials.

112.4e Zinc Oxide. Like the various white lead pigments, zinc oxide combines with the acid components in paint to form zinc soaps, which contribute several helpful characteristics, particularly to exterior house paints.

First, the zinc soap improves film hardness and helps retard chalking; it also adds brilliance to white paints and helps retain color integrity, because it absorbs a considerable part of the sun's more damaging ultraviolet rays. It adds body to a paint and aids in grinding and mixing, and, finally, it is effective in combatting mildew.

Extensive tests have shown the ability of zinc oxide to prevent yellowing and to maintain film integrity—both in linseed oil house paints, which are solvent-thinned, and in latex house paints, which are water-reducible.

Nonetheless, until about 1970, few latex house paints contained zinc oxide, partly because it formed harmful soaps, which broke down emulsions and ruined the paint. Years of research overcame the problem. Then, testing by the most prestigious paint companies under the most rigorous standards proved to their satisfaction that house paints with zinc oxide impart superior durability and color retention. Now many companies are offering them, although their raw material costs are somewhat higher.

112.4f Calcium Molybdate, White. A non-toxic, anti-corrosive pigment, this product came along in the mid 1970's just as lead and chrome pigments came into question by environmentalists and toxicologists.

Because of poor hiding, moly white—as it is known—has relatively little effect on the color of primers, making it easy to cover them with one topcoat.

Moly white is believed to protect iron via a complex process by which simple corroded iron is converted to a protective ferrous-molybdate compound. As the process continues, the entire surface becomes a protective barrier. One form of moly white is intended for water-reducible coatings; the other is for solvent-reduced.

112.4g Antimony Oxide. Classification of antimony oxide as a reactive pigment is questionable because it doesn't react in the true sense of the word, inasmuch as it does not serve as a component of a new compound, as do lead and zinc oxide.

It does serve a role other than hiding, which it does very poorly. Antimony oxide apparently acts as a catalyst, causing other chemicals to have desirable functions; for example, it was once used in exterior paints and automotive enamels to improve chalk resistance. Now, its main pigmentary use is as a catalyst in certain fire resistant paints used almost exclusively by the U.S. Navy. The antimony oxide is believed to catalyze the chlorine in the special alkyd used for this purpose to delay

What Goes in a Coating

ignition of the coating and the surface.

112.5 Colored Pigments. To simplify a very involved subject, the arrangement of colored pigments used for coatings will be divided into organic and inorganic types, and then each color will be arranged to afford comparisons of individual materials by performance:

Inorganic

(Lead-Containing)		(Lead-Free)
Chrome Yellows	Iron Oxides	Siennas
Chrome Oranges	Cadmiums	Chrome Oxides
Chrome Green	Iron Blue	Nickel Titanate
Molybdate Chrome Orange	Ultramarine Blue	Zinc Yellow
Red Lead		
Basic Silicon Lead Chromate		

Organic

Lithols	Benzidine Yellow	PTA (phototungstic)
Para Reds	Nickel Azo Yellow	PMA (phosphomolybdic)
Toluidine Red	Quinacridones	Indanthrones
BON	Arylide Red	Phthalocyanines
Hansa Yellow	Thioindigoids	Anthraquinones

112.5a Advantages of Organic and Inorganic Pigments.

Inorganic
1. Nonbleeding in organic solvents
2. Superior heat resistance
3. Less costly per pound
4. Opaque

Organic
1. Some slightly soluble in organic solvents
2. Strong (tends to offset high cost)
3. Transparent
4. More intense

112.5b Disadvantages of Organic and Inorganic Pigments.

Inorganic
1. Less heat resistant

Organic
1. Low tinting strength
2. More costly per pound

112.5c Yellow Pigments.

Chrome Yellow:
(Lead chromate)

Advantages—Clean, crisp color, low cost, high hiding, nonbleeding. Good durability, declining with lightness.

Disadvantages—Poor alkali and soap resistance, tendency to darken on exposure, contains lead; discolors in presence of sulfides.

Zinc Yellow:
(Zinc potassium chromate)

Advantages—Prevents deterioration of aluminum and magnesium; yields low-priced, clean green shades with phthalocyanine blue; rust inhibitive; light fast.

Disadvantages—Greenish yellow; highly reactive in some resin systems and must be compounded carefully to avoid livering; mainly suitable for primers where topcoat is applied early because of its slight water solubility; poor hiding.

Iron Oxide:

Advantages—Lightfast; nonbleeding; nontoxic; high hiding; good chemical resistance; nonreactive; available in low-hiding grades for transparent and translucent films.

Disadvantages—Poor gloss retention; dirtyish colors.

Nickel Titanate Yellow:

Advantages—Lightfast; exterior durability; chemical resistance; withstands high temperatures.

Disadvantages—Tendency to chalk; weak in tinting strength.

Hansa Yellow:
(Toluidine Yellow)

Advantages—Available from primrose to medium yellow; lightfast masstones; alkali and acid resistant; lead-free.

What Goes in a Coating

Disadvantages—Poor hiding; bleeds badly; poor heat resistance; high price; poor lightfastness in tints.

Benzidine Yellows: *Advantages*—Twice the tinting strength of Hansa yellow; lead free; good masstone strength and lightfastness; good chemical and bleed resistance.

Disadvantages—High cost; low hiding power; and high oil absorption.

Flavanthrone Yellow: (Red Shade) *Advantages*—Excellent lightfastness, good resistance to bleeding and chemicals.

Disadvantages—High cost; and low color intensity in metal finishes.

Anthrapyrimidine Yellow: (Green Shade) Same as flavanthrone, except has a green cast and is transparent.

112.5d Orange Pigments.

Chrome Orange: *Advantages*—Low cost; nonbleeding; lightfast; rust-inhibitive.

Disadvantages—Contains lead; darkens in masstone; poor hiding and soap resistance compared with molybdate chrome orange.

Molybdate Chrome Orange: *Advantages*—Low cost; nonbleeding; combines with organic reds to give low-cost, high-intensity, durable finishes of various shades of red.

Disadvantages—Contains lead; poor alkali and soap resistance; tends to darken in masstones on exposure.

Benzidine Orange: *Advantages*—High color intensity; high tinting strength; nontoxic; good chemical resistance.

Disadvantages—High cost; poor lightfastness; poor bleed resistance; low hiding power.

112.5e Green Pigments.

Chrome Greens: *Advantages*—Low cost; wide variety of shades, since they are made of inorganic iron blue and chrome yellow, which can be varied; good hiding; nonbleeding.

Disadvantages—Poor alkali resistance; contain lead; moderately lightfast; tend to flood and float; rarely used in water systems.

Chrome Oxide: *Advantages*—Nonreactive; good heat, light and chemical stability; high hiding.

Disadvantages—Olive green shade; poor gloss retention on exterior exposure; low tint strength, about one-fourth that of chrome green.

Hydrated Chrome Oxide: *Advantages*—Cleaner, brighter green than chrome oxide; excellent exterior durability.

Disadvantages—Low tinting strength; poor blister resistance.

Phthalocyanine Green: *Advantages*—Clean color; high transparency in metallics; lightfast; resistant to chemicals and bleeding; tinting strength.

Disadvantages—High cost; tendency to bronze on exposure.

What Goes in a Coating

112.5f Blue Pigments.

Iron Blues:	*Advantages*—Low cost; high tint strength; good durability in masstone and dark tints; nonbronzing.
	Disadvantages—Poor lightfastness in light tints; poor can storage stability and alkali resistance, which is overcome somewhat by blending with phthalocyanine blue.
Ultramarine Blue:	*Advantages*—Reddish tone; brilliant; low cost; alkali resistant; heat resistant; good infra-red reflectancy.
	Disadvantages—Low hiding; very poor acid resistance; poor tinting strength; poor outdoor durability.
Phthalocyanine Blue:	*Advantages*—Lightfast; heat-stable; chemical resistant; high in tint strength and lightfast; bleed resistant.
	Disadvantages—High cost; tends to bronze in dark shades, which can be overcome by blending with iron blue.

112.5g Red Pigments.

Red Iron Oxide:	*Advantages*—Low cost; high hiding; lightfast; good chemical resistance.
	Disadvantages—Dirty color; poor gloss retention; low tint strength.
Red Lead:	*Advantages*—Excellent film integrity for exterior paints; high rating for rust-inhibition in primers.

Disadvantages—Low in hiding; useable only for primers, since it fades; actually is orange rather than red.

Toluidine Reds.
Advantages—Bright, relatively low-priced organics; outdoor durability; lightfast in deep tones; excellent acid and alkali resistance.

Disadvantages—Poor bleed resistance.

Para Red:
Advantages—Low cost; good hiding; good acid and alkali resistance; lightfast in deep tones.

Disadvantages—Bleed badly when used for tinting and have poor lightfastness in tints.

BON Reds and Maroons:
Advantages—Masstone lightfastness; excellent bleed resistance.

Disadvantages—Poor alkali and soap resistance and poor tint lightfastness; not useable in coating systems that are highly acid or alkaline, since the manganese component will cause wrinkling.

Lithol Rubine:
Advantages—Good bleed resistance; combined with molybdate chrome orange to make low cost reds; low cost; deep clean colors; bluish tone.

Disadvantages—Poor hiding; poor lightfastness and alkali and soap resistance.

Chlorinated Para Reds:
Advantages—Low cost; good hiding; good acid and alkali resistance; good deeptone lightfastness; better tinting lightfastness than para red; brilliant tones.

What Goes in a Coating 67

Quinacridone:
Disadvantages—Bleeds.
Advantages—Best durability; nonbleeding; tint lightfastness; exceptional brightness.

112.5h Metallic Pigments.

Aluminum:
Advantages—Good hiding, light and heat reflectance; and resistance to water permeability; light weight; excellent acid and salt spray resistance; leafing effects.

Disadvantages—Careful formulating required to avoid loss of reflectance because of deleafing on the dried coating's surface.

Bronze:
Advantages—Variety of simulated gold colors produced by varying amounts of constituents: copper, zinc, antimony, and tin; leafing grades.

Disadvantages—Higher cost than aluminum; poor hiding; heavy.

Zinc Dust:
Advantages—Sacrifices itself in an anticorrosive paint on steel, thereby providing protection even when pinholes or small cracks in the coating permit atmospheric water to reach the substrate.

Disadvantages—Very heavy; use is limited to prime coats, so must be topcoated; paint must be carefully made so that zinc dust is in contact with metal substrate to permit the so-called galvanic action whereby a minute electric charge is set up when the zinc dust is wet, sacrificing the zinc instead of corroding the metal substrate.

112.6 Extender Pigments.

112.6a General. These are called pigments because they have the proper physical credentials, except for their virtual inability to hide although some of these materials can even hide under certain circumstances.

In our discussion of white hiding pigments, we saw that the refractive index of a pigment gives the key to its hiding ability. We said that hiding ability is directly related to the spread between the refractive index of the pigment and that of the binder. Titanium dioxide, the king of the hiding pigments, has a refractive index of 2.76, which is a long way from the R.I. of an alkyd resin, which is 1.50 to 1.60, so it hides well. See Table 112.6a.

China clay, on the other hand, has an R.I. of 1.56, which is

Table 112.6a. *Space Filled by 100 lbs. of Extender Pigments.*

	Gal./100 lbs.
Silicates	
Diatomaceous silica	6.15-5.0
Clay, natural	4.6
Silica (quartz)	4.6
Mica (phlogopite)	4.37
Talc	4.21
Wollastonite	4.13
Calcium Sulfate	
Gypsum	5.11
Anhydrite	4.07
Precipitated	4.07
Calcium Carbonate	
Precipitated types	
Colloidal	4.48
High oil absorption	4.48
Low oil absorption	4.48
Surface-treated	4.53
Natural types, calcite	
Water ground	4.43
Dry ground	4.43
Limestone	4.43
Imported chalk	4.43
Barium sulfate	
Ground barytes	2.65
Blanc fixe	2.75

What Goes in a Coating

practically the same as an alkyd, so it provides little or no hiding. The other extender pigments are just about as deficient in hiding power since their R.I. is very close to those of alkyds and other commonly used resin binders.

Extenders, since they are used by the thousands of tons each year by coating manufacturers, obviously have other virtues beside hiding. They are mostly low-cost materials; and they can do cheaply one of the tasks of a pigment, which is to provide body and bulk to a surface coating.

Each extender pigment has some principal characteristic that contributes to coating performance.

If we merely added just enough white hiding pigment to a binder so that we could spread that binder over a surface and hide it, we would be robbing that surface of desirable characteristics. First of all, the coating probably would be runny and would provide uneven protection because it had no body. The coating would look saggy and ugly. Then, if it had just enough white hiding pigment to hide the substrate, a large proportion of the binder would be "free binder," which means that it would not be held in tow by the tiny pigment particles and would be able to reflect large quantities of light. This free binder would cause the dried film to be very glossy; very often this is not desirable.

In that circumstance, too, all of the free binder could cause deep, wasteful penetration into certain porous surfaces like gypsum board, or soft woods, or plaster. Thus, some sort of pigment material beyond what is needed for white hiding is also needed to tie up the "free binder" so that it will not make the surface glossy or so that it won't penetrate certain surfaces so deeply and wastefully.

Simply put, titanium dioxide—the best hiding pigment but which is quite expensive—is used only to the extent necessary to provide opacity; the rest of the pigmentary duty is carried out by the extenders (except for tinting with inorganic and organic color pigments and metal soap formation by zinc or lead reactive pigments).

Since the integrity of a protective coating often depends on the formulator's judicious use of extender pigments, the following description of these materials is presented so the specifier will understand why one or another is selected for a particular coating.

112.6b Calcium Carbonate. This common material, known as whiting, calcite, ground limestone, or chalk, is used in many kinds of paints because it imparts tint retention, and a degree of mildew resistance to an exterior coating, plus crisp brightness—as distinguished from gloss, or sheen—to all paints in which it is used.

A variety of grades from natural to synthetics are available, which offers the formulator a selection of inexpensive materials to meet his requirements whether it be for an extender that absorbs much oil or practically none. One hundred pounds of one grade of calcium carbonate will tie up 9 lbs. of linseed oil, while 100 lbs. of some other grades will be able to tie up as much as 55 lbs. This has an important bearing on spreadability and penetrability. (See Table 112.6b.)

Table 112.6b. *Oil-Holding Ability of Extender Pigments.*

	Lbs. of Oil per 100 lbs. of Extender
Silicates	
Diatomaceous	150.0
Mica, muscovite	47.5
Clay, natural	36.0
Clay, calcined	
Talc	27.0
Silica	25.0
Wollastonite, fibrous	40.0
Wollastonite, non-fibrous	20.0
Calcium Carbonate	
Precipitated types	
Colloidal	55.0
High oil absorption	40.0
Low oil absorption	17.0
Surface-treated	15.0
Natural types, calcite	
Water ground	15.0-16.5
Dry ground	9.0
Limestone	10.5
Imported chalk	12.5
Calcium sulfate	
Precipitated	50.0
Anhydrite	25.0
Gypsum	21.0
Barium Sulfate	
Blanc fixe	14.0
Ground barytes	6.0

What Goes in a Coating

Calcium carbonate, because of its low cost, is extensively used for calks, putties, and sealants, as well as all types of paints.

112.6c Clay (Aluminum Silicate). Good dispersibility in water has made clay an important extender in latex paints. This, plus its low specific gravity and consequent high bulking value, has made it a big favorite of formulators.

Two types of clay are used: (1) natural clay, with the characteristics described above, and (2) calcined clay with a fine particle size that imparts what is described erroneously as *high oil absorption,* which really means that it has a lot of receptivity to the liquids in which it is to be dispersed, whether oil, resin, or water.

One natural clay, bentonite, has such a fine particle size that it is used as a thickener and sag control agent.

112.6d Talc (Magnesium Silicate). The ability to retard the cracking that could occur if small particle-size or reactive pigments are used, has made talc an important extender. It also plays a role in flatting. Talc is used in large quantities in white house paints, but because it yields a white chalk it is not used in tints.

Talc's definite improvement of paints justifies its use. Other extenders have far better bulking value, or ability to fill up a lot of paint with relatively small poundage. In fact, talc has the lowest bulking value of all the extenders.

112.6e Silica. Finely divided quartz offers high bulking and is used in metal finishes to reduce the gloss and in wood fillers and undercoaters to give good sanding characteristics.

Another silica made by fuming a silica salt, is available in microscopic-sized particles and is a flatting and thickening agent. It is used in small quantities as a supplement to other extenders.

A third silica is diatomaceous, or derived from one-celled plants that petrified centuries ago on old river and sea bottoms. Diatomaceous silica has low weight per unit of volume, which means that a pound goes a long way in a gallon container. It also has a high rate of oil absorption and, thus, ties down the binder used in a formulation. This tying-down function cuts the binders' ability to reflect light and helps make the paint flat rather than glossy, a characteristic it would have if there is much free binder.

Because it has high bulking value (low weight per unit of volume) and a high rate of oil absorption, small quantities are used for thickening and flatting.

112.6f. Wollastonite (Calcium Metasilicate). A fibrous form of Wollastonite, a non-toxic silicate, has replaced asbestos in many texture paints. A non-fibrous form is used because its high alkalinity buffers acidic resins such as polyvinyl acetate. Wollastonite's needle-like form aids film strength, scrub resistance, and color and retards burnishing.

112.6g. Sulfate Extenders. Barium sulfate and calcium sulfate are two sulfates that have been used to a greater or lesser degree for many years in the coatings field. The barium salt is found in nature as a crude ore that can be used with minimum treatment or that can be bleached by an acid treatment.

Synthetic barium sulfate is also offered. Aside from its use as an extender for zinc sulfide in the making of lithopone, which is no longer widely used as a white hiding pigment, barium sulfate is now used mainly in formulating sanding primers, because they have low binder demand and are soft.

Calcium sulfate, or gypsum, has not been used extensively by itself. However, a form of calcium sulfate had been co-precipitated with titanium dioxide to a combined weight of some 100,000,000 pounds a year as an economical white hiding pigment for interior alkyd flats and semigloss enamels, as well as some latex and traffic paints.

Unfortunately, this co-precipitated grade has proven uneconomical to the single manufacturer producing it. Production has been phased out, and paint manufacturers will have had to reformulate around it. To maintain competitive pricing, they have sought to replace the calcium sulfate component of all the composite pigment with an extender.

Fine-particle size clay, talc, and calcium carbonate have been found satisfactory replacements.

113 Solvents.

113.1 General. Most binders used in protective coatings are so viscous that they would be unable to adhere to a surface, even

What Goes in a Coating

if they could be spread. To make them spreadable, sprayable, and manageable, they must be dispersed or dissolved. Latex emulsions, as we have seen, are particles dispersed in water; this means the particles and the water are, more or less, attracted to a third material, an emulsifier; and the particles, the third material, and the water are intimately mixed to the extent that precipitation is avoided. A solvent differs in that it doesn't need a third material to help it mix intimately with the resin or oil binder to be dissolved in it.

No really definitive theory has been enunciated about the physics of solutions. Most of the high-flown attempts at an explanation boil down to a simple statement that certain molecules of matter consist of atoms and combinations of atoms that just do not seem to stay electrically balanced when they are placed in most liquids; as a result, they tend to turn in on themselves and precipitate in these inhospitable liquids or to form clusters and then drop out as globs, or agglomerates. Certain other liquids seem more able to team up with the lightly charged ends of these over-balanced molecules and balance them out so that they can circulate in the liquid and avoid precipitation.

These are solvents. Each solvent will work to a greater or lesser degree with certain materials and may not work at all with others. The larger the molecule to be dissolved, the more difficult it is for a solvent to have any effect on its disbalance; and this inability to influence the balance of outsize molecules limits what can be dissolved in a particular solvent. Water, for example, can dissolve sugar. It cannot dissolve polyvinyl latex, which is a relatively huge molecule. But, toluol can dissolve polyvinyl latex. It cannot dissolve Teflon®, which is much larger still.

Since it is desirable to use water as a means of reducing latex paint, some means of reducing polyvinyl acetate other than by dissolving it had to be utilized. So emulsification was devised. This is the use of a material that literally holds the hand of water on one of its ends and holds the tail of polyvinyl acetate on the other. The net effect is the dispersion of PVA in water. So when the term emulsifying is used, we are referring to a means of dispersion.

The alkyd molecule and the epoxy molecule are examples of two different resins requiring different solvents. They are not often reduced with water (although breakthroughs in this direction have been made and are in development stages). Alkyds that dry at room temperature are usually thinned with mineral spirits, a common hydrocarbon of a type known as aliphatic. Epoxies are much more difficult to dissolve and require "solvent systems," consisting of various combinations of aliphatics and higher-boiling, stronger aromatic hydrocarbons; and, in some instances, alcohols must be added to the scales to achieve our hypothetical solvency-balance.

Solvents do more than merely thin coatings. They also help wet the surface and contribute to adhesion by penetrating into the pores and crevices and taking the paint with them. In the can, during storage, they provide a "head" of vapor that helps retard the skinning caused by oxidation.

They also play a role in laying down uniformly thick films. This is known as leveling. If a solvent evaporates too quickly, the film may dry too rapidly causing it to harden before it can spread out evenly. As a result the paint brush marks may be frozen into the surface instead of smoothing out.

Too rapid evaporation can also cause a crust of dried paint to form on the surface. Materials remaining below the crust are soft, and the end result is an easily-damaged finish.

For these reasons, and others too complicated to justify explanation here, specifiers should treat the solvent contents of coatings with respect. When significant, solvent details will be included in Performance Comparative Charts in Chapters 7 and 8 and in the Detailed Specification Charts at the end of the book.

As a general rule, paint formulators are almost obliged to stick by uniform principles in selecting solvents. Regional and national manufacturers with established reputations can be counted on to furnish suitable solvents, but it never hurts to make certain.

Architectural specifiers are not likely to be affected to any great extent by air pollution regulations since paint manufacturers are obliged by law to conform and must be responsible. Products specified for factory finishing may have been influenced by changes in solvents to conform to air pollution regulations, most of them influenced by Los Angeles' famous Rule 66, which aims at restricting the release of smog-forming, photo-sensitive chemicals into the atmosphere.

What Goes in a Coating 75

Specifiers may rest assured that coatings applied at the factory and which have had to be reformulated to Rule 66 standards are as good as they were before the regulation.

113.2 Types of Solvents. The following paragraphs describe the types of solvents used in architectural finishes or in finishes likely to be used on materials specified by architects and builders, and finished at the factory.

113.2a Hydrocarbons. More hydrocarbon solvents are used than all others combined. They are derived mainly from petroleum, although some are still obtained by the distillation of coal tar.

Two types of hydrocarbons are used: (1) *aliphatics,* which, in general, evaporate more slowly and have less solvent power than the aromatics; (2) the *aromatics,* which often require extra care in handling because their vapors will ignite more easily, since they usually have lower flash points than the aliphatics. On the other hand, the low flash points and the lower range of distillation temperatures usually signify stronger solvency than that found in the aliphatics with lower temperature ranges.

113.2b Mineral Spirits. These are the most common solvents in run-of-the-mill architectural paints, other than water-reducibles. They are described as *standard,* which have higher solvency, and *odorless,* which are free of the slight pungency of standard mineral spirits, but are only about two-thirds as powerful a solvent. The odorless are likely to replace the standard in many cities with air pollution regulations, since they are less photosensitive when they evaporate.

113.2c Painters (VM&P) Naphtha. Varnish makers and painters naphtha preceded mineral spirits as the favored petroleum thinner, but its low flame-point and its high-speed evaporation cut into its popularity. Safer mineral spirits took over, but VM&P naphtha is still used in spray paints and traffic paints where rapid evaporation is important.

113.2d Solvent 100 or High Flash Naphtha. This solvent is about twice as strong a solvent as mineral spirits, but it is twice as expensive. It is more toxic and evaporates more quickly, hence it is more likely to leave brush marks.

113.2e Solvent 150. This solvent, which gets its name from its flash point, evaporates more slowly than other commonly used solvents. It could be called a retarder-solvent since it serves primarily to slow down evaporation in order to permit the coating to level. In hot weather, when standard mineral spirits evaporate too quickly, painters would have a hard time overcoming the drag of their viscous paints. Solvent 150 maintains flowing properties and eases their job. It is rarely used as the sole solvent.

113.2f Turpentine. Turpentine, derived from pine gum, has been replaced as a solvent by mineral spirits, although the latter has only half its solvent power. Toxicity, high cost, and strong odor proved the undoing of turpentine.

Dipentene, another pine gum derivative, has greater solvency and is used to a limited extent in some formulations to slow down evaporation and as an anti-skinning agent.

113.2g Esters and Acetates. The acetates are primarily used as lacquer solvents, and have limited interest to architectural specifiers.

Glycol esters are used as solvents for urethanes and as additives (ethylene glycol monoethyl ether acetate is an additive that improves flow or blush resistance). Their main interest to the construction fields, however, is in their use as coalescing agents in latex paints.

Two glycols—*ethylene glycol monobutyl ether acetate* and *diethylene glycol monoethyl ether acetate*—play an interesting role in latexes. They could be called latent, or dormant solvents, because they appear to idle along in the can of paint until the film is spread and most of the water has evaporated. At that time, for just an instant, the sleeping glycol ether acetate flashes awake and pulls the tiny particles of latex together and dissolves them, tying them together in a continuous film just long enough for complete drying to take place. That is the generally accepted rationale for film coalescence in latex paints.

What Goes in a Coating

113.2h Alcohols. Alcohols, except for cutting shellac, are not encountered very often in structural paints. One use is as the sole solvent for *polyvinyl butyral resins,* which serve as wash primers for factory-primed metal.

113.2i Glycol Ethers. These are mainly used for factory-applied finishes. *Ethylene glycol monoethyl ether* and *ethylene glycol monobutyl ether* are used to a limited extent with epoxies. The latter improves flow and application properties. The first ether is also used in phenolic varnishes.

113.2j Ketones. The main use of the available ketones are in factory-applied finishes. *Methyl isoamyl ketone* has some use in brush-applied finishes. Diacetone alcohol, a crossbreed, is sometimes used as a modifying solvent to improve flow and leveling of brush-applied coatings. Its slow evaporation and low toxicity fit it for this function.

114 Additives.

114.1 General. A long time ago, in the Antediluvian age of the paint industry—meaning the 1940s—the journeyman painter used surface coatings that consisted of lead pigment, either red or white, and linseed oil, or perhaps alkyds. Of course, he had a few organic and inorganic colors to help satisfy the requests of his customer.

The lead pigment not only provided opacity, it also supplemented the oxidation of linseed oil fatty acids to assure the through drying of the protective film.

Much has happened since those days "before-the-flood" when water paints and sophisticated solvent-thinned coatings arrived. Most importantly, surface coatings have been substantially improved. A 20-year paint job is a possibility, not a dream. Five-year and 10-year life is common. (The old lead-linseed paint could be counted on for three years at best.)

The marvels of polymer chemistry have made this possible. As it happens, very little comes easily. Better surface coatings are no exception. Years of experimentation by stubborn dreamers, who knew that polymer chemistry could be tailored to improve coatings, led to trial upon trial of supplementary materials to make the promising polymers work.

Those materials that were tried and found useful are included among the invaluable additives used in coatings. They range from alcohols and hydrocarbons, to keep water-thinned paints from foaming, to complicated methoxy phenols and methyl ethyl ketoximes, to prevent skinning.

Some additives, like grinding aids, merely ease the job of manufacturing and don't concern the construction specifier. But chemicals to prevent fungus on the outside of a house are important. Also, ultraviolet absorbers are needed for tint retention in most paints; and if journeymen painters, with their high hourly wages, are not to bleed the builder, the right amount of flow agents are needed to help them spread paint at an economic pace.

The following paragraphs outline the most common additives that can affect coatings performance.

114.2 Antiflooding Agents. Antiflooding agents, or deflocculators, are included in many color formulations to keep the white hiding pigments from separating from the color particles. Silicone oils and fatty acid esters are used for this purpose in solvent-reduced coatings; and anionic and nonionic emulsifiers are used in latex. Without them, colors may separate either in the can or after application during drying.

114.3 Anti-Gelling Agents. Anti-gelling agents prevent gelling-in-the-can, which sometimes occurs to chlorinated rubber. Epichlorohydrin eliminates this.

114.4 Anti-Skinning Agents. Anti-skinning agents are needed for paints that dry by oxidation, particularly quick-drying materials like floor enamels or four-hour enamels. These agents have to retard oxidation in the can but must disappear when the coating is applied. Proper formulating achieves this, but unskilled formulators can cause trouble here. The most popular anti-skinning agents are butyroldaxime and methyl ethyl ketoxime, because they are so volatile in open air that they do not hinder drying, and because they have no effect on color.

114.5 Coalescing Agents. Coalescing agents serve, more or less, as a quickly disappearing mucilage to bind together the tiny particles of latex polymers used in water-thinned paints just as they are about to dry. As the water evaporates, these latex polymer particles begin to pack together to form a film. To assure a uniform packing rather than clusters, the coalescing agents hold them in a soft, tacky embrace for the instant needed to get them to blanket the surface uniformly. Typical coalescing agents are ethylene glycol monobutyl ether, diethylene glycol monoethyl ether, and diethylene glycol monoethyl ether acetate.

114.6 Driers. Driers are really catalysts, or chemicals that play a role in a chemical reaction without actually entering the reaction. In this instance, driers speed up the reaction of oxygen in the atmosphere with such binders as alkyds and vegetable oils, so they will dry in a reasonable time. A typical alkyd wall paint, for example, would probably take a week to dry thoroughly; but with driers, they dry overnight. Even some latex, or water-thinned paints, oddly enough, contain small amounts of drier. That is because latex house paints often have quantities of vegetable oil or alkyd resin added to aid adhesion over old, chalky paint; and these solvent-materials need driers.

Driers are, mainly, metal soaps formed by reaction of metal-bearing substances and either naphthenic acid, tall oil fatty acids, or octoic acid; recently, neo-acids derived from petroleum are being used to make highly concentrated driers. Cobalt and manganese driers are called primary driers since they are effective by themselves. Cobalt is the most effective. Manganese has one big drawback: It tends to discolor white paint, so it should be avoided in whites. Lead, calcium, and zirconium are auxiliaries of the primary driers and promote uniform through-dry. Lead has recently come into disfavor because of its toxicity and will be banned from households.

114.7 Defoamers. Defoamers prevent foam-formation in the manufacture of latex paints, which could delay manufacture and result in under-filled cans. Moreover, on application foam in the can can lead to bubbled-surfaces, or cratering, which is most undesirable since each bubble is a potential weak spot. Typical defoamers reduce the surface tension, which causes the foam. These are mineral spirits, pine oil, octyl alcohol, and various esters of fatty acids.

114.8 Emulsifiers. Emulsifiers are added to compensate for the emulsifiers in latex resins that are lost, or tied up in production of paint. These latex resins are literally born in a fluid of emulsion because they are made by *emulsion polymerization,* a process in which the major component of the polymer is emulsified so that the final product is compounded in an emulsion. An additive emulsifier is often used because the pigments in the paint extract emulsifier from the mixture; and if none were added, the emulsion might break, causing seeding or gelling. Common emulsifiers are amine soaps and anionic or nonionic surface active agents.

114.9 Freeze-Thaw Stabilizers. These stabilizers lower the freezing point of a paint to reduce the likelihood of its freezing in storage. Glycols—ethylene, diethylene, or propylene—are used most frequently.

114.10 Pigment Suspension Aids. Pigment suspension aids prevent hard-settling during storage, which can cause serious problems of redispersion, particularly if the precipitates cake on settling. These suspension aids lead to light gel-like structures that hold the pigment in tow; and even when they settle, they are softer and easier to stir into suitable condition. Materials used are the same as those used for pigment wetting agents.

114.11 Pigment Wetting Agents. Pigment wetting agents help binders coat each particle of pigment, which means they must aid in the replacement of air around the pigment. Estimates put the amount of space consumed by air in a bag of pigment at 80 percent. Not all the pigment is surrounded by this air; some is lumped together in agglomerates, which must be broken up in the usual milling process. Pigment wetting agents speed the separation of these ground particles and help to keep them separated. Without them, dispersion would be inadequate, with consequent inconsistencies in hiding, gloss, color, and texture. For solvent paints, lecithin and esters of fatty acids are used. Water-thinned paints use potassium tripolyphosphate and tetra potassium pyrophosphate.

114.12 Preservatives. Preservatives eliminate bacterial contamination, which leads to enzyme degradation of cellulosic thickeners and other organic matter in paint. Phenyl mercury salts have been used for many years, but lately various substitutes have been introduced in anticipation of Federal laws banning mercury in paint.

114.13 Thixotropic Agents. Thixotropic agents provide body to a paint so that it appears to have a creamy, easily spread consistency. When pressure is applied, either by dipping the brush in the can or brushing on a surface, the paint thins as pressure is applied; then the creamy consistency returns when pressure ceases. This thick-thin ability is imparted by so-called thixotropic agents, or false-body donors. A paint containing proper thixotropic agents requires little stirring after storage, because its body holds the pigments and other materials in place. The result is a material that is easy to spread and that won't sag on the surface since it regains its body after spreading. These materials also function as viscosity control agents.

114.14 Viscosity Control Agents. Viscosity control agents permit a paint formulator to hold his product at a definite range of thickness during storage in the can. They are indistinguishable from the thixotropic agents described above and are listed separately only for classification purposes. Solvent-thinned paints are controlled by amine-treated clays, usually a type called bentonite; or by waxy materials derived from hydrogenated castor oil. Some viscosity control of solvent paints is obtained from certain colloidal silicas or aluminum stearate. Latex paints commonly use cellulosic thickeners, such as hydroxy ethyl cellulose, methyl cellulose, or carboxy methyl cellulose. These must be preserved during in-can storage by mercurial preservatives, which have come under attack from environmentalists. If satisfactory substitutes for mercury salts are not accepted, it is possible that the cellulosics will be replaced by fumed silicas or other colloidal silicas, or by aminated clays.

HELPFUL ADDITIONAL READING

1. Hay, J. Kirk, and Garmond G. Schurr, "Moisture Diffusion Phenomena in Practical Paint Systems," *Journal of Paint Technology*, Vol. 43, No. 556 (June, 1971), pp. 63-72.

2. Martens, C. R., *Technology of Paints, Varnishes and Lacquers* (Interscience Publishers: New York, 1965), Chaps. 2-22, pp. 12-390.
3. Parker, D. H., *Principles of Surface Coating Technology* (Interscience Publishers: New York, 1965), Chaps. 4-18.
4. Solomon, D. H., *The Chemistry of Organic Film Formers* (John Wiley & Sons: New York, 1967), Chap. 8, pp. 202-18.
5. Stieg, Fred B., "Titanium Dioxide or Is It?" *Journal of Paint Technology*, Vol. 43, No. 561 (November, 1971), pp. 36-43.

chapter 2

Testing—Or the Art of Making Sure

200 General. Actual performance tests on specified coatings are rarely needed because reputable manufacturers usually carry these out in their quality control departments. Tests may be desirable, however, in those instances where performance is critical and the failure of a coating may result in expensive shutdowns of equipment or in extensive damage to installations. Even in these instances, certification by reputable manufacturers that specifications have been met should be satisfactory.

To acquaint architects and builders with the tests used to check out performance characteristics, a brief summary of test methods used in paint plants and in private testing laboratories on behalf of architects and engineering firms is presented here. Chapter 9 discusses tests for fire retardant coatings; see especially Fig. 902.3.

TYPES OF TESTS

In the methods outlined in the following paragraphs, simple, informal tests will be described, wherever possible, in addition to the more formal tests requiring special equipment.

201 Drying Time. An informal test as old as the pyramids—use a dry finger and feel if the surface is dry to the touch in the indicated time.

Actual, recognized stages of drying are described in Federal Specification TTP 141b, Method 406.1, if anyone is interested; and for the more serious, the American Society for Testing

Materials has a specification ASTMD154-52, which was developed along with hundreds of other highly technical and exacting specifications by Committee D-1 of ASTM, which meets twice a year and whose members spend thousands of man-hours each year developing satisfactory methods for testing coatings.

Drying tests, in general, can be split into five parts dealing with dust-free stage; tack-free stage; through-dry stage; full hardness; and recoat time.

These tests are all dependent on the opinion of the man or men conducting the tests and should, of course, be done with clocks to permit notations of time.

Testing laboratories and many paint companies have instruments that automatically and continuously apply sand, or lint, or some sort of indenting device to the surface. The ability of the sand to stick to the drying surface at certain intervals and the depth of the indenting device indicate rates of drying.

201.1 Dust-Free Stage. The *dust-free* stage is determined by noting if the finger will be free of paint when brushed lightly over the surface. In place of a finger, cotton fibers can be used. Since everyone has a different idea what constitutes a light touch, this is a somewhat subjective test.

201.2 Tack-Free Stage. The *tack-free* stage is reached when a slight pressure applied with the finger leaves no mark on the surface. A piece of paper may be pressed on the coated surface; and if it is applied with a specified weight for a definite time, then the coating is said to be tack-free if the paper will drop off when the panel is turned upside down.

201.3 Through-Dry Stage. *Through-dry* is reached when considerable pressure is exerted by the finger in a rotating motion without distortion of the film. That indicates that the entire depth of the film is as dry and hard as the surface.

201.4 Full-Hardness Stage. *Full-hardness* requires that a fingernail scratch into the surface without removing the coating, except with great difficulty.

201.5 Recoat Time. *Recoat time* is the period required for the film to dry so thoroughly that application of a second

Fig. 201.5. Shore Durometer.
A spring-loaded indenter protrudes through this device. When pressed into a freshly applied film, the indenter is forced back in relation to the hardness of the film. The test is made at intervals following application to determine how soon the film reaches the specified Shore hardness rating, indicating that it has dried hard enough to receive a second coat.

coat is possible without lifting or loss of adhesion of the first coat and without causing any other film malfunction. See Fig. 201.5.

202 Gloss. Tests to rate a coating for gloss depend on the ability of surfaces to bounce back varying amounts of light beamed on them.

We can understand how these tests work if we recall that a flat mirror reflects virtually all the light beamed on it. That's because it is almost perfectly smooth. Surface coatings, even the smoothest, have some degree of roughness. As it approaches perfect smoothness, it has more gloss. As its roughness increases, even if this requires a microscope to detect, the lower its gloss becomes. In effect, a gloss reading shows the relative reflectibility of the coated surface as compared with a smooth flat mirror.

What happens is this: a light beam applied to a coated surface is to some extent absorbed by passing into the minute imperfections; some light is diffused by surface imperfections, or roughness, and a small amount is converted to heat. The remaining portion is bounced off as *incident light*. If the beam that is played on the surface is held at a certain angle—let's say 80 degrees, then the incident light will bounce off at a determinable angle. Photocells are placed at the point where the incident light should be reflected, and the reading for a given angle is said to be the *specular gloss* for that angle.

A reading between 5 and 15 at 85 degrees, specular gloss is required for flat latex paints in federal specifications; specular gloss at 60 degrees for a semigloss must be between 25 and 55; and a high gloss enamel requires no less than 70. The use of 85 degrees for flat paints permits greater accuracy at low gloss.

The numerical readings for the various tests are based on the readings of a high-gloss, black-glass panel as a standard of 100. If the photo-cell reading for the test panel is 70 percent of the reading for the black-glass standard, then the coating on the test panel is said to have a rating of 70. A typical instrument used to test for gloss is shown in Fig. 202.

Not all companies use instruments for checking gloss. They may establish standard panels for gloss, semigloss, and flat coatings. They may then check gloss at various grazing angles and then match the results against the standard panels, which may be numbered in accordance with their rating; i.e., the panel numbered 100 may be high gloss; and so on down to zero for dead flat.

203 Flexibility. To see how much "give" is in a coating, two official tests are available.

203.1 Mandrel Test. This test uses mandrels, or steel cylinders of various diameters. Metal panels are coated with the finish to be tested and are bent double over the mandrel. This test method is described in Federal Specification 622.1. Many federal specifications designate the diameter of the mandrel over which the test panel is to be bent without cracking. As the specification gets tougher, the mandrel diameter becomes smaller.

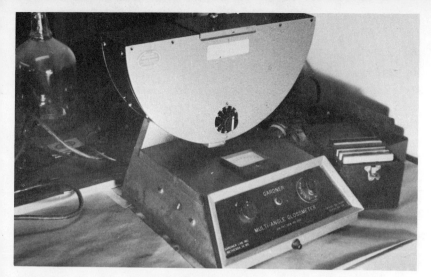

Fig. 202. Gardner Multi-Angle Glossmeter.
Coatings rated as high gloss, or semigloss, should be checked by this multi-angle gloss indicator. Angle at which light falls is shown on the meter, and the degree of gloss is compared with standards held in the box at right.

203.2 Conical Mandrel Test. The second flexibility test was developed by the American Society for Testing Materials and is called the Conical Mandrel Test, ASTM D522-41. The panel is clamped to the mandrel while a draw bar bends it. Percentage elongation at the point of cracking determines the product's flexibility rating. See Fig. 203.2.

204 Hardness. Two tests are commonly used for rating film hardness.

204.1 Pencil Hardness Test. This test utilizes pencil with leads of varying hardness. The hardest pencil, usually 6H, is pushed into the film. Successively softer pencils are pushed in. The first pencil whose point breaks determines the Pencil Hardness of the film. In effect, this system rates a coating's hardness by which pencil's point is crumbled when it tries to penetrate the coating's surface. If the coating is hard enough to break the point of a 6H (or hardest) pencil point, then it rates 6H, and is a hard surface indeed. If it can be penetrated by all the pencils harder than 6B (the softest), which it causes to disintegrate, then it is a soft coating.

Fig. 203.2. G.E. Heavy Duty Conical Mandrel.
Flexibility is measured by the percentage elongation of the coating at the point of cracking over the cone.

204.2 Sward Rocker Hardness Test. This test is more scientific. The Sward Rocker has two 4-inch metal wheels attached to each other in parallel. A weight is placed on one side of the arrangement; and starting at a fixed standard position, the rocker teeters back and forth until it comes to equilibrium with the weighted portion on the bottom. The softer the coating the more the rocker will be damped and the less it will oscillate. Conversely, the harder the coating the greater the oscillations of the rocker.

The tester counts the number of oscillations of the rocker on the coating under test. Regular plate glass is used as a standard. It has a Sward Hardness of 50.

204.3 Impact Resistance. Rapid deformation of a coating because of sudden, sharp impacts can lead to film failure. ASTM D 2794-69, "Resistance of Organic Coatings to the Effects of Rapid Deformation," calls for a coated test panel to be placed over a hole on a base plate. An impactor over the panel is hit by a 2- or 4-lb. weight elevated to a height prescribed for the test. Specifications range up to 160 inch lbs. After the impact test has been concluded, the surface, under low-power magnification, must show no evidence of flaking, cracking, or deformation. See Fig. 204.3.

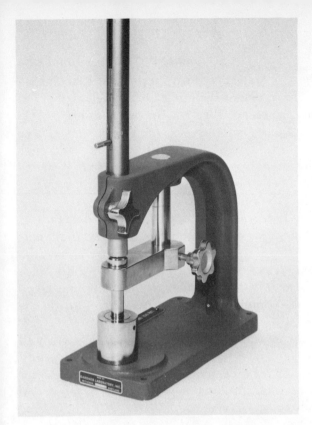

Fig. 204.3. Gardner Heavy Duty Variable Impact Tester.
A 2- or 4-lb. weight is dropped through a rod at a prescribed height to exert from 0 to 160 inch lbs. of impact against a coated test panel.

205 Lightfastness (indoors). Even with supposedly lightfast colors, fading occurs under some conditions. Ordinary interior lighting can lead to yellowing and darkening of whites, and can cause fading and color changes in tints and solid colors.

Actual, long drawn-out exposure tests, naturally, are the most accurate; but in the absence of time for these exposures, dependable accelerated tests are used. These involve an instrument called the Fade-Ometer, which uses a special carbon arc to subject coatings to days of trial that are accepted in place of months of normal exposures. ASTM and the federal government accept lightfastness tests based on this system.

206 Adhesion. Parallel scratches made into a coated surface in a systematic manner are used to rate adhesion of the material to its substrate.

In one test, the General Electric Adhesion Test, an instrument utilizing 11 razor blades clamped in parallel, with 1/32 inch between them, scratches through the test surface. Across the lines thus scratched, another set of 11 lines is drawn at right angles, thus producing 100 sections, each 1/32 inch square. The numerical rating of the film's hardness is stated in terms of the number of such squares that remain attached to the substrate after the razor work.

If the test specifier wants to make it more demanding, he can require that masking tape be placed over the intact squares (that is, if the razors have failed to loosen any) and then have the tape removed with a sharp, rapid pull. This test, very likely necessary for hard epoxy and silicone alkyd coatings, can be rated numerically in accordance with the number of squares still holding tight to the surface.

The adhesion for a particular coating on a soft wood will not apply with respect to a hard wood; and certainly it will be invalid for metal or for a cementitious surface. Performance specifications for a single product take this into account and usually state which substrate was used.

207 Cold Checking and Cracking. A sharp, sudden drop in temperature causes contraction of both the surface coating and its substrate. Because these rarely, if ever, contract at the same rate, stresses are set up in the coating, which may cause it to form tiny checks and cracks. Sometimes this is merely harmful to appearance, but if the initial adhesion is not good, flaking and peeling may remove surface protection.

Federal Specification 622.3, Flexibility (cold cracking) describes mandrel-bending tests on metal panels alternately heated in an oven and cooled in ice water.

Many cold-check tests are made by seeing how many heating and cooling cycles can be run before checking develops without any mandrel bending at all. Specifiers can develop their own tests to conform to probable conditions on the job. A coating for a Florida hotel isn't likely to have much of a problem, while a New England warehouse will need a set of tough tests. A realistic series of tests would seem to call for rapid drops from 80° F to zero degrees or below and back to 80°, repeated after reasonable intervals. This could be changed in accordance with the need of the coating.

Fig. 208.2. Humidity and Salt Fog Cabinet.
Coated metal panels are hung from glass rods in this cabinet and are checked for resistance to salt fog and plain moisture.

208 Resistance Tests. A series of tests is available for measuring resistance to various detrimental conditions and materials. These include: water, salt spray, solvent, chemicals, and bleeding.

208.1 Water Resistance. The ability of coatings to resist water is usually determined by immersion tests. These will be described in Section 209.

208.2 Salt Spray Resistance. A continuous fog of salt solution is wafted from a nozzle in a special corrosion-resistant chamber to test the ability of coatings to balk corrosion of metal. The coated metal panels are hung from glass rods; and to prevent corrosion from electrolysis set up between adjoining panels, they are separated from each other. See Fig. 208.2.

Coatings intended to prevent metal corrosion have a lot expected of them: they must also resist the spread of under-film corrosion in tests in which ugly crosses are cut across the coating and down to the metal. This tests the coating system's passivating ability. A system is said to passivate when it discourages or stops the electrolytic action that causes corrosion when moisture gets down to metal. The number of hours that a panel survives a salt-spray test measures its effectiveness.

208.3 Solvent Resistance. Specifiers may establish their own requirements here in accordance with circumstances to be encountered in use. Coatings required for the metal pole supporting a sign near a gasoline pump in a service station may require harm-free 24-hour immersion in gasoline. A cosmetic factory may require coatings able to withstand exposure to acetone for fingernail polish. The test for a service station finish, naturally, will differ from that used for the cosmetic plant coating.

After suitable periods of immersion in the solvent or solvents likely to be encountered by the coating, visual inspection is made to determine softening of the film, or separation from the substrate, which may take the form of wrinkling or blistering. Results are rated by number, with 10 indicating virtually no effect, and 0 indicating complete failure. In-between scores are subjective.

Immersion tests may be too stringent, and the specifier may call for application of the solvent by cloth or sponge if that comes closer to actual exposure conditions.

208.4 Chemical Resistance. Like solvent resistance tests, those for chemical resistance should be tailored to the particular conditions likely to be experienced by the coating. The chemical to be tested should be described as to strength, and the conditions relating to time of exposure and temperature, and other variables should be provided in detail.

208.5 Bleed Resistance. Solvents in a topcoat may pick up colors from the undercoat or from colored chemicals in wood substrates. These picked-up colors may ruin the appearance of the topcoat and can be a source of great dismay to homeowners and plant maintenance personnel.

Certain types of pigments are more likely than others to be subject to bleeding. One way to avoid this situation is to specify nonbleeding colors. When some questionable pigment combinations are necessary because of such factors as cost or because certain tints must absolutely be used because of an adamant homeowner or decorator, then bleed-resistance tests should be conducted.

Similarly, if dark resins such as asphalts or bituminous coatings must be used on a substrate because of a waterproofing condition, for example, then tests should be run to assure that the solvent used in the topcoat will not cause the black material to bleed into the topcoat. Knots and streaks caused in wood by sap deposits must be sealed before coating to avoid solvent pick-up and resulting discoloration of the finish.

Whenever bleeding is feared, a preliminary test coating should be made on a small area under circumstances to be duplicated later on the entire job. The safest specification is to call for a bleed-proof system. That puts the burden on the paint manufacturer to provide a primer, or sealer if necessary, and a topcoat or topcoats that have the correct combination of solvents and compatible colors to prevent bleeding.

209 Water Immersion. For testing a coating's ability to withstand immersion in water, or any specified liquid, Federal Specification Standard 141-A, Method 6011, provides detailed instructions for putting the coating through its paces, even down to liquid temperature, but the specifier has to state what he's looking for. And then, the permissible degree of harm has to be described and must be interpreted subjectively by the inspector.

Similarly, ASTM D 870-54—Water Immersion Test of Organic Coatings on Steel—leaves it up to the specifier. Both test methods list blistering, wrinkling, roughening, disintegration, loss of adhesion, and color change, and invite the specifier to add his own trials if he can think of any. These may include rust formation or development of fungus after immersion has been terminated.

The specifier must state how many coats he wants on the test panel and how many mils the system is to deposit on the surface. He should also state how long he wants the film to age before immersion and how long it should be immersed; or if it should be immersed until certain conditions appear. He should also give the temperature of the liquid, the time after the immersed panel has been removed before inspection, and the percent of permissible weight loss.

ASTM's method is more specific than the federal procedure. ASTM calls for a temperature of 37.8° C, plus or minus 1°; panels should be uniformly thick, and no less than one-half inch should separate the panels in a test tank in order to avoid cathodic deterioration. ASTM requires that the test liquid be changed every 72 hours to avoid contamination and that inspections be made of panels that have been removed and dried after 2 hours' immersion and after 24 hours, or as the specifier may desire. Comparisons at those times and after the test has been completed should be made with control panels that have not been immersed. Considerable range is thus afforded the specifier as to the requirements he wants to set up for the behavior of a coating during or following a test for immersion, either in water or some other liquid.

210 Moisture Vapor Permeability. This test measures an important characteristic of a coating: its ability to let moisture vapor pass through it from one side where moisture vapor is high, to the other side where it is low or nonexistent. This ability is important on exterior walls because excess moisture vapor in some rooms will build up so much pressure that it will try to leave any way it can. Sometimes it will get in the air space between inner and outer walls. The exterior of wood or cement may have enough tiny voids or capillaries to permit the moisture vapor to pass through it. However, if the coating will not allow the vapor to pass, the pressure will build up against it and cause blistering or peeling.

Federal Specification Test Method Standard 141-A, Method 6171, requires that a moisture differential be set up across a test specimen, using an apparatus that permits weighing the moisture that has permeated through the film under test.

The apparatus consists of a Permeability Cup, which permits 25 sq.cm. of the film to be exposed to a condition of high humidity on one of its sides, with a completely dry area on the other side of the cup. Moisture can pass into the dry side only through the film, which has been sprayed on a test panel that is capable of permitting moisture to pass. The material is air-dried or baked for a required time and at a specified temperature, depending on the circumstances under which it will be used. The coated side of the panel should be exposed to the wet side of the Permeability Cup, and

Fig. 210. Gardner-Park Permeability Cups.
A coated test panel is placed over the opening of this cup and located in a humid area. Weighings are made periodically to determine the specific permeability of the film, as indicated by the amount of moisture passing through.

the uncoated side to the dry area. Periodically, weighings are made of the cup to determine how much moisture has passed. The milligrams of water that have passed at a constant rate per square centimeter of film (1 mil thick) at 25° C with 100 percent humidity in the wet side and an initially desiccated atmosphere in the other is declared the Specific Permeability of that film. See Fig. 210.

211 Abrasion. The ability of a coating to withstand the effects of scraping and scratching action is measured by its resistance to the wearing effect of sand, one of the hardest materials to which a coating is likely to be exposed.

211.1 Taber Abrader. The most commonly used test device for abrasion resistance is the Taber Abrader, which is recognized by Federal Specification Method 619.2, *Abrasion Resistance by the Taber Abrader*. Over the years tests by this instrument have shown excellent correlation with results in service. Two carborundum wheels on the machine rub in opposite directions: one rubs from the center toward the outside, and the other from the outside toward the center of the test specimen. Applied to the carborundum wheels are loads that can be varied from

250 to 500 to 1,000 grams. Results of the test can be stated in terms of weight loss per given number of cycles of the wheels at a stated load, or it may be the number of cycles required to wear through the film.

211.2 Other Abrasion Tests. Other tests available are the *Falling Sand Method, Air Blast Abrasion Method,* and the *Armstrong Sandpaper Abrasion Machine.* The Falling Sand Method uses Ottawa Sand graded through an 840-micron sieve and retained on a 590 micron sieve. Federal Specification TTP 141-B, Method 619.1 and ASTM D968-51 and ASTM D658-44 cover the methods whereby sand is either projected onto the surface by an air blast of a given strength or simply dropped, falling by gravity. In all instances the sand is directed at a limited area and at an angle of 45 degrees. See Fig. 211.2. The air blast method uses a Bell Laboratory Abrasiometer to impel carborundum sand against the test panel at an air flow of 2.5 cu.ft. per minute.

The *Armstrong Sandpaper Abrasion Machine* was developed to eliminate abrasive surface-clogging, one of the weaknesses of the Taber Abrader. With this system, test specimens are rotated on an endless belt in an opposite direction to a rotating wheel with sandpaper pulled over its perimeter. The sandpaper is re-rolled after each contact with a test panel, thus preventing clogging. Results of the test are therefore believed more dependable than even the highly regarded Taber test.

Test results are correlated with control zinc panels to compare the abrasion of the coating under test with the abrasion of the zinc standard. Some tests use acrylic plastic sheet as the standard. The test is considered complete when the control panel, either zinc or acrylic sheet, has lost between 0.50 and 0.55 gms.

212 Hiding Power. The capability of a paint to cover a surface when applied at a certain rate of spread per gallon is stated as its *hiding power,* or its *spread rate.* Both federal specifications and ASTM tests use black and white substrates to test ability of a paint to obliterate. ASTM's test is to furnish indications of a paint's ability to hide in comparison with another. The federal test determines what is known as Contrast Ratio. This is an interesting test carried out by painting over two polished

Fig. 211.2. Gardner Falling Sand Abrader.
Abrasive sand is dropped through the 36-in. guide tube on to the coated test panel until the substrate is exposed. The number of liters of sand needed to expose 5/32nd of an inch of the surface is the basis for rating.

glass panels, one white and the other black. Controlled film thicknesses are obtained by special paint-spreading implements such as a doctor blade. Wet-film thickness gauges are utilized to assure accurate film thickness readings.

Unless the black-glass surface is virtually obliterated, light reflected from the paint covering it will be different from that reflected from the painted white glass panel. When the film thickness on the black glass has been increased to the point that obliteration has occurred, or that the reading of the reflected light is within 98 percent of the painted white panel, then the paint is regarded as having achieved solid hiding.

The hiding power in square feet can be calculated by simply dividing the film thickness into the volume of paint, or by using Table 212, which is based on the percentage of resin solids in the gallon of paint and the Pigment Volume Concentration of the material; since this table is based on 2-mil coatings, a mathematical adjustment is necessary to convert its figures to the film thickness determined in the test.

In ASTM D344-39, the relative dry hiding power of paints is determined by weighing the amount of paint needed to hide

Table 212. *Hiding Power Per Gallon in Sq.Ft. as a Function of Resin Volume and PVC.**

Pigment Volume Concentration	Percent Resin Solids by Volume							
	15	20	25	30	35	40	45	50
0	120	160	200	240	280	320	360	400
10	133	178	222	267	311	356	400	444
15	141	188	235	282	329	376	423	470
20	150	200	250	300	350	400	450	500
25	160	213	267	320	373	427	480	533
30	171	229	286	343	400	457	514	571
35	185	246	308	369	431	492	553	615
40	200	267	333	400	467	533	600	667
45	218	291	364	436	509	582	654	727
50	240	320	400	480	560	640	720	800

Source: Dow Technical Data Report 35-0 (Feb., 1970).

*These calculations are based on a 2-mil coating.

Note: Wherever possible in the specifications provided in the section on surfaces, PVC and percent resin solids by volume will be furnished to permit use of this table for figuring coverage to be obtained. Some specifications will list coverage requirements at a specific film thickness. Use of the hiding tests outlined above should also be a way to prove that claims for PVC and resin solids are accurate.

Testing—Or the Art of Making Sure

black and white squares provided in hiding-power charts furnished by the Morest Co. A syringe is used to measure the paint to be used on a specified area. With this and the weight of a gallon of the tested paint, it is simple to calculate the hiding area in square feet per gallon.

Specifiers should be aware that high hiding is not necessarily a standard for selecting a coating. Low-priced paints, because they usually have less resin (which is clear and expensive) may hide better than some expensive paints—but will probably deteriorate far sooner. On the chart above it will be noticed that materials with pigment volume concentration (PVC) of 50 will need only 25 percent resin solids by volume to cover 400 sq.ft.; but when the PVC is only 30, the resin solids must be 35 percent; and if the PVC is 20, then resin solids must be 40 percent to get coverage of 400 sq.ft. With higher resin solids we can rightly expect better durability, stain resistance, and scrubbability.

213 Washability. A scrub-tester applies soapy water with a brush-type device to the coating being tested. The coating is applied over a glass panel primed with an alkyd flat paint or it may be put on a contrast-paper panel. After drying 24 hours, it is alternately washed and allowed to dry until signs of adverse effects begin to appear. The number of washings is then noted as an index of washability.

214 Scrubbability. A device similar to the one used for washability, but with a brush able to apply pressure, is run over the test panel. The number of scrubs required to reach the substrate is a measure of scrubbability. This is often considered a test of durability.

215 Stain Removal. Coated panels are stained with various materials such as lead pencil, lipstick, crayon, mercurochrome, black ink, and grease. The harder the surface, according to the theory, the less the stain will penetrate it, and the fewer washes should be required to remove the stain. If too many scrubs are required for removal, this means that the process will cause excessive wearing of the protective surface whenever stains are removed. Ideally, stains should be readily wiped off.

A high volume of pigment in relation to resin means a soft finish able to absorb stains into the film. Rubbing the stain out of the film may seem easy, but it comes out at the expense of the surface. Less penetration of a harder film based on a lot of resin may be more superficial, but it may cause more difficulty in stain removal because of the stubbornness of the resin. Thus, easy removal of many stains is desirable, but it could indicate that difficult stains will only be removed by hard scrubbing.

Thus, we have the anomaly of a tough, scrub-resistant coating that may give up stains stubbornly because those stains that are able to penetrate it are held in a very firm embrace; while, softer, less durable coatings may give up their stains more easily because, although penetration is probably deeper, the protective coating is so much softer that it can be eroded by scrubbing; and in the erosion, the stain is worn off along with the soft, pigment-loaded finish. Ideally, the surface will be hard enough to resist penetration so that wiping is all that is necessary for stain removal.

216 Weathering. Tests to determine the ability of coating materials to withstand the rigors of actual weather exposure are conducted under circumstances duplicating actual conditions as closely as possible. For this reason test areas are usually scattered throughout various climatic zones. See Fig. 216.

Florida test areas check performance under conditions of considerable sun and rain and relatively uniform temperature. Los Angeles tests show performance in fairly dry, warm climates. A single test area in New England, Middle Atlantic states, or the Midwest will be satisfactory for all these sections. If tests are to show performance where moisture causes severe mildew conditions, Louisiana is likely to be selected, while North Texas would be chosen for the opposite situation, very dry. Oregon is the choice for a very wet climate, and North Dakota or Montana would be the choice if performance tests under extreme temperature fluctuations are wanted.

216.1 House Paints. Special panels are used for exposure tests, and they are set out in a certain manner to duplicate conditions as closely as possible to those on actual houses. For example, they are set out vertical to the ground with panels exposed

Fig. 216. Weathering Test.
Test fences are used in various climates to check the ability of coatings to withstand sun, water, condensation, and blowing sand before marketing.

on both north and south sides of the test rack. Southern exposure yields more varied and significant information, such as checking, cracking, flaking, mildew, and dirt collection, and the fading of tinted colors; but while northern exposure is sun-free it does show performance under conditions affecting gloss retention, mildew resistance, and dirt collection. An instrument used to test for chalking is shown in Fig. 216.1.

Test panels for outdoor exposure are 3 ft. long, 6 in. wide, and start on one edge at a thickness of 1/2 in., thinning down to 1/4 in. at the other. Three kinds of wood are normally used: Western red cedar, representing fine-grained woods,

Fig. 216.1. Jacobsen Chalk Tester.
A felt tape is pressed against a weathered painted surface at a given rate of pressure. The adhering chalk is rated by visual reference to photographic standards on the Jacobsen Chalk Rating Chart (10-0 for low-chalk to high-chalk level). This test is for erosion under actual weathering conditions.

which are best for holding paint; white pine, typical of clapboard siding, and medium in paint-holding ability; and Southern yellow pine, with its poor paint-holding power.

Panel selection is covered by ASTM standards also. The reference is ASTM D358-51T, *Wood To Be Used as Panels in Weathering Tests of Paints and Varnishes.*

Exposures are made by individual paint companies and raw material firms as well as by independent test fence operators in various sections of the country. A great deal of testing is constantly underway, mainly at the research facilities of raw material firms. Scores of tomorrow's house paint formulations are on racks now undergoing three and five years of testing for paint companies whose technical staffs have developed new products that previously passed accelerated weathering tests and thus became candidates for tests under actual naked conditions of sun, storm, wind, rain, sleet, and all the deterioration that natural microbiological enemies can impose.

All coatings manufacturers with any pride in the integrity of their products—and that covers most—are zealous in their weathering tests and have data to sustain their claims.

216.2 Structural Steel. Testing the protective coating of steel is complicated by the need for strict duplication of surface conditions on the test materials. Panels that have been wire-brushed first and exposed for even a relatively short time are likely to have oxidized more than panels prepared just before paint application. This can affect performance because meticulous preparation of steel surfaces is frequently vital to optimum coatings performance.

The ASTM provision for Conducting Exterior Exposure Tests of Paints on Steel, ASTM D1014-51, calls for using three test surfaces: one is hot-rolled steel angles, 4 in. wide, 1 ft. long, and 3/8 in. thick. Before testing, one-half of each flange on each side should be 50 percent wire-brushed but rusty, and 50 percent natural mill scale.

The second consists of hot-rolled steel plates of a type used in tank construction. These should be 12" × 12" × 1/4" and should also be one-half wire-brushed and rusty and one-half mill scale.

The third panel should be sheet steel completely free of mill scale and rust.

The first two represent reality, because that, unfortunately, is how most steel appears when painted in the field. The third material is how it should be if the painting contractor selected for the job is conscientious, which is another way of saying that his foremen are properly trained, or if the architect or prime contractor involved has eagle-eyed inspectors.

The accuracy of weathering tests for structural steel depends on the care of the testing organization in controlling film thickness and in getting comparable amounts of rust and mill scale on the test panels. The former is somewhat easy; the latter requires that the specifier ask what laboratory did the testing. Special laboratories doing nothing but coatings work have gained acceptance of government agencies, and where structural steel protection is sufficiently important, their tests may be necessary.

To overcome the subjective factors involved in panel preparation, at least four panels of each type should be used, and their results averaged. Photographic standards for degrees of rusting are provided in ASTM D610-43 and Ld-1-39. Quick determinations of the comparative resistance of different formulations to corrosion can be obtained on test fences where southern exposure at 45-degree angles are made—near ocean salt spray.

216.3 Accelerated Weathering. Technical men have long sought accelerated dependable testing methods, but only recently have techniques been developed that offer a fair degree of dependability.

Weather-Ometers, consisting of carbon arc lights to simulate the sun's deteriorating ultraviolet rays and water jets, intermittently emitted, have been correlated with actual performance of some materials; and thus, they have some validity for exterior performance of enamels, lacquers, and clears. For house paint and structural steel, they may aid in the early elimination of clearly inferior coatings, but they have little or no standing as performance indicators. Even as screening devices for house paint they are open to question, because some products they have condemned turned out to be worthwhile on actual exposure tests.

To overcome the objections to accelerated testing, ingenious technologists have varied Weather-Ometer procedures, seeking to approach the timing of daylight and night. Water cycles have been varied, arc lights have been switched and alternated, and in some instances, water has been sprayed constantly even when lights were turned off for periods, to simulate the diurnal change.

217 Film Thickness. Effective protection often depends on adequate film thickness. Painters are not scientists, nor are many of them concerned with end results; hence, their indifference can lead to films that fail to do the job. That means that all the skillful selection of a coating and meticulous surface preparation can be all for naught because the coating is applied in too thin a flow. Film thickness gages are available, and inspectors should use them.

217.1 Wet Film Gages. These are described in ASTM D1212-54 and are of two types. The first is in the form of a spool with eccentric barrels on flanges. As the barrel is rolled on the film, a change in clearance occurs between the surface of the wet film and the eccentric barrel as the flanges cut through to the substrate. That point where the barrel makes

Fig. 217.2A. Tinsley Thickness Gage.
Magnetic attraction is measured over coated ferrous surfaces by pulling the glass tube away from a spring-held magnet in its nose. Thicker coatings decrease magnetic pull on the magnet; thinner coats increase it.

contact with the paint film can be noted directly as the wet-film thickness in mils.

The second ASTM-approved gage is the Pfund gage, which uses a convex glass lens that is pressed into the wet film until it hits the substrate. A wet spot remains on the lens, and its diameter is the depth of the film. The Pfund gage is often preferred in the field because it leaves a smaller mark on the film than the barrel type.

217.2 Dry Film Gages. These are described in ASTM specifications that are available for instruments used to measure nonmagnetic coatings on magnetic substrates (ASTM D1186-53), and for electrically nonconductive dry films on nonmagnetic metal substrates. A third ASTM specification describes a small, drill like instrument that penetrates the film, measuring the distance from the film surface to the substrate. Typical gages include the Aminco-Brenner Magne-Gage, and the American Instrument Company's Filmeter. See Figs. 217.2A and 217.2B.

A common method for measuring dry-film thickness relies upon microscopic examination. The film is cut at a small angle with a sharp knife. The microscope is focused on the top of the cut, and the fine adjustment wheel's position is marked down. The focus then shifts to the exposed substrate surface. The film thickness is calculated from the number of turns needed to focus from one spot to the other.

Fig. 217.2B. Inspector Thickness Gage.
A dry-film gage for field work, the Inspector works with nonmagnetic films over ferrous surfaces by measuring magnetic attraction as it changes with coating thickness.

218 Moisture Testing. Moisture meters are available to determine if excess moisture is present in wood or cementitious materials, on the surface of metal or in the atmosphere to such an extent that it would interfere with drying of the painted surface.

Fig. 218.1. Delmhorst Moisture Detector.
Current passing between the electrode probes tells when moisture in wood is excessive. This pocket meter may also be used for plaster and for asbestos shingles.

218.1 The Delmhorst Detector. This meter uses a probe that can determine water content of wood or cement. See Fig. 218.1.

218.2 Portable Moisture Meters. Portable moisture meters are available for field detection of excess moisture in the air. Considerable numbers of specifiers are now writing into their instructions prohibitions against painting when moisture content of the air rises above preestablished levels.

HELPFUL ADDITIONAL READINGS

1. Burns, R. M., *Protective Coatings for Metals*, 3rd ed. (New York: Reinhold Publishing, 1967), pp. 314-75.
2. Chatfield, H. W., *The Science of Surface Coatings* (Princeton, N.J.: D. Van Nostrand Co., 1962), pp. 442-75.
3. Nylen, Paul, and Edward Sunderland, *Modern Surface Coatings* (New York: Interscience Publishers, 1965), pp. 586-608, 697.
4. Taylor, C. J. A., editor, *Paint Technology Manuals*, Vol. 5 (New York: Reinhold Publishing Co., 1965), pp. 25-163.

chapter 3
Surface Preparation

300 General. Teams of skilled chemists have succeeded in providing the construction industry with a variety of excellent coatings that can perform just about any reasonably difficult protective task demanded of them—if they are used on suitably prepared surfaces.

When we get to the subject of know-how in surface preparation, we part from the chemist and have to lean on the acquired experience of maintenance engineers, science-minded painting contractors, and dedicated sales service engineers working with paint companies and equipment firms. Paint specifiers often ignore the lessons learned by these practical men and simply specify materials and let human nature run its course in regard to surface preparation procedures. Too many clients, as a result, run high risk of suffering severe disappointment. Even the finest coating is a prime candidate for failure or, at best, short life if it is applied on an improperly prepared substrate.

One important lesson that has been learned by analytic consumers of large quantities of paint is that the cost of paint used on a job verges on being insignificant; the big cost is in surface preparation and application; and preparation is more often the major expense.

In a study of 16 paint systems used on industrial structures, where paint costs are high, it was found that even here the cost of material comprised only 6 percent of the total cost of painting and maintaining a building surface. Application and

Surface Preparation

preparation made up 94 percent. Preparation ranged from 15¢ per sq.ft. for wood or plastic flooring to 19¢ for hand brushing of exterior wood or steel surfaces, and up to 30¢ for sandblasting exterior steel surfaces for service in corrosive atmosphere. By 1977, these costs had increased by 75 percent or more. Regardless of the cost—within reason of course—suitable surface preparation is indispensable if maximum life and finest appearance are to be obtained.

In the last decade an entirely new calling has developed —coating systems engineering, to develop information and devise plans for obtaining sustained surface protection for buildings, ships, tanks, highway bridges, or whatever structure is subject to environmental deterioration. The task of these men is to decide how to get maximum protection for a minimum long-range investment.

Further details of the work of these men, together with the return on invested dollar-per-square-foot-per-year is reported in Chapter 10. Suffice it to say here that money invested in surface preparation has been well spent, according to these professionals.

It has long been acknowledged that such sensitive substrates as iron and steel require special treatment prior to the application of coatings; but, by and large, many specifiers rely heavily on supplier-company representatives and painting subcontractors to determine what preparation is needed— even where iron and steel are involved.

When it comes to cementitious surfaces and wood, very little is usually specified as to condition of surface prior to painting.

This section will be devoted to methods of preparation as well as the need.

301 Four General Guidelines for Surface Preparation. Four general rules or guidelines should be carefully observed:

1) Clean off loose material that doesn't belong on the surface, such as dust, rust, old paint, or anything that can be removed by hand with the aid of a brush, broom, chisel, scraper, sandpaper, or steel wool.
2) Utilize artificial means such as power equipment or chemicals to eliminate tight, stubborn material on the surface, such as grease, chemicals, gums, exudates, and

anything else that may lift later and take off the surface coating with it or that may prevent adhesion of the coating.
3) Enhance adhesion where necessary by roughing the surface through chemical etching, sandblasting, or some method of abrading to provide an anchor pattern.
4) Modify the surface chemically, where necessary, to create a substrate that will be receptive to the coating when the original material is inhospitable.

302 Implementing the Four Surface Preparation Guidelines.
Hand cleaning is usually adequate for carrying out the first rule or guideline, which deals mainly with loose or easily removed material. See Fig. 302.

The second rule requires power tools or chemicals. Included under power tools are a special category of machinery that directs abrasives to surfaces with stubborn contaminants.

The third rule sometimes requires abrasive blasting but more often merely simple hand-abrading with sandpaper or steel wool. Some required roughing may be accomplished with chemical etching.

The fourth rule requires chemicals that will react with substances in the surface material itself to provide a hospitable synthetic substrate for the coating, a substrate that also may help prevent harmful chemical reactions between the actual substrate itself and chemicals, or moisture, in the atmosphere.

Techniques listed above will be described briefly in the next paragraphs. Detailed treatment of each of these techniques will be found later in this chapter.

303 Hand Cleaning. Hand cleaning includes use of water and detergents and solvents on brushes or rags to wipe off foreign matter. Hand cleaning may be used under each of the following circumstances:
1) For long-oil alkyd, oil, or phenolic paints, which are good wetting paints and more adaptable than most to surface problems.
2) For even more sophisticated coating systems when areas are small and don't justify the use of heavy equipment, or where abrasive dust and other foreign matter likely

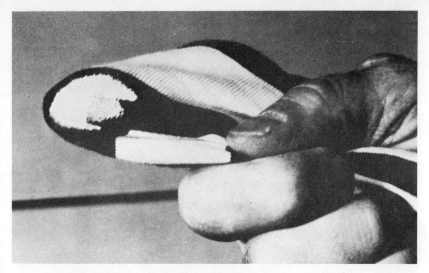

Fig. 302. Hand Cleaning of Chalked Surfaces.
Hand cleaning can be used for removing chalk prior to repainting. A test for chalk is to wipe a rag over a suspected surface. Oleoresinous primers are preferred for chalky surfaces, but latex may be used with an additive to improve wetting.

to result from mechanical preparation will harm adjacent surfaces or machinery.
3) When time does not permit waiting for mechanical equipment.
4) Where surface is in a tight space and equipment cannot be brought in.

It should be mentioned that equipment is available to facilitate the surface cleaning task even when requirements are relatively light. Shown in Fig. 303 is a low pressure detergent cleaner which can take the place of the traditional sponge, detergent, and "elbow grease."

304 Power Cleaning. Power tools include such devices as wire brushes, disks, scalers, scrapers, needle guns, and tiny hammers. The following points are relevant:
1) Power tools may provide a quicker way to do some jobs that are done too slowly with hand tools.
2) They may be used to remove tightly formed oxides on metal, which if not removed will eventually cause paints to lift, making way for moisture and then corrosion.
3) They provide adequate roughening for some surfaces where the profile doesn't have to be too deep or definite.

Fig. 303. Low Pressure Detergent Cleaner.
This piece of equipment aids in speeding up a job which could have been done by hand.

304.1 Needle Guns. Needle guns permit easy conforming to irregular surfaces such as seams and corners and welds and bolts.

304.2 Wire Brushes. When specifying the use of wire brushes it is important to require that surfaces are roughened rather than polished. Moreover, rotary tools sometimes generate enough heat to cause tiny particles of metal to fuse; and these can accelerate rust and corroding action.

304.3 Water Blasting. Water blasting is usually inadequate to remove tight mill scale, which forms on oxidizable metal, but it will remove loose rust or dirt. This is often the best method available when abrasive sand would damage nearby equipment

Fig. 304.3. Water Blasting.
Water blasting equipment is being used to clean an exterior industrial surface.

or sensitive instruments or material. It can encourage rust, however. See Fig. 304.3.

305 Abrasive Blasting. Natural sand and flint are used as well as ground walnut shells and corn cobs. The latter two are used mainly for delicate surfaces and will not remove mill scale or tight rust. Manufactured abrasives are available, such as copper or lead slag, which are fast cutting, sharp, and needle-like. Care must be taken that they do not get stuck in steel surfaces and cause trouble later on. Also available are nonmetallics such as silicon carbide and aluminum oxide, which can be reused.

Typical particle sizes used are 16 to 40 mesh or 30 to 50 mesh; finer sizes are faster but larger ones cut deeper. Sand should be of selected grades and uniform. Suppliers should be required to furnish angular shaped particles for cutting and round ones for a polishing effect, when that is desired.

Fig. 305A. Abrasive Blast Cleaning.
Shown here is a vacuum-type sand blasting unit by Clementina, Ltd.

Degrees of blast cleaning as established by the Steel Structures Painting Council are described in Section 309. Abrasive blast cleaning equipment is shown in Figs. 305A and 305B.

306 Chemical and Solvent Treatment. The five types of chemical and solvent treatment are described below.

306.1 Solvent Wiping and Degreasing. Solvents capable of removing oil, grease, dirt, paint stripper, and any other foreign matter are utilized. Ordinary mineral spirits will dissolve and dilute most contaminants sufficiently to permit them to be wiped or washed off. First, dry materials should be removed with a wire brush. Next brushes or rags should be saturated

Fig. 305B. Abrasive Blast Cleaning.
Abrasive cleaning units are available with portable features for ease in use.

with solvent and worked over the surface. Several rinses may be necessary to clean adequately. Care should be taken to prohibit use of solvents with low flash points or that are toxic at utilized concentrations. Solvents likely to cause fumes harmful to nearby personnel, equipment, or materials should also be avoided.

306.2 Alkali Cleaning. Alkali cleaners such as trisodium phosphate, caustic sodas, and silicated alkalis are considered more efficient, and less likely to be harmful than solvents. They require more work because they are usually heated in water to somewhat high temperatures (150° to 200° F). The higher the temperature the more efficient the operation. *These alkali cleaners should never be used on aluminum or stainless steel.*

Alkalis go right to work on oils and grease, setting them up as soaps that wash away in water. Other ingredients in the cleaners aid in eliminating dirt and in preventing mildew.

306.2a Rinses. It is very important that clear water rinses follow use of these alkalis because their presence on the surface could be worse than the original contaminants. Hot water rinses, preferably under pressure, should be used.

306.2b Test for Alkali Removal. To determine if the alkali cleaner has been thoroughly removed, Universal pH test paper should be placed against wet steel to check for the presence of free alkali. As a precaution, the specifier should require that the cleaners contain 0.1 percent chromic acid or potassium dichromate to prevent corrosion.

306.3 Steam Cleaning. Pressurized steam or hot water, preferably with a detergent added, removes oil and grease by liquefying and diluting them. Wire brushing or brush-off blast cleaning, when used, should come after the steam-cleaning process.

306.4 Acid Cleaning. Iron, steel, concrete, and masonry may be cleaned with acid solutions. (See details in Sections 307 and 309.)

Iron and steel are treated with solutions of phosphoric acid with small amounts of solvent, detergent, and wetting agent. This removes oil, grease, dirt, and miscellaneous contaminants. Phosphoric acid has an advantage over alkali cleaners because it removes light rust and provides a faint etch to the surface, thereby aiding adhesion of applied coatings. Four types of acid cleaners are usually obtainable, each to be used for a particular method of use:

1) *Wash-off,* whereby the cleaner is applied and allowed to act for a specified time, rinsed, and the surface dried.
2) *Wipe-off,* used where rinsing is impractical. The cleaner is applied, allowed to act for a specified time, then is wiped off with clean, damp cloths in successive wipes until thoroughly clean and then the surface is dried prior to painting.
3) *Hot-dip,* where the surface can be immersed in hot cleaner, rinsed in hot or cold water with successive rinses in weaker cleaning solution, dried, and painted.
4) *Spray,* using pressurized equipment for the wash-off method.

Surface Preparation

306.5 Paint Removers (Solvents). Architects working on remodeling projects, and maintenance department heads are likely to need specification information for paint removal if their roles in surface preparation are to be carried out.

Paint removers, classified as chemical cleaners because they contain solvents, are divided into four types. These four types are discussed in the following subsections in order of increasing effectiveness.

306.5a Nonchlorinated Solvents. For removing oleoresinous finishes and relatively easy-to-remove ones, low-cost strippers based on nonchlorinated solvents are available. They usually consist of lacquer-type solvents classified as aromatic and having high solvency ratings. Ingredients include benzene and toluene, with methanol and some ketone solvent often added as co-solvents. Thickeners and waxes are usually present to hold the removers on the surface and to aid in lifting the final product from the surface.

306.5b Nitric Acid Solvents. For removing tough coatings verging on the hardness of porcelain, such as epoxies, a solution of nitric acid in dimethyl sulfoxide is used at a temperature range of 120°-130° F.

306.5c Chlorinated Hydrocarbon Solvents. The most effective paint removers, other than those using expensive dimethyl sulfoxide, mentioned above, are chlorinated hydrocarbons, of which methylene chloride is the most efficient and the least toxic. However, public antipathy for chlorinated hydrocarbons may eventually halt the use of this material except under carefully controlled conditions.

306.5d Other Solvents. Several uncommon solvents are effective for paint removal, in addition to dimethyl sulfoxide, and should be investigated if all else fails. They are 2-nitropropane; tetrahydrofuran; 1,1,2-trimethoxyethane; and dimethyl formamide, which, like dimethyl sulfoxide, is among the richest substances in hydrogen double-bonds. This means that these solvents are the most effective agents available for penetrating into materials hitherto regarded as impenetrable. That accounts for their ability to cut right into epoxies.

306.6 Chemical Modification of Surfaces. Four types of chemical modification may be identified.

306.6a Hot Phosphate Treatments. Hot phosphate treatments, because they require heat, are mainly used for shop application to architectural iron or steel that will be used in corrosive atmospheres and where adhesion of coatings will be problematic without such treatment. Zinc or iron phosphate solutions are used, the latter when high gloss topcoats are required. (Details of this treatment and cold phosphate treatment will be found in Section 309.)

306.6b Cold Phosphate Treatment. Although not as good as hot phosphatizing, cold phosphate treatment has the advantage of suitability for field application. The cold method uses phosphoric acid, a wetting agent, and a water-miscible solvent with water.

306.6c Chromate Treatment. Chromate treatment is a factory process mainly used for aluminum, zinc, tin, magnesium and their alloys. It involves dipping the metal in a hot solution of sodium chromate and sodium carbonate, yielding a coating of hydrated chromium oxide plus an oxide of the treated metal, which is a good base for paint and protects against atmospheric corrodants. A second process uses chromic, phosphoric, and hydrofluoric acids for the dip and gives a grey-green to grey-brown adherent surface film which, chemically, is a combination of chromium phosphate, metalic oxide, and metalic phosphate, depending on the metal treated.

306.6d Wash Primers. These are really a form of cold phosphatizing, because phosphoric acid is the active ingredient. A film-forming synthetic resin, usually polyvinyl butyral, is added to the acid, along with zinc chromate. The material leaves such a thin film, about 0.0004 in., that it cannot be called a primer, but it does aid adhesion of subsequent coats. It also permits delay in application because the surface is protected against atmospheric contaminants. Wash primers are particularly helpful when metal is to be used underwater.

307 Preparing Cementitious Surfaces. Clean, dry surfaces are,

Surface Preparation

of course, necessary when cementitious surfaces are to be painted. In addition, however, the surfaces must be nonreactive, which means that the alkali chemicals in the mixes used to produce the surface must be neutralized; construction defects also must be corrected to assure that moisture will not seep through from behind and eventually break the bond of adhesion between surface and coating. Moreover, any mildew on the surfaces must be removed by scrubbing with a solution of properly reduced sodium hypochlorite, which is common household bleach.

307.1 Types of Cementitious Surfaces. Types of cementitious surfaces include stone, brick, and concrete.

307.1a Stone. Washing or brushing should remove dirt and loose stone particles, but if the face of the stone is shiny, indicating extreme hardness and probable imperviousness, blasting may be necessary to provide sufficient roughness for coating adhesion.

307.1b Brick. Few problems should be encountered in cleaning brick. Grease, dust, or other foreign matter should be removed with a household detergent or solvent. Efflorescence—a white, powdery deposit—and laitance on the cement grout, which is a fine cement powder that sometimes floats to the top of drying cement, should be removed in accordance with information contained in Section 307.3.

307.1c Concrete, Cement, and Mortar. Because the liquid portion of wet portland cement is alkaline until the cement reacts with air to form carbonates, most paint suppliers recommend at least three months of weathering before new concrete, mortar, or stucco is coated. When circumstances require earlier painting, or before excess alkalinity has dissipated, neutralization can sometimes be accomplished by using a wash consisting of phosphoric acid and zinc chloride. Some paints may not be compatible with this wash, and latex paints, particularly, should be checked for suitability. Some paint companies suggest a zinc sulfate wash when oil-based paints are to be used. When wood or metal forms are used to shape the concrete surface, oils are often used to free the forms, and these oils are generally nondrying. They, too, must be weathered for dissipation or be removed, or else the adhesion of coatings will be affected.

The outer surface of concrete (at least 1/16 inch deep) should be dry before painting. When in doubt, a moisture meter should be utilized. Prior to painting, a check should be made to see if any holes have been made on the surface, in which case they should be opened up sufficiently to permit proper patching with approved patching compounds.

307.2 Preliminary Cleaning. Hand scrubbing with detergent solutions may be needed to remove oily or greasy contaminants from a cementitious surface. If this is not adequate, steam and detergent combinations or solvents may be used. When acidic contaminants are present, an alkali cleaner may be needed; while if alkali contaminants are present, an acidic cleaner may be required. In any event, before using acid etching, sandblasting, or power grinding, these contaminants should be removed to avoid having them ground into the surface by these preparation methods.

307.3 Removing Efflorescence, Laitance and Glaze. Concrete and masonry surfaces must be free of efflorescence (soluble salts leached out of these materials by water or other liquids and which end up on the surface as an unattractive, white dewy film) and also should be free of light glaze. Concrete and masonry surfaces also may have laitance, a fine cement powder that for various physical reasons floats to the surface after concrete is poured and while it is hardening. This too must be removed, because, like efflorescence, it will later cause paint to lift. Laitance and efflorescence should be wire brushed or scraped. Any grease or oil on the surface should be removed by solvent wiping or by steam or alkali cleaning before the acid-cleaning treatment is carried out.

307.3a Acid Cleaning. Acid cleaning is done in the following steps:

1) Wet the surface thoroughly with clean water.
2) Scrub the surface with a 5 percent solution, by weight, of muriatic acid, or 10 percent in extreme cases of

Surface Preparation

efflorescence or laitance or where the surface is strongly glazed and the muriatic acid is to serve as an etch.
3) The acid solution must not remain on the surface more than five minutes before scrubbing; otherwise, it can lead to the surface formation of salts, which are difficult to remove.
4) Without delay, when scrubbing has ceased, sponge or rinse away the solution with clean water.
5) Work should be done on small areas, no more than four sq.ft. in size.

307.3b Glaze Removal. If steel-troweled concrete appears so dense, or even glazed, that coatings may not be able to penetrate, a simple test should be made by placing a few drops of water on the surface. If the water is not quickly absorbed, then acid etching (described above) is necessary to eat into the glaze.

1) If the acid etch is not adequate to permit passing the water-absorption test, removing the glaze will require either abrasion by a hard, rough stone or by light sandblast. Another method is to allow it to weather for 6 to 12 months.
2) Still another method for handling glaze is to treat the surface with a solution of 3 percent zinc chloride and 2 percent phosphoric acid. This is allowed to dry on the surface and acts as a tie-coat for the primer coat of paint.
3) Some coatings are sensitive to the alkali in concrete, notably alkyds; and when these are to be used, acid-cleaning methods are needed to eliminate the free alkali likely to be present on the surface. To assure that all free alkali will be removed, specifiers should require application of dampened pH testing paper to several places.

307.4 Sandblasting. Blasting is needed, particularly when poured concrete or precast concrete forms are to be subjected to heavy service, such as frequent abrasion, continuous immersion, or even frequent spillage. This is due to the air pockets that almost always form in these kinds of concrete, which must be opened sufficiently to free the air and permit filling, so that a smooth coating, entirely free of tiny holes, will result. If these air pockets are not opened, the air will eventually force its way out, causing breaks in the coating, permitting moisture to enter the concrete, and eventually causing the coating and concrete to disintegrate. Neither acid etching nor hand or power abrading opens the surface sufficiently to free the air trapped in the pockets.

307.5 Hand or Power Tools. Loose material can be removed efficiently enough by wire brushing, but this will not be effective in opening air pockets. Impact tools are too slow for economic usage. Power grinders are also considered too slow, but may be used for small areas or for finishing off incomplete sandblasted jobs.

307.6 Acid Etching. After first removing loose concrete and greasy residues, if any, etching may be done with a solution of 1 part muriatic acid, 30 percent, with 2 parts of water by volume. Application is usually recommended at a rate of 50 to 75 sq.ft. per gallon. The acid solution should be kept on the surface only two to three minutes and then rinsed off completely—to prevent salt formation.

307.7 Sacking. New cement may be prepared for paint by hand-rubbing a mixture of cement, sand, and mortar into the surface. This has been found effective in sealing voids and bubbles and leaves a surface receptive to a coating. These so-called sack coats should not be used with thick applications of paint, which tend to shrink and pull the sack coat away from the underlying concrete. Use should be limited to surfaces on which thin, flexible, nonshrinking protective coatings will be applied.

307.8 Specifying Information. For heavy-duty service, see data on sandblasting in Section 307.4. For moderate service, hand- or power tool-roughening is adequate if abrasion is likely to be light and if liquid spillage is expected to be infrequent. Voids need not be completely widened, but dust trapped in them should be drawn out by vacuum or air blasting. For horizontal surfaces subjected to moderate service, acid etching

Surface Preparation

is adequate; on a vertical surface, however, the etchant will not remain on the surface the necessary 2 to 3 minutes. Light sandblasting may be necessary for vertical surfaces. On green concrete, a sack coating may be desirable.

307.9 Special Types of Cementitious Surfaces.

307.9a Concrete Block. Because of its rough surface, concrete block is not troubled to any extent by air pockets so that the removal of dust and dirt is all that is needed unless free alkali or grease is present, in which case chemical cleaning may be necessary.

For the most efficient paint usage, as well as appearance, block fillers should be used prior to painting. These can be low priced, consisting of conventional polymers that do moderately good jobs, or higher priced, consisting of polyesters or low-viscosity members of the epoxy family, or synthetic rubbers. The filler should have high pigment content and low solvent to minimize shrinkage. If surface voids are large, the use of two coats is advised. Allow 24 hours for the final coat to dry before painting.

307.9b Sprayed Concrete. Sprayed concrete is hard and dense and usually has few air pockets. Because of these characteristics hand cleaning is all that is necessary prior to painting, unless the surface is very smooth, in which case acid etching or sand blasting may be needed.

307.9c Concrete Floors. Steel-troweled floors may be too smooth to provide adequate adhesion for paint so acid etching or sandblasting may be needed; however, troweled surfaces usually have no air pockets, so roughening is all that has to be accomplished.

307.9d Stucco. This consists of portland cement, mortar, sand, plus coloring matter on some occasions. Thus, it requires weathering, as does concrete, to cut alkalinity.

307.9e Plaster. About one month is usually needed for plaster to dry thoroughly. If high performance coatings, with great cohesion are to be used—particularly if the plaster is old or very smooth—some degree of nondestructive roughening will be necessary prior to application.

307.10 Repairing Cracks and Openings. When filling cracks, holes, or other gaps in cementitious materials or wallboard areas to be filled should be dampened with clear water and then the compounds should be applied with a putty knife or small trowel. If possible a slightly convex shape should be given to the treated surface to allow for shrinkage. If portland cement grout is used, the repaired cracks should be kept damp for several days after application.

The following materials may be used:

1) *Patching plaster*—For large gaps in plaster, the patching material is virtually ordinary plaster except that it has a component to accelerate hardening.
2) *Spackling compound*—Small holes and cracks in wallboard and plaster are repaired with soft powder that is mixed with water, or which sometimes comes in paste form. Spackling compound is also used as a joint cement in conjunction with perforated tape to level off joints between sections of wallboard. For this purpose the compound is spread over the joint with a spackling knife, covering adjacent nail heads. Perforated tape is laid on this and then covered with the compound.
3) *Portland cement grout*—Portland cement grout and brick grout are used for cracks in concrete and masonry.

308 Preparing Wood Surfaces. Since wood is porous, care must be taken that contaminants are removed entirely from the surface, or isolated if they have penetrated into the wood cells and pores. Oil, grease, and waxy substances are especially difficult to remove from the cells and pores below the surface. Saturating the wood usually will not remove them. Shellac, aluminum paint, or knot sealers, are able to isolate them.

A common contaminant frequently responsible for adhesion problems is moisture. Wood is usually kiln dried to remove moisture, but often some water remains. Specifiers, to assure themselves that moisture content is low enough, should require painters to check lumber with a moisture meter before applying wood primers or sealers.

Surface Preparation

Safe moisture levels are 9–14 percent for exterior woodwork; 5–10 percent for interior woodwork; and 6–9 percent for wood flooring.

Even in acceptably sound wood, some unexpected knots and discolored sap wood may turn up. These can lead to irregular or sap-dyed finishes when the rosin or natural dyes in the wood are absorbed by the solvents in the applied paint. When unsound wood is met, painters should be instructed to remove it and fill in the space with putty or plastic wood.

Where fresh sap or natural dyes in the wood or knot threaten to discolor the coating, which is usually receptive to these natural wood chemicals, it is necessary to seal the threatening surface with a material that is nonreactive with the discolorant and which, at the same time, can provide a stable substrate for the paint. For this, an alcohol solution of a phenolic resin and polyvinyl butyral are used.

Plywoods to be used for interior or exterior surfaces should be selected from among resin-laminated grades if they are to be painted. Casein or animal glue adhesives when used to bind the plies may permit moisture penetration, which will cause loss of adhesion, peeling, and blistering. Some so-called interior plywoods are not sufficiently moisture resisting and should be avoided.

Essential steps in preparing wood surfaces include sanding, sealing pores, roughening, treating mildew, removing troublesome finishes, flame cleaning, and sealing joints.

308.1 Sanding. Where coarse sanding is necessary, it should be done at an angle to the grain. Medium sanding should follow on the grain. Fine sanding should also be with the grain until the surface is able to reflect light at roughly a 45-degree angle. Rough spots can be detected and smoothed.

If a curved area is to be sanded, the sand paper should first be flexed over a curved surface to shape it for abrading.

Small depressions in wood needing to be sanded can usually be smoothed first by wetting the area with hot water since the depressions are usually caused by compression of the wood cells, which can regain their former bulk by reabsorbing the moisture that had been pressed out of them. One way to do this effectively is to place a drop or so of water in the depression and apply a hot wire to boil the water and swell the compressed cells. In any method, sufficient time must be allowed to let the treated area dry thoroughly before sanding and painting.

For varnishing and lacquering, where the decorative grain is to be visible, sanding requires extra care. Coarse sanding should be at an angle to the grain, but medium and fine-grained sanding which follow should be with it. Small irregularities in the sanded surface can be observed in the light and smoothed over after a coat of hard varnish or shellac has been applied to facilitate sanding.

When sanding has been completed, mineral spirits should be used to wipe off the wood dust.

308.2 Sealing Pores. Sealing of pores is necessary in open-pore wood. (See Table 308.2.) A filler made of fine silica in oil or varnish is used. It may be colored. Its consistency depends on the porosity of the wood and is adjusted by adding mineral spirits. A stiff brush is used to apply it, and excess sealer is wiped off with a dry rag—across the grain. The final step is application of varnish or lacquer.

Before painting, nail heads and screws should be driven below the surface of the wood. The resulting small holes should be filled, after priming, with putty or plastic wood. Small cracks and other surface defects should also be filled using these materials before priming could cause the oils and solvents in the primer to be absorbed in the wood pores or cells. The absorption would ruin the fillers. The dried filler has to be sanded flat before the entire area is painted.

When oak or other open-pored hardwood is to be painted, a thin version of putty is applied to close the pores. The dried, sealed surface should be sandpapered prior to painting.

End pieces of wood that will be exposed, such as window sills or frames, or wood that will be built into masonry, should be given a heavy primer sealer to prevent water absorption into the pores.

308.3 Roughing. Previously painted wood must be prepared by roughing to receive a new coat. Loose paint must be removed by a wire brush or some form of abrasive. Stubborn paint may be flame-cleaned, if safe to do so. (See Section 308.6.) Hard wood should be stripped with paint removers only, since flame may singe the surface pores. (See Section 306.5.)

Surface Preparation

Table 308.2. *Wood Classification According to Openness of Pores.*

Name of Wood	Soft Wood	Hard Wood	Open Pore	Closed Pore	Notes
Ash		x	x		Needs filler
Alder	x			x	Stains well
Aspen		x		x	Paints well
Basswood		x		x	Paints well
Beech		x		x	Varnishes well, paints poorly
Birch		x		x	Paints and varnishes well
Cedar	x			x	Paints and varnishes well
Cherry		x		x	Varnishes well
Chestnut		x	x		Requires filler, paints poorly
Cottonwood		x		x	Paints well
Cypress		x		x	Paints and varnishes well
Elm		x	x		Requires filler, paints poorly
Fir	x			x	Paints poorly
Gum		x		x	Varnishes well
Hemlock	x			x	Paints fairly well
Hickory		x	x		Needs filler
Mahogany		x	x		Needs filler
Maple		x		x	Varnishes well
Oak		x	x		Needs filler
Pine	x			x	Variable
Teak		x	x		Needs filler
Walnut		x	x		Needs filler
Redwood	x				Paints well

308.4 Treating Mildew. Mildew, even if only suspected, should be treated by a wash consisting of bleaching compound and cleaners. A suggested formulation that can be specified for this purpose comes from the National Paint and Coating Association.* It calls for:

    2/3 cup trisodium phosphate
    1/3 cup detergent powder
    3 quarts of warm water
    1 quart of 5 percent sodium hypochlorite (bleach)

---

*Preventing Mildew (Washington, D.C.: National Paint and Coating Association, 1969).

This makes a little more than one gallon. It should be scrubbed on full strength and then thoroughly rinsed off.

308.5 Removing Troublesome Finishes. Seemingly uncoated exterior or interior plywood may have a finish on it that can cause adhesion problems. Some of these materials are overlaid with resin-impregnated fiber and some have overlays of varied density. The supplier should be asked to furnish informaton about the finish, if any, with suggestions for painting.

When information is not available it is advisable to clean the surface with steel wool and then follow with mineral spirits or if necessary paint-removing solvents.

308.6 Flame Cleaning. Passing reference was made to flame removal of paint from old wood surfaces. If the specifier believes the painting contractor will use this method, it is important to require that an experienced operator carry out the cleaning, because charring of the surface leads to poor adhesion of applied paint. Erratic char patterns on the substrate may cause corresponding irregularities in the finished surface.

To avoid this, a specification should be given that the flame pass over the surface to be cleaned at a consistent speed, and that a stripping knife follow behind it to remove the hot, fluid paint that rises as the flame passes over it.

It is also important that the knots encountered in the flame-cleaning process be "boiled up" or heated sufficiently so that all remaining resin is flushed out by the flame and then removed by knife. Otherwise, exudations may occur later. If a knot exudes excessively, it should be reamed or drilled out, with a wood plug used to replace it unless the resulting hole is small enough to be filled with plastic wood or oleoresinous filler.

308.7 Sealing Joints. Suitable sealants should be selected wherever wood is to be joined to other materials. This is especially necessary where the juncture is with cementitious surfaces that may transfer water to the wood. Sealants are also required where some degree of movement is possible at the joint. Calking compounds are needed to prevent entry of moisture where wooden window ledges, trim, or molding meets other materials. See Section 400.

Surface Preparation
309 Preparation of Metal Surfaces.

309.1 General. Any one of a number of surface contaminants or by-products of metal manufacture—if not removed—can cause paint failure. These include grease, dust, oil, and—trouble of troubles—mill scale, which is a combination of several kinds of oxidized iron clinging tenaciously on the surface of steel where it is formed as a scarcely visible rash under the fiery influence of the rolling mill. If any of these impediments to good surface coating remain on the surface, the prospects for paint durability fall off sharply. They all contribute to poor adhesion and diminish the likelihood of perfect bonds between coating and substrate. When the bond inevitably ruptures because of the presence of one or more of these impediments, the metal substrate becomes vulnerable to water—the big enemy—which enters, sets up its liaison with atmospheric oxygen and hydrogen, and provides proper chemical and electrolytic conditions for the breakdown of iron, steel, or other metal surfaces, a condition we call *corrosion.*

To rid the surface of these unwelcome visitors, brushes, power needles, sand blasters, centrifugal blasters, acids, alkalis, and pickling solutions are needed to the same degree that paint is needed for the final job of covering the surface.

The proper preparation of metal surfaces has often provided dramatic proof of its benefits. J. C. Hudson, head of the famous Corrosion Laboratory of the British Iron and Steel Research Association, made a thorough study of the effect of surface preparation on life expectancy of an applied protective system.[*] He found that preparation by sandblasting increased life expectancy more than four times over that of a steel surface that had been merely weathered and wirebrushed before painting.

He also found that sandblasting was more effective when followed with two coats of paint than a four-coat system used on inadequately prepared metal.

Table 309.1A, based on Dr. Hudson's work, shows that intact mill scale, under ideal conditions, was found to be an effective substrate but still not as effective as sandblasting. (Remember that intact mill scale is rare in real-life conditions.)

---

[*]*Steel Structures Painting Manual,* Vol. 1 (Pittsburgh: 1969), pp. 110-15.

Table 309.1A. *Surface Preparation and Durability of Coatings on Metal Surfaces (in years of life).*

Surface Preparation	4-Coat Scheme (2 of red lead paint & 2 of red iron oxide paint)	2-Coat Scheme (both red iron oxide paint)
Intact mill scale	8.2°	3.0
Weathered & wirebrushed	2.3	1.2
Pickled	9.5°	4.6
Sandblasted	10.4	6.3

°Indicates that some test specimens survived when test had been concluded.

Even moderate weathering, as the tests show, can lead to rust formation and future trouble. Each preparation method listed was used on several test specimens. The figures given are averages.

Researchers were able to show conclusively that the easy way—weathering and wirebrushing—is the poorest way and that sandblasting, or pickling with low-priced sulfuric acid or other acids to remove mill scale and rust, yield substantial economic dividends, which, incidentally, are not indicated here but which will be discussed in more detail in Chapter 10.

Dr. Hudson published another study, which, spread over 15 years, showed that weathering alone is bad business and that specifying organizations should go to great lengths to see that applicators are zealous in their surface preparation. Here micaceous iron ore, a natural form of iron oxide, which only recently has become generally available in the U.S., was used effectively with pickling as the surface preparation, while weathering, or weathering plus surface heating, led to failure in about seven months. In Table 309.1B the third column refers to surface preparation that included heating the weathered surface to 200° C and wirebrushing.

Similar tests conducted in this country for the American Society for Testing Materials and the National Association of Corrosion Engineers confirm the findings of Dr. Hudson and his co-workers.

Since mill scale figures so prominently in the strategy of metal preparation, a better understanding of what it is may help specifiers know why it is so important to have it removed.

One of the best explanations appears in the *Steel Structures*

Surface Preparation

Table 309.1B. *Method of Surface Preparation for Painting vs. Durability.*

Painting Scheme (Two Coats)	Over Weathered Surface (years)	Over Pickled Surface (years)	Over Weathered and Heated Surface (years)
Black bituminous	2.6	13.1°	2.2
Lead chromate	2.7	14.5°	7.5
Micaceous iron ore	0.6	15.0	0.6
Red lead	1.1	13.1°	5.9
Red oxide	1.8	8.1°	2.8
Red oxide and zinc chromate	3.7	9.6°	5.3
White lead	2.2	7.0	1.7
Average	2.1	11.5	3.7

°Back surface had not failed after fifteen years' exposure.

Painting Manual, Vol. 1. In a section entitled "Simplified Theory of Corrosion," F. N. Speller writes that clean, metallic iron, even at normal temperatures, forms a thin, invisible oxide film by direct chemical combination with atmospheric oxygen. As temperatures increase in iron processing and tempering, the thickness of this oxide film grows until it becomes visible as it takes on color. These colors, known as tempering colors, range on carbon steel from a pale straw at 400° F to deep blue at red heat.

Mill scale formed on steel during rolling depends on the type of operation and the temperature. Of the three layers of iron oxide that make up mill scale (and which, incidentally, are magnetic), the layer adjoining the steel surface is ferrous oxide, FeO, which is unstable and easily oxidized to ferric iron. This is soon changed to ferric oxide, or rust. This rust expands as it forms; and, in expanding, it may loosen the mill scale. If paint has been applied over the scale, the loosening basically caused by the underlying rust undermines the paint.

An odd fact about mill scale is that it resembles a "noble" metal in relation to steel. That means that if water sneaks to an interface where mill scale and steel meet and sets up a tiny electric current—the kind that causes erosion of one of the surfaces—the surface that will erode is the steel.

We have seen in Table 309.1A that perfectly intact mill

scale under paint is an effective barrier against moisture and helps prevent corrosion of steel; but in the imperfect world of construction, a real mill scale surface is likely to have defects that can let in moisture. Because mill scale is more "noble" than mere steel it will be intact while "plebeian" steel is sacrificed in its behalf. That explains why, if a coating over mill scale is scratched, as so often happens, the mill scale beneath provides a perfect setup for a tiny corrosion cell at the crack, which insidiously spreads under the scale.

When such cracks do start, oxygen can reach down to the unstable bottom layer of the scale (remember, there are three layers, the bottom one easily convertible to rust) and start a corrosion process that causes the scale to lift and take the paint with it. See Chapter 7, Fig. 703.1.

Thus, we have two destructive processes possible when mill scale is allowed to remain on the metal surface to be painted.

Nonetheless, blasting or chemical treatment to remove scale is fairly costly, and some circumstances may not justify the expense. The specifier can choose among several cleaning methods, so he has scope for making a decision.

309.2 Mill Scale Removal. After metal surfaces have been cleaned of oil, grease, and dirt, they are ready for final preparation, which may be accomplished by: (1) mechanical treatment, (2) acid pickling, or (3) chemical modification. See Fig. 309.2.

309.2a Mechanical Surface Preparation. Mechanical treatment is divided into four classes, discussed below in increasing order of effectiveness.

Class 1. The least effective method of cleaning relies upon hand or power tools. This approach is effective where the corrosive situation is mild or normal, and the coatings used have sufficient wetting ability to adhere to tight contaminants or residues normally likely to remain on surfaces following cleaning.

Hand cleaning or power cleaning with simple tools is satisfactory for certain situations where thorough cleanup is either unnecessary or impractical. These measures are adequately

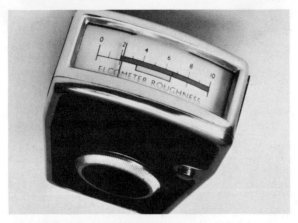

Fig. 309.2. Elcometer Roughness Gage.
Special magnetic contacts slide over a rough, sand, or shot blasted surface indicating changes in magnetism as peaks and valleys are passed over. This gives a surface profile to show the metal surface is ready for primer application.

described in the earlier part of this section where general methods are discussed.

Class 2. Flame cleaning or brush-off blasting is used to improve surface preparation when the environment is a little worse than normal.

Flame cleaning. The surface is rapidly heated by means of an acetylene or oxyhydrogen torch. Mechanical strains are set up in the mill scale which cause it to flake off; and the old rust present, because of its water content, literally explodes off. The surface is warm and should be painted at once. Flame cleaning is not considered as good as blast cleaning for removing rust and dirt. See Table 309.2a for gas consumption rates for flame cleaning.

Brush-off blasting (Steel Structures Painting Council SP-7). A low-cost mechanical method that is hardly more than a way to save time by supplanting hand tool and hand power tool labor. This method is used to remove old finishes that are in poor condition, loose rust, and loose mill scale. It is worthless for tight mill scale. It is useful for removing loose or deteriorated paint from masonry.

Table 309.2a. *Cleaning and Gas Consumption Rates for Flame Cleaning.*

Condition of Steel Surface	Net Linear Speed (ft./min.)	Surface Area Flame Cleaned (sq.ft. per hr., 6-in. head)	Oxygen or Acetylene Consumed (cu.ft./sq.ft.)
Relatively clean, flat surface	35	1000	.13
Slight amount of rust and loose scale	28	800	.17
Moderate amount of rust and loose scale	21	600	.23
Relatively clean riveted section	18	500	.26
Heavily rusted flat surface	14	400	.33
Heavily rusted riveted section	7	200	.65

Class 3. Commercial blast cleaning may be used for surfaces in moderately corrosive atmospheres. This includes water immersion and exposure to industrial or marine environments. Commercial blast cleaning is considered adequate to the performance of most paint systems under normal conditions of exposure. Specifications are covered by Steel Structures Painting Council SP-6.

This degree of blasting calls for removal of all loose scale, rust and other surface contaminants. Commercial blast, where the atmosphere is not likely to be corrosive, offers a big advantage over the finer methods—white metal and near-white metal blasting; this advantage is speed. The list of cleaning rates shown in Table 309.2a1 illustrates this advantage.

Class 4. This level of cleaning would be used for surfaces to be exposed to, or in direct contact with, strong chemicals

Table 309.2a1. *Rate of Cleaning.* *

Type	Sq.Ft.
White metal	100
Near-white	175
Commercial	370
Brush-off	870

*These are approximate cleaning rates at 100 psi with a 5/16-in. nozzle, per hour.

Surface Preparation 135

where any degree of rust formation on metal surfaces would be intolerable or where circumstances justify the highest degree of surface preparation.

White metal blast. Used for coatings that must stand exposure in very corrosive atmospheres and under such circumstances the high cost of exacting surface preparation is justified. The term "blast cleaning to white metal" means removal of all rust, mill scale, and all contaminants. Maximum paint performance on metal is materially aided when this degree of blast-cleaning is carried out.

Near-white metal blast. (Steel Structures Painting Council SP 10) Used in moderately severe situations and costs about 10 to 35 percent less than white metal blasting. Complete cleaning must be judged visually and is somewhat subjective. It is achieved when the blasted surface shows shadows, streaks, and discolorations across the general surface area rather than as concentrated manifestations in localized areas of the surface.

309.2b Selection of Abrasives. High-speed, low-profile blasting results from using steel grit able to provide dry-honing, and Ottawa sand, from very fine to fine grades. Higher profiles start with use of medium-grade Ottawa sand, or fine iron grit, or medium iron shot.

Profiles higher than 2 mils, obtained with medium iron shot, fine iron grit, or medium Ottawa sand, require care because of the danger of too-deep valleys.

A caution about the use of iron shot should be observed. Heavy shot causes impaction of the surface, and this can influence the adhesion of the primer coating and, consequently, the entire system.

Coarse and medium iron grit and coarse Ottawa sand are particularly valuable—despite their general tendency to produce high peaks—in those situations where scale and slag must be removed from weld seams with their stubborn, irregular deposits. Care must be observed to refrain from proceeding too far in cleaning weld-joints; otherwise, the weld can be worn down and weakened, or even pitted. The latter condition can cause thin-paint areas at high points. There bare spots will appear, and, later, rust.

Further data useful for specifying is provided in Table 309.2b.

Table 309.2b. *Specifying Abrasives* and Surface Profile.*

	Average Mesh	Rate (sq.ft./hr.)	Max. Hit of Profile (mils)
Steel grit (dry honing)	100	120	1.5
Ottawa Sand—Very Fine	80	175	1.5
Ottawa Sand—Fine	40	150	1.5
Ottawa Sand—Medium	18	115	2.0
Ottawa Sand—Coarse	12	90	2.8
Iron Grit—Fine	50	100	2.0
Iron Grit—Medium	25	65	3.2
Iron Grit—Coarse	16	60	4.5
Iron Shot—Medium	30	110	2.0
Iron Shot—Coarse	18	75	2.5

*At 80 psi using 5/16-in. Venturi nozzle at 18 to 24 inches from mill scale-covered mild steel plate.

Note: Care must be required to see that extreme peaks and valleys are not left on the surface, because these require additional buildup of paint; and if too high, these peaks can lead to early paint failure because they provide thin spots from which coatings can easily be worn, leading to bare spots where moisture can find a toehole to start the corrosion process.

309.2c Abrasive Blasting Techniques. Blasting equipment works on one of three principles—pressure, suction, or centrifugal force. Selection of the system can influence performance or cause harm to surrounding areas. Pressure equipment is the most common for field work. In this method, the abrasive is carried in either a high-pressure stream of air or water or a combination of them. In the suction method, the abrasive is shot under pressure and then picked up by vacuum set up in a cone over the blasted surface. The centrifugal method causes the abrasive to be hurled from the blades of a rotating paddle wheel.

Selection of the suction or vacuum method, where suitable, helps cut down the dust hazard and permits reclaiming the abrasive since the material is drawn into a vacuum bag. The method loses efficiency on very irregular surfaces, because the suction head cannot make complete contact with the surface, diminishing the pull of the vacuum. In some instances, where the surface is badly contaminated with rust, algae, etc., the vacuum cone may become clogged. In this case, the cone

Surface Preparation

can be held just above the surface and operated as a semi-open device.

Wet blasting combines abrasive and water in the pressure method. It almost eliminates blasting-dust and the hazards to nearby equipment associated with it. Specifying this method, however, can lead to problems of troublesome cleanup if the structure to be cleaned has a fairly large number of ledges and horizontal girders as well as upturned angles. Wet material trapped there is more troublesome to remove than dry dust because it forms a sludge that must be eliminated by rinsing, brushing, or by the force of compressed air. Another serious limitation of this type of blasting is the risk of rusting, although rust inhibitors are added to the rinsing water. Surfaces must be dried thoroughly before applying paint.

Centrifugal blasting utilizes vaned wheels rotating at controlled speed. Onto these, abrasive grit is dropped and spun at great force against the steel plates to be cleaned. This is particularly useful for preparing steel plates prior to erection and just before they are to be coated.

General precautions for blast cleaning. The following precautions should be observed:

1) Before blasting, require that grease or oil be removed by solvent cleaning.
2) Require that dry blasting be done only if the temperature of steel is less than 5 degrees above the dew point to prevent moisture condensation on the surface during painting, an almost certain prelude to rust.

309.3 Acid Pickling. An effective and relatively inexpensive way to remove mill scale and rust from steel and iron, and metallic oxides from zinc, aluminum, copper, and brass is by a process known as pickling.

Sulfuric acid is used for most of these materials because it is cheap and has a higher boiling point than hydrochloric, nitric, hydrofluoric, and phosphoric acids, the other acids used. For aluminum, acids are never used for pickling, except for a final dilute rinse after caustic soda treatment is used.

When the plate or parts to be pickled are small enough

they are usually immersed in a bath; but the acid may be sprayed or washed on the surface.

Speed of cleaning is fairly good, but enough time must be allowed to permit the acid to work beneath the surface through cracks in the harder, scaly outer layer. Once at the softer, under-layers of mill scale, the acid eats rapidly down to the bare metal. Inhibitors are added to the acid solutions so the bare metal won't be eaten away too rapidly in those portions where the acid reaches first. Surfaces to be pickled must first be solvent- or alkali-cleaned to remove oil and grease. After pickling has been carried out, several clear-water rinses are needed to remove all acids and salts; and the final rinse should have a weak alkali in it to retard rusting in the time elapsing between rinses and surface coating.

Caution: Do not pickle aluminum in acid. Do not pickle stainless steel.

309.3a Specifying Information for Pickling. A typical pickling process is covered in Steel Structures Painting Council SP-8. To remove scale and iron oxide, a solution of 5 percent (by volume) sulfuric acid is used at a temperature of 170° F. When descaling has been accomplished, the steel is rinsed for two minutes in 170° F water to remove the acid solution. It is then given an inhibitive treatment in a weak dichromate-phosphoric acid bath at 190° F. The parts are hot, and they dry quickly, permitting the shop to prime them quickly. This should be specified to avoid any likelihood of rusting prior to priming.

The English Footner process has the steel pickled in sulfuric acid, then rinsed with water before being plunged in a hot solution (175° F) of 2 percent phosphoric acid and 0.3 to 0.5 percent iron, thus forming a thin layer of iron phosphate to prevent rusting.

309.3b Pickling for Particular Types of Metals. *Zinc* pickling is accomplished in cold 2 to 20 percent sulfuric or nitric acid. *Aluminum* is pickled in dilute caustic soda for a few seconds, or in soda solution for a few minutes. Following this, the aluminum is washed with water and bathed briefly in dilute nitric acid. *Copper* and *brass* pickling is done in sulfuric acid.

309.4 Passivation or Metal Pre-Treatment by Chemical Modification.
Susceptible metal surfaces are sometimes protected against corrosion by modifying their chemical composition. In this process, inorganic liquids are applied to the metal surface where they form metal compounds that react with water in a different way than the original metal.

What happens is that the resulting compounds act passively when water reaches them. The original metal—iron, for example—would act forthrightly with water, and would end up as rust. The compounds that make metal *passive* in the presence of water are, not surprisingly, called *passivators.* An interesting phenomenon occurs when a surface is properly passivated. A large scratch, instead of expanding angrily, stays almost intact or spreads its rust very gradually.

Earlier we saw that rust results from the reaction of water with iron on the surface. The rust is iron oxide, which is more noble than plain iron, or steel, which means that whenever water invades the interface between the rust and the steel or iron below, the nobler rust disdains the water and the less-noble iron or steel reacts with it, forming more rust.

When the steel or iron is passivated, the new composition, on its surface, which is an iron salt, has a higher degree of nobility than the original metal. It does not combine so readily with the oxygen brought in with water; and when some rust does form because of damage to the passivating film, the higher nobility of the new iron salt composition slows down, or even halts, expansion of the small amount of rust that forms.

Using these inorganic solutions is almost like painting, so it may be difficult to regard these chemical modifications as pre-treatments. They differ from paint in several ways. First, they combine chemically with the substrate, which paint does not do; they are applied in coats of minimum thickness; and they offer little actual physical protection by themselves.

Passivating, or the chemical modification of metallic surfaces, is performed most frequently on iron and steel. They are treated with hot or cold phosphatizing materials. The result is a thin surface of a phosphate salt, probably a complex of iron phosphate plus other metallic phosphates.

Aluminum, zinc, magnesium, tin, and their alloys are pretreated with chromates. (See Section 309.4b.)

309.4a Treating Iron and Steel. Iron and steel are treated with hot phosphate and cold phosphate.

Hot phosphate treatment. Because heat is needed, this method is mainly suitable for shop treatment and may be specified where architectural steel must meet certain rigid standards. Zinc or iron phosphate solutions are used in this process. Zinc phosphate gives better results; but if a high gloss is desired in the topcoat, iron phosphate is preferred because it is deposited in a finer crystalline structure and in a thinner film; hence, it does not absorb as much of the binder in the topcoat. If a gloss coat is desired with zinc phosphate treatments, thicker paint overcoats may be used to compensate for the binder given up to the zinc salt.

Care must be given to time and temperature to control desirable thickness of the phosphate. Thicker deposits are more brittle, but they also aid adhesion and rust prevention of the subsequent coats. At one time, treatment for one hour at about 200° F was necessary, but now with the addition of certain nitrates as accelerators, an effective treatment can be carried out in as little as 2 or 3 minutes at about 100° F. Because of the reduced time needed, it is now possible to spray the phosphate. Following the phosphating treatment, a water-rinse is needed, and the treatment is completed with a rinse in water containing 0.05 percent chromic acid, which must be used before the final water rinse has dried.

Specifying information. Thickness of coatings desired for zinc phosphate is 1 to 10 micrometers and for iron phosphate it is 0.1 to 0.5 micrometers. Or, by weight, 100 to 4,000 milligrams per square foot may be specified. It is important to require that the phosphated metal be painted within 24 hours of treatment, and sooner where possible. The reason is that the phosphated layer is not absolutely continuous and contains pores, which may permit corrosion to start if either painting or impregnation is not carried out within the time limit. Phosphatizing has been found compatible with alkyd and alkyd amino finishes and with epoxies, water-soluble coatings, and with thermosetting and thermoplastic acrylics, all of them used with architectural metals.

Cold phosphate treatment. Paint adhesion to surfaces treated with cold phosphatizing is not as good as with hot phosphatizing; however, the former can be applied in the field. In cold phosphatizing, phosphoric acid, with a wetting agent, a water-miscible solvent, and water are applied. An acid concentration of about 5 to 7 percent (by weight) is used.

Cold phosphate should not be used in high summer temperatures or in high winds or direct sunlight—all of which hasten evaporation and lead to high acid concentration.

Specifying information. Proper reaction and dilution has occurred when a dry, grayish-white powder has formed. This usually happens a few minutes after the phosphate has been applied. A dark color and sticky feel indicates the acid is too concentrated.

309.4b Treating Nonferrous Metals.

Chromate treatment and wash primers are the two most frequently used methods.

Chromate treatment. Used mainly on aluminum and the other metals cited above.

In the MBV method, aluminum is dipped for a minute or so in a hot solution consisting of 15 gm. sodium chromate and 50 to 80 gm. of sodium carbonate per litre. This yields a protective layer of 75 percent hydrated aluminum oxide and 25 percent hydrated chromium oxide. The thin film provides a good base for painting and offers protection against deterioration in harmful environments.

In the *Alodine Process,* a combination of chromic acid, phosphoric acid, and hydrofluoric acid is used instead of the alkaline bath used in the first method. The Alodine Process gives a grey-green to grey-brown coating, which consists of aluminum phosphate, chromium phosphate, and aluminum oxide in a thickness running between 0.5 and 5 micro meters.

Wash primers. These have been found more efficient than the method previously described as cold phosphatizing.

Actually, wash priming is a form of cold phosphatizing since phosphoric acid is a key ingredient. These materials are not primers in the usual sense since they are so thin (the usual film thickness is 0.0005 inch, a half-mil). It was found that by adding a film-forming synthetic resin, polyvinyl butyral, and a corrosion-inhibitive pigment, zinc chromate, to phosphoric acid, better results were obtained than if either component or a combination of two were used together. These wash primers provide good adhesion to surfaces that are otherwise difficult to paint, such as galvanized or stainless steel and aluminum. By putting them on immediately after blast cleaning or pickling they provide a protective coat against rust and mill scale and permit delay in applying final coats. Wherever metal is to be used under water, they are particularly desirable.

HELPFUL ADDITIONAL READINGS

1. Burns, R. M., and W. W. Bradley, *Protective Coatings for Metals*, ACS Monograph No. 163, 4th ed. (New York: Reinhold Publishing, Inc., 1967), pp. 27-54, 64.
2. Levinson, Sidney, and Saul Spindel, *Paints and Protective Coatings* (Washington, D.C.: Departments of the Army, Navy, and Air Force, 1969).
3. Martens, Charles R., *Technology of Paints, Varnishes, and Lacquers* (New York: Reinhold Publishing, Inc., 1968), pp. 619-42.
4. Nylen, Paul, and Edward Sunderland, *Modern Surface Coatings* (New York, 1965), pp. 611-32.

chapter 4
Calks and Sealants

400 General. Frequent reference appears throughout the sections on surface preparation and surface defects to the importance of sealing construction surfaces from atmospheric water. Because water can work its way through many materials, particularly porous wood and cementitious surfaces, care must be exercised to seal off juncture points so that water will not be able to reach absorbent edges and then penetrate throughout the surface where it can disturb or destroy the adhesion of the coating. See Fig. 400.

Calks and sealants are soft moisture-impervious compounds which, under modest pressure, can be made to conform to crevices, cracks, and joints of various sizes and shapes. Calks are used for fixed joints, or those with slight movement; sealants are mainly used for joints where some movement is anticipated.

401 Calks. A calking compound consists of an oil or resin binder with a filler, which may be any one of a number of extender pigments, such as calcium carbonate, clay, mica, talc, or asbestos, and other materials including colorants, thickeners, and dispersing agents. They are used in cracks and crevices to provide a smooth surface for paint, and they are used to seal off fixed joints, or joints with extremely limited movement. Because they are made of binders that do not dry hard right away, calking compounds during their effective life remain soft and adhesive beneath the surface, although they are rather hard on top, where oxygen readily reaches them.

Fig. 400. Calking of Brick-Wood Interface.
A brick-wood interface requires calking compound to prevent passage of moisture to rear of wood.

401.1 Low-Priced Calks. Low-priced calking compounds, all too often used by painters of the "quick and dirty" school, are made with raw linseed oil or fish oil, depending on the state of the market or the pride of the manufacturer. These bargain calks may cost less than $0.75 per tube. They have a life expectancy of one or two years; yet, unless specifications are given and enforced, they may be used to seal joints of wood to wood, wood to metal, or wood to cement, where paints costing as much as $20 per gallon have been specified.

Conscience-less builders may regard the homeowner as a prime victim and let contractors get away with cheap calks, since such contractors are probably already permitting painters to substitute low-quality paint for the ones originally specified by the architect. However, if the architect is given authority to demand suitably durable products and the right to approve subcontractors, he should specify high-grade calking compounds and approve only painting contractors with pro-

Calks and Sealants

ven reputations for meeting specifications.

401.2 High-Grade Calks. High-grade calks for durably painted surfaces are available. These include butyl-rubber and acrylic-latex calks.

401.2a Butyl Rubber Calks. Butyl-rubber calks with a life expectancy of three to four years, for example, can be bought for relatively little more than oil-based calks.

401.2b Acrylic Latex Calks. Next above the butyls in life expectancy are acrylic latex calks. Outdoor exposure tests, lasting five years, show that acrylic calks last much longer than oil and butyl calks. Their useful life is at least 5 years, and very possibly ten years.

Acrylic latex calks cost about 25 percent more than butyl and twice as much as oil calks.

Other advantages of acrylic calks, in addition to longer life, are:

1) Since they require no solvent, acrylic calks are odorless.
2) They flow easily.
3) Their latex binder enables them to be applied to damp surfaces.
4) They can be painted within 30 minutes of application.
5) They have greater mildew resistance than oil types.
6) Cleanup can be accomplished with soap and water.

Specifiers calling for acrylic calks, however, must observe a few precautions. Storage should be at temperatures above freezing. Application should be made only when temperatures are above 40° F. Two other recommendations when acrylic calks are used are to clean old, chalky surfaces and scrape off old calks before applying the new calking.

401.2c Special Acrylic Qualities. As shown in Table 401.2c, the calk 12-DHR-123 is a high-density material suitable for gun and squeeze tube use. In the technical literature in which this table appeared, three variations of this basic formulation were characterized as to ultraviolet's effect on color. One of the three contained a *plasticiser* combination consisting of Flexol (R) Plasticisers TCP and TBF. The synergizing effect

Table 401.2c. *Typical Performance Properties of Calks and Sealants Based on Acrylic Latex.** *

Properties Tested	Calk 12-DHR-123	Sealant 12-DHR-117
Total solids, percent by weight	83.2	79.5
Pounds per gallon	13.49 (a)	12.26 (a)
Color	White	White
Package stability (30 days at 50° C)	Pass	Pass
Freeze-thaw stability (5 cycles)	Pass	Pass
Consistency (grams per minute) (b)	221	1694
After 4 days at 158° F	218	1624
Tack-free time (minutes)	<30	<30
Cure-through time (days)	2–3	2–3
Bleeding or staining (c)		
78° F	None	None
158° F	None	None
Slump		
Channel	None	None
Right angle (glass, aluminum and wood)	None	None
Gunability after 7 days,		
(32° F—81% R.H.)	Good	Good
(30° F—26% R.H.)	Good	Good
Shore A hardness,		
14 days at (73° F—50% R.H.)	11	20
1092 hours in Weather-Ometer	23	33
Specific volume (ml/gm)	0.65	0.62
Color and cracking, Weather-Ometer,		
300 hours, aluminum channel	None	None
300 hours, wood channel	None	None
300 hours, glass channel	None	None
Low temperature flexibility,		
1/16 to 1/8-inch film on Teflon,		
1/2-inch rod,—30° C	Passed	Passed
1/4 inch film, 1092 hours in		
Weather Ometer,—10° C	—	Passed
Tensile psi, 2 weeks at (73° F—50% R.H.)	6.2	8.0
Percent elongation, 2 weeks at (73° F—50% R.H.)	300–400	650–850
180° peel adhesion (lbs./in.) (d)		
Yellow pinewood	7.0	32.0
Stainless steel	9.0	30.0
Aluminum	9.0	21.5
Glass	7.0	28.5
Glazed ceramic tile	6.0	25.0

(a) This value will vary slightly with the type of plasticizer used.
(b) Semco Calk Gun, 50 psi air pressure, through a 1/8-in.-diameter nozzle orifice.
(c) Canadian Specification 19-GP-5a, Section 6.24.
(d) All cohesive failure on sealants (12-DHR-117); adhesive and cohesive failure on calks (12-DHR-123): Canadian Specification 19-GP-5a 6.2.6.2.

*UCAR® 153, Union Carbide Corp.

Source: American Paint Journal.

Calks and Sealants 147

of the combination is instrumental in achieving good color stability. Where color retention is a factor, it may be necessary to specify that the acrylic latex calk contain these plasticisers.

The high-density 12-DHR-123 has several significant differences in its formula from that of Sealant 12-DHR-117, a low-density sealant whose characteristics are described in the same table. The calk contains 4.34 lbs. of acrylic latex (55 percent solids) per gallon, while the low-density sealant has 5.8 lbs. per gallon. The calk also has 7 lbs. of calcium carbonate per gallon, and the sealant has 5.5 lbs.

401.2d Pigment Volume Concentration. If we want to use PVC as an indication of the quality of the two formulations, we have to translate these weight indicators to gallons; and we find, after simple arithmetic, that the PVC of the calk is roughly 39, and the sealant has a much lower PVC of 27. Since the lower PVC means that it has more resin, which we have already learned, it also means that the sealant is harder, less dense, has far better adhesion characteristics, and twice the elongation possibilities, which is vital for a sealant and not needed in a calk. Reference to these closely related formulations permits us to see what distinguishes a calk from a sealant, and this leads us right into the subject of sealants.

402 Sealants. Flexibility is the major distinguishing characteristic of sealants. Their primary use is to seal joints where movement from any one of a number of causes is likely, such as contraction and expansion due to weather, impact, vibration, or any kind of motion requiring elasticity to maintain the integrity of a joint.

Sealants are of two basic types: (1) *elastomers,* or stretchable polymers, which set up and firm by the release of solvent, and (2) those which set up because a chemical change of some kind takes place in the applied material to cure the binder.

402.1 Elastomeric sealants. Elastomeric, solvent-release type sealants are made with the following binders: (a) butyl rubber, (b) Neoprene® (polychloroprene), (c) Hypalon® (chlorosulfonated polyethylene), and (d) acrylics.*

---

*Neoprene® and Hypalon® are registered trade names of E. I. duPont de Nemours Company. Plexiglass® is the registered trade name of Rohm and Haas, Inc.

402.1a Butyl Rubber. Butyl rubber is a copolymer of isobutylene and isoprene, which are chemically related to the synthetic rubber used in tire making. Because it sets up by solvent release, butyl rubber would have to contain more solvent than is acceptable in a sealant and would shrink excessively. To avoid this a permanent plasticiser, polybutene, is used with the solvent to limit the fluency of the material and to restrict shrinkage. Other permanent plasticisers are sometimes used, such as oils, resins, and asphalts. Since butyl rubber and polybutene maintain tackiness, a drying oil, such as bodied castor oil, is often added so that the surface will oxidize and remain impermeable and tack-free.

Extenders. Butyl sealants may be filled with any one or more of the common extenders, but calcium carbonate is most often used. Butyl sealants can take almost unlimited colors. Mineral spirits is the most usual solvent.

Identifiable properties of butyl-based sealant. The following characteristics of butyl-based sealants should be kept in mind by the specifier:

Curing-type—Solvent evaporation from surface inward.

Tack-free time—30 minutes to 6 hours.

Life expectancy—10 to 20 years.

Application temperature—41° to 122° F.

Service temperature—22° to 203° F.

Hardness when aged—15 to 40 Shore A.

Weather resistance—Remains flexible beneath surface despite reasonable ozone, oxygen, and moisture exposure.

Permissible elongation—25 to 50 percent.

Limitations—Probability of shrinkage with low-solids material; strength and stretch less than rubber-based compounds that cure; inadequate recovery after stretch.

Advantages—Easy working, low cost, easy cleanup, weathers fairly well.

Suitable usage—General wood, masonry, metal sealing; glazing; sealing stable joints or those with slight movement.

Calks and Sealants

402.1b Neoprene. Neoprene,* like other members of the polychloroprene family, is a soft solid that can be made into fluids of various viscosities by addition of plasticisers. Three types of sealant-grade Neoprenes are offered. By judicious blending of these grades, and/or by the proper selection of plasticisers, a broad line of products for a variety of temperature ranges and service conditions can be offered. An inexpensive plasticiser, *butyl oleate,* improves low-temperature properties and ozone resistance at the expense of fungus and flame resistance, which under some circumstances may not be important, especially if additives are used to compensate.

Specifiers, knowing the circumstances in which the sealant will function, can be assured that suppliers will have or will provide, if quantities justify, Neoprene sealants to meet their particular needs.

Modification with *acrylonitrile* improves the heat, petroleum oil, and gasoline resistance of Neoprene. From 18 to 50 percent of this nitrile is used. The more nitrile used the better the resistance to oil, gasoline, and heat; but as the modifier is increased, a reduction occurs in resilience, flexibility at sub-zero temperatures, and ability to resist heat buildup.

Neoprene sealants may be filled with reinforcing extenders, carbon black, and the following metallic oxides, which regulate the cure rate of the binder and tie up hydrogen chloride that may be released during aging: zinc oxide, red lead or litharge, and magnesium oxide. If water resistance is important, red lead or litharge should be specified. The usual combination in sealants and calks made of Neoprene is 10 parts of zinc oxide and 20 parts of lead oxide, either as red lead or litharge, to 100 parts of binder solids.

These sealants are available as two-component or one-component products. The one-component form usually requires a slightly elevated temperature for curing. The two-component cures at room temperature, or with heat.

Identifiable properties of neoprene sealants. The following characteristics should be kept in mind:

Curing type—Solvent evaporation.

Dry-to-touch—About 4 hours; less for 2-component grades.

Life expectancy—30 years.

---

*Neoprene® and Hypalon® are registered trade names of E. I. duPont de Nemours Company.

Curing time—3-7 days; less for 2 component.

Application temperature—41° to 122° F.

Service temperature—30° to 248° F.

Weather resistance—Slight chalking.

Elongation, 77° F—500 percent; 320 to 400 percent for 2-component grades.

Elongation, 32° F—300 percent.

Fire resistance—good.

Tensile strength—1,800 psi; 560 to 920 psi for 2-component grades.

Taber Abrasion resistance, CS-17, 1000 revolutions—18 gms wt. loss.

Water immersion—2.6 percent weight gain.

Resistance to oil and gasoline—Good.

Hardness—38 to 44 Shore A.

Limitations—Stretch exceeds recovery; lengthy cure; shrinkage; poor color stability; black only; cures slowly for 2 component; other limitations same as 1 component.

Advantages—Resists deterioration by weather; 2 component is solvent-free; nonshrinking; sets fast.

Suitable usage—Sealing, both 1 and 2 pkg., seams and laps in metal; binding shingles; repairing asphalt cracks.

402.1c Acrylics. Acrylics, already encountered in the calking section, offer several advantages as sealants. Usually, no primer is required because of its outstanding surface adhesion. Its flexibility, color stability, and resealing ability offset some of its disadvantages, which are odor, soft-setting and somewhat slow-cure, plus its need for heat in many instances where best results are desired.

Precautions to be specified when these sealants are used are restrictions against applying when temperatures are below 41° F; or when the surface is to be continuously under water. Applicators must be required to remove all dust, oil, grease, wax, moisture, or dirt.

Calks and Sealants

In addition to the performance characteristics listed under calks, the following characteristics should be noted when acrylics are used as sealants:

Curing type—Solvent evaporation.

Life expectancy—20-year minimum.

Tack characteristics—Requires anti-tack agent.

Curing time—14 days at 77° to 122° F.

Application temperature—Gunnable at 70° F.

Service temperature (NAMM specif.)—30° to 200° F.

Accelerated aging—No adhesive or cohesive failure or oil exuding after 1,000 hours.

Salt spray resistance—No failure after 200 hours at 40° F.

Ultraviolet through glass—Excellent resistance.

Elongation—250 percent at 45° F.

Weight loss after heat aging—5.19 percent, average.

Hardness—35 Shore A.

Resealing—More than 75 percent positive contact.

Water immersion cohesion—At least 150 percent extensibility for 33 hours in accord with TTS 00230.

Color stability—Good to excellent.

Limitations of acrylics, in general, for sealants—Soft setting and somewhat slow cure; odor; requires heat for convenient application (but heating and pressure devices are available); dirt pickup probable; no continuous water immersion.

Advantages—Self-resealing; outstanding adhesion to most surfaces without primer; no staining; wide color choice and stability; fine extensibility and flexibility.

402.1d Hypalon. Hypalon (chlorosulfonated polyethylene) offers outstanding weather resistance, and its ability to withstand ozone exposure is better than neoprene and butyl rubber. In fact, one researcher found that ozone exposure equal to

200 times the amount in the atmosphere caused so little erosion that he concluded that the Hypalon joint would easily outlast just about any structure.

These durable sealants, made by treating polyethylene in a solution with chlorine and sulfur dioxide, are often a blend of three binders, each with its own strengths and weaknesses. They are known as Hypalon 20, 30, and 40. The characteristics of the three types are shown below:

Properties	Hypalon 20	Hypalon 30	Hypalon 40
Hardness range, Shore A	45 to 95		
Tensile strength, psi			
Carbon black stock	to 3,000		
Gum stock	to 1,500		
Color stability	Excellent		
Strength at tear	Fairly good		
Abrasion resistance	Outstanding	Outstanding	Outstanding
Chemical resistance	Excellent	Satisfactory	Excellent
Compression resistance	Fair	Poor	Good
Flame resistance	Fair	Very good	Good
Ozone resistance	Excellent	Excellent	Excellent
Petroleum oil resistance	Fair	Excellent	Good
Weathering	Excellent	Excellent	Excellent
Low-temperature behavior	Good	Poor	Good

Curing systems achieve cross-linking, or vulcanization. Litharge used as a curing agent aids in water and chemical resistance. When magnesia is also added, litharge imparts heat resistance, but water and chemical resistance is diminished. Magnesia-pentaerythritol curing compounds are recommended when discoloration is not significant. Other curing systems used in proprietary products are epoxy resins, amines, phenols, and sulfonates.

402.1e Helpful Notes on Solvent-Release Elastomeric Sealants. While sealants of the solvent-release type are substantially superior to oil and alkyd versions, which they have, in the main, replaced, and while their high molecular weight provides excellent flexibility as well as resistance to oil, chemicals, ozone, and heat, they still suffer from varying degrees of shrinkage caused by solvent evaporation and, when elongated, they have only partial recovery.

Calks and Sealants

If technical requirements do not justify paying the high cost of complex sealants that cure through chemical reactions, these solvent-release materials are more than merely adequate.

However, when "Tiffany" products are indicated, where long life, nonshrinking, hardness, stretch-recovery, and resistance to chemicals and solvents is important enough to ignore cost, then selection must be limited to chemical-curing sealants.

402.2 Chemical-Curing Sealants. These sealants are made with the following binders: (a) Polysulfides; (b) Polyurethanes; (c) Silicones.

402.2a Polysulfides. Polysulfides are prepared for one-package sale as well as two-package and have been proven by many years of successful use in notable structures throughout the nation. Life expectancy is in excess of 20 years. Its elongation, ranging from 150 to 300 percent, is followed by excellent recovery.

Formulations can be developed with incredible elasticity and temperature range. However, as in most chemical compositions, other properties would be sacrificed. Polysulfide sealants are available to exceed requirements of Federal Specification TTS 00230 (one-package elastomeric sealant) and TTS 00227 (two-package elastomeric sealant), which also can be met by formulations based on other materials. See Table 402.2a.

Manufacturers claim that two-component polysulfide sealants are capable of sealing any conceivable joint where movement is expected to be up to 50 percent, in joints wider than 3/4 inch.

Surfaces must be very carefully cleaned before use, and porous surfaces or surfaces to be immersed in water must be primed. Minimum joint size is 1/4 inch by 1/4 inch. For joints wider than 3/4 inch, depth should not exceed one-half of width, if two component sealers are used. One-component sealants are suitable for movements expected to be up to 35 percent. For the one-component sealant, maximum joint size is 3/4 inch by 1/2 inch deep, although deeper ones can be sealed if a closed cell polyethylene back-up plug is used.

Primers are frequently required for polysulfide sealants. Marble, glass, white concrete, and plexiglass surfaces at joints,

Table 402.2a. One-Component Elastomeric Sealant Compound—Federal Spec. TTS 00230

	Type I Flow Type for Horizontal Joints	Type II Gun Grade for Vertical Joints	Commercial Product #1 Flow Type for Vertical Joints
Material	Unspecified	Unspecified	Polysulfide
Curing type			Moisture in air
Application temperature			35° to 120° F
Application life	20 secs. at standard temp. & R.H.	45 secs.	Meets TTS 0030
Service temperature			−40° to 220° F
Hardness—21 days standard	15 to 50 Shore A	15 to 50 Shore A	25
Sag or flow	N.A.	Not more than 3/16"	Good resistance
Weight loss after 7 days at 158° F	Not over 10%	Not over 10%	Meets TTS 0030
Tack-free	Not over 72 hrs.	Not over 72 hrs.	2 to 48 hrs.
Bond cohesion after 3 test cycles on mortar, plate glass, and aluminum	Under 1 sq.in.	Under 1 sq.in.	Meets TTS 0030
Adhesion in peel from plate glass and aluminum alloy after 21 days' cure	Not over 25% adhesive loss	Not over 25% adhesive loss	Meets TTS 0030
Shrinkage, percent			11
Resistance to ozone, ultra-violet, weather			Exceptional
Water impermeability			Good
Flexibility at 0° F			Good
Recovery from 100% elongation			100%

*Courtesy Gibson-Homans Co.

Calks and Sealants 155

according to specifications of one major polysulfide sealant supplier, require priming with silane material.

An amino silane primer is recommended for white concrete and marble surfaces, and a modified silane is recommended for priming glass and plexiglass at joints, after all loose material, oil, dirt, grease, and oil have been removed.

402.2b Polyurethanes. Polyurethanes offer high tensile and tear strength, and an ability to resist deformation under considerable stress. Abrasion resistance is an outstanding feature of sealants based on polyurethane. In this characteristic, it is superior to commonly used polymers. For this reason, it is usually selected for floors, highways, parking lots, and any place where there is heavy traffic.

Its ability to oppose deformation also suits it for joints where small objects such as rocks may be thrown against it.

Urethane sealants adhere to many unprimed surfaces but, generally, experts recommend using primers based on amino silanes, modified silanes, or silicones. Improvement of bond strength through use of primers is definite.

Sealant suppliers should recommend suitable primers. At least one important manufacturer offers a system of company inspection and a guarantee, provided that all directions are followed.

Some sealant formulations include an amino silane additive, thus obviating the need for a primer coat.

A study showed that 2 percent amino silane added to a urethane sealant used on glass increased the adhesive strength from 205 to 7,160 grams of pull per centimeter of width. On steel, the control had 1,460 grams strength; and when amino silane was added, adhesive strength rose to 4,117. On aluminum, the control had 626 grams, and silane-modified sealant could withstand 2,528 grams pull. After two months' exposure, the controls had lost appreciable strength, but the silane-modified sealants had lost little adhesive strength. The modified sealant used with aluminum had actually increased its bond. Performance comparisons for one-component urethane sealants, gun grade and pour grade, are shown on page 158.

Table 402.2b. Two-Component Elastomeric Sealing Compound—Federal Spec. TTS 227.

Type	TTS 227 Type I Flow Type for Horiz. Joints	TTS 227 Type II For Vertical Joints	Commercial Product #1 Flow Type	Commercial Product #1 For Vertical Joints	Commercial Product #2 Flow Type	Commercial Product #2 For Vertical Joints	Commercial Product #3 Flow Type and Vertical Joints	Commercial Product #4 Flow Type and Vertical Joints
Material			Polysulfide Chemical	Polysulfide Chemical	Polysulfide Chemical	Polysulfide Chemical	Polyurethane Chemical	Chemical
Curing type	Not specified	Not specified						
Application temperature	40° F+	40–122° F	35–122° F	35–122° F	40–122° F	40–122° F		40–122° F
Pot Life	3 hrs. +	3 hrs. +	3 hrs.	1–6 hrs.	3 hrs.	4 hrs.		45 min.
Service temperature			−60–250° F	−60–250° F	−60–250° F	−60–250° F	−48–220° F	−45–200° F
Hardness—14 Days, Standard Shore A	15–50	15–50	45	25			40 ± 2	40–50
Hardness—7 Days Standard +21 at 158° F	Not over 60	Not over 50	Meets TTS 227	Meets TTS 227	Meets TTS 227	Meets TTS 227	Meets TTS 227	Meets TTS 227
Joint movement, total	Class A 50% Class B 25%	Class A 50% Class B 25%	Meets TTS 227	Meets TTS 227	Meets TTS 227	Meets TTS 227	Meets TTS 227	Meets TTS 227

Calks and Sealants

Property						
Weight loss after heat aging	Not over 10% original wt.			Meets TTS 227	Meets TTS 227	Meets TTS 227
Adhesion in Peel	Not less than 5 lbs.		Meets TTS 227	Meets TTS 227	Meets TTS 227	Meets TTS 227
Tack-free time	Within 72 hrs.	12–48 hrs.	6–48 hrs.	12–15 hrs. (75 F + 50% R.H.)	12–24 hrs. (75 F + 50% R.H.)	24 hrs.
Complete cure		3–7 days	3–7 days	7 days	7 days	Light Traffic 72 hrs. heavy
Vertical sag			No sagging or slumping at 40–122° F Meets TTC 598		Meets TTS 227	
Shrinkage		Negligible	10%	N.A.	N.A.	
Percentage recovery after 100% elongation		100%	100%	N.A.	N.A.	75–85%
Weather-Ometer—300 hrs.					No Change	Very good
Compression set					3%	

Performance, One-Component Urethane Sealants

	Gun grade	Pour grade for horizontal
Tack-free, hours	1 to 24	1 to 24
Cure time, days	5 to 10	4 to 8
Tensile strength, psi	300 to 400	250 to 500
Ultimate elongation, percent	500 to 600	100 to 200
Method of cure	Chemical reaction with moisture in air	
Shore A hardness, initial	20 to 30	20 to 30
one to six months	35 to 50	35 to 50
Water immersion	Good	Good
Aging	Excellent	Excellent
Service temperature	−40° to 250° F	−40° to 250° F
Recover after 100 percent elongation	90	85
Maximum joint movement	10 to 15 percent	10 to 15 percent

Specifications—Federal Specification TTS 230

For performance data of typical two-package polyurethane sealants, see Table 402.2b for chemically-cured two-component sealants meeting Federal Specification TTS 227.

402.2c Silicones. Silicone sealants, like one-package urethanes, cure by chemical reaction with moisture in the air. They have the important ability to be able to leave a calking gun at about the same viscosity whether the temperature is −40° or over 200° F. Because it is not affected by high temperatures, the sealant does not sag in the joint when used at elevated temperatures.

Tack-free time for silicones is shortest of all for sealants, taking place in one hour or less depending on the humidity. A skin of cured silicone rubber begins to form almost immediately.

Silicone sealants are provided by General Electric and Dow Corning, both of whom provide surface primers where needed. G.E. claims its product adheres well to many surfaces, while Dow Corning says that all surfaces should be primed. Performance characteristics of a typical silicone sealant are provided below:

Tack-free time (40 to 80% R.H.) 50° to 80° F—1 hour or less

Hardness, Shore A at 75° F—30 to 35
    Aged 1 to 6 months   35 to 45

Service temperature—65° to 250° F

Resiliency—Very high

Resistance to extension—High

Recommend maximum joint movement—20 to 25 percent

Ultraviolet resistance
 Direct—Excellent
 Through glass—Good to excellent

Life expectancy—20 years plus.

HELPFUL ADDITIONAL READINGS

1. Damusis, Adolfas, editor, *Sealants* (New York: Reinhold Publishing, Inc., 1967), pp. 92-270.
2. *Architectural Reference Manual,* Toch Bros., Division of Carboline Co.

chapter 5

Application of Coatings

500 General. Putting paint on a surface may seem quite a simple job, suited to the very limited talents of the least endowed intelligence of our society. That kind of thinking has caused some of the most disappointing paint jobs ever encountered by trusting property owners. True, a high degree of formal education is not required to wield a paint brush, but it helps if the painter is under the guidance of a well-qualified construction or painting supervisor.

Fortunately, architects or builders looking for painting contractors able to apply highly sophisticated coatings to potentially troublesome surfaces can select from a considerable number of firms with trained chemical engineers on the staff, or college-trained managers with their eyes intently scanning painter performance because they want to get their jobs done efficiently and with the sort of results that will bring future business.

In some areas, or for reasons of economy, however, it may be necessary to depend on lesser companies; and in the case of builders and industrial maintenance teams, it may even be necessary to employ painters directly and to set up supervisory systems.

For these reasons, it appears worthwhile to discuss the details of paint application so that it can be determined if the jobs are being done with satisfactory efficiency and in a manner to assure durable service.

First of all, paint materials have to be checked to be sure that the containers are full and that the paint is in good

Application of Coatings

condition. Coatings must be mixed properly so that separated components are put into a suitable state for application.

Containers smaller than five gallons should be stirred by hand, the thick material separated if necessary so that lumpy portions can be pressed out with the stirring stick against the side and incorporated into the more liquid portion. When mixed properly, personnel should then follow manufacturers' directions for adding and stirring in thinner. Paint to be sprayed should, of course, have thinner added to compensate for the solvent almost certain to be lost between spray gun and surface.

501 Application Methods. Four important methods of application may be identified: (1) brush, (2) roller, (3) flat-pad applicator, and (4) spray gun.

501.1 Brush. Brushes should be selected:

1) For most primers, particularly where surfaces have tiny irregularities that may be missed by rollers or spray or where penetration is especially important
2) For corners, edges, and odd shapes
3) For trim and molding
4) For small areas where masking for spray applications is not worthwhile.

501.2 Rollers. Rollers should be selected:

1) For topcoats where the stippled effect produced by rollers does not matter
2) For jobs when skilled brush painters are not available
3) For fairly large areas where adjacent surfaces requiring masking are so extensive that spraying would be too costly
4) For large flat areas where spraying would create a fire hazard or might endanger nearby equipment or personnel.

501.3 Flat-pad applicators. This method may be chosen:

1) For jobs that could be done by brush, except for those jobs requiring considerable penetration
2) For jobs that could be done with rollers.

501.4 Spray Guns. Spray guns may be selected:

1) For extensive surfaces where time lost in masking is not great enough to offset the benefits of spray application speed
2) For irregular or round surfaces if brush application would be inefficient or if brush marks must be avoided
3) For the application of quick-drying lacquer-type coatings
4) For jobs requiring very smooth finishes.

502 General Rules for Coatings Applications. It is possible to set forth certain general rules that apply to all four methods of application:

1) Paint only when surface and ambient temperatures are between 50° F and 90° F* when using a water-thinned coating; and between 45° and 95° F for other types of coatings.
2) Maintain coatings in container at a temperature range of 65–85° F at all times on the job.
3) Paint only when the temperature is expected to stay above freezing in the period that the coating is to dry.
4) Paint only when wind velocity is below 15 mph.
5) Paint only when relative humidity is below 80 percent.
6) Observe the recommended spread rate for each kind of coating.
7) Tint each coat differently if the same paint is to be used for successive coats in a system to assure complete hiding.
8) Allow sufficient time for each coat to dry before applying another.
9) Allow adequate time for the topcoat to dry before permitting service to be resumed.

To acquaint specifiers with advantages of the various systems, rates of coverage of brush, roller, air spray, and airless spray methods are presented in the following table:

---

*It is claimed that some emulsion latex paints can be used between 36° and 40° F, if certain coalescing agents are used. However, the temperature range cited above is recommended in official, U.S.-endorsed manuals.

COVERAGE PER DAY

Method	Square Feet
Brush	1000
Roller	2000–4000
Air Spray	4000—8000
Airless Spray	8000–12000

503 Brush Application. This section discusses the specific considerations involved in brush application. Other sections later in this chapter discuss specific considerations for roller application, flat-pad application, and spray application. Because of its special nature, a final section is devoted to electrostatic spray application.

503.1 Materials. A brush is a collection of bristles attached to a handle by means of a plastic setting compound and protected at the base by a metal ferrule. Bristles should be checked carefully for quality. Cheap bristles are to be avoided; inferior bristles may break off, ruining the appearance of the painted surface. High quality *natural* or *synthetic* bristles are available. The best natural fibers are made of *hog bristles* from China. Lesser-quality hog bristles come from Poland. (The Chinese bristles are better because Chinese producers allow the hogs to live longer so the hair is longer, stronger, and more resilient.) The virtue of hog bristles is a natural forking at the end, called "flagging," which enables the brush to hold more paint than those made of other natural bristles. They deposit films with finer wet brush marks, which have a better chance of leveling and disappearing as they set prior to drying.

Lower quality bristles, usually *horse hair,* do not flag and in use soon become limp. They hold less paint than hog bristles and leave coarser brush marks. Some brushes blend horse hair and hog bristles with results depending on the proportion of hog bristles used.

Badger brushes are used for varnish application; and squirrel and sable-hair brushes have special applications for striping, lettering, and more-or-less art uses.

Synthetic fibers, notably nylon, are flagged and curled to simulate natural hog bristle, enabling them to load paint and leave relatively thin brush marks. They should never be used with lacquer, shellac, or any coating diluted with a strong solvent, because the fibers may be damaged. Since water doesn't swell these synthetics, they are often preferred for water-thinnable paints. Just before the rapproachement of the U.S. and China made Chinese hog bristles available once again in this country, a major synthetic fiber manufacturer introduced a new synthetic fiber that is equally good with water or oil-reducible coatings.

503.2 Brush Types. Brushes may be classified into four main types: (1) wall brushes, (2) sash and trim brushes, (3) enameling and varnish brushes, and (4) stucco and masonry brushes.

503.2a Wall Brushes. These brushes are flat and range in size from 3 to 6 inches. They are used primarily for large surfaces, and are available in a wide variety of natural and synthetic bristles.

503.2b Sash and Trim Brushes. These are available in four common types: round, oval, flat with a square edge, and flat with an angle edge. They range from 1-1/2 inches to 3 inches in width, and from 1/2 to 2 inches in diameter. For very fine work, chisel-shaped edges can be purchased.

503.2c Enameling and Varnish Brushes. These brushes have shorter but finer bristles than other brushes so they can lay down high-viscosity finishes smoothly and evenly. They are from 2 inches to 3 inches in width and have flat square edges or chisel-edges.

503.2d Stucco and Masonry Brushes. Stucco and masonry brushes are wide and look like flat wall brushes. Nylon is preferred for rough surfaces, but hog or other natural bristles may be used. Quality is not critical.

503.3 Helpful Notes. The following suggestions should be heeded with respect to brush application:

503.3a Brushes. Synthetic fiber brushes can be used immediately after purchase, but natural-bristle brushes will give better service if soaked before use. Soak the bristle portion of the brush in linseed oil for 48 hours prior to use. This gives the bristles added flexibility and swells them in the ferrule

Application of Coatings

so they will be less likely to pull out during the application process.

503.3b Surface Discontinuities. Painted surfaces which at intervals seem to have irregular dull areas indicate that the painter has been covering too much surface before coming back to his edges. If edges of a wall area handled by a particular painter are allowed to dry out too much before he gets back to expand the area and thus extend the edge, the newly applied paint cannot blend in with the old. This happens because the latter has progressed too far in the hardening process. The result is a break in the smoothness of the surface. This is evidence of either an unskilled painter or the use of paint that may be drying more rapidly than paints the workman is accustomed to using. One way to help eliminate this condition is to specify coatings that are slower drying. In this case, however, there is risk of more dirt pickup. The other way to prevent this surface discontinuity is to require each painter to work within the bounds of an imaginary vertical line that is from 5 to 6 feet high. If he holds his width to about 2 feet, edges will still be wet when he extends them.

When two painters work on the same large surface, one should work from right to left of the area, and the other from left to right. As each finishes an individual 6′ × 2′ area, he then completes the vertical portion of the wall or other surface above it, and then comes back to the still-wet edge that was covered first. Finally, the two painters meet.

The circumstances will determine the actual method to be used. The desired purpose is to avoid breaks in smoothness.

503.3c Uneven Spreading. Paint that has spread out unevenly has undoubtedly been applied by a neophyte. This means that the paint has been allowed to run and that the painter hasn't brushed out the resulting irregularity. Sanding down the high spots is then necessary, followed by repainting. See Fig. 503.3c. If these runs are observed while the paint is still quite tacky, it may be possible to brush them out, using considerable pressure, and then finish the coating off in the usual direction. Care must be taken that the adjoining area is brushed as well to avoid breaks in continuity.

Fig. 503.3c. Uneven Paint Spreading.
Uneven spreading of paint can be seen directly below the brush where a run of paint is visible.

503.3d Enamels. Enamels are more difficult to brush, as a general rule, and the use of partially worn brushes is advisable. Their shorter, worn bristles tend to spread the more fluid enamel more rapidly and reduce the likelihood of runs or curtains. A new, longer-bristled brush may tend to deposit enamel without working it into the surface.

503.3e Painting Order. In general, large walls are painted from the top downward, starting with a corner or any vertical divider that seems logical. Trim, doors, windows, and molding other than those with narrow underedges should be painted after major surfaces have been finished. An exception, of course, is where scaffolding is required; in that instance, everything within reach of the scaffold should be painted before moving it.

Experienced painters will know that moldings and clapboard, or anything with narrow underedges, should be

Application of Coatings

painted before large continuous surfaces. Corners, as well as edges, should be painted in strokes that sweep off the edge rather than from the edge inward.

504 Roller Application. The second major method of applying a coating is by means of rollers. Basically, a paint roller is a fabric cylinder capable of being saturated with a liquid coating. The cylinder is mounted on a simple spoked metal frame that rotates about a rod attached to a handle.

Rollers function by picking up paint from a traylike metal container and then spreading it onto a surface. As the paint-saturated roller passes over the surface, its swollen fibers deposit their individual tiny loads in a process of expansion and contraction as pressure is imposed and released during rotation of the roller.

504.1 Roller Selection. Rollers differ with respect to width, nap length, and cover material.

504.1a Width. Roller widths range from 1-1/2 inches to 18 inches, but these extremes are considered special. Most common widths are 3 inch, 4 inch, 7 inch, and 9 inch. Inside diameters run from 1-1/2 inches to 2-1/4 inches.

504.1b Nap Length. The nap length of a roller cover influences roller performance. Smooth surfaces require short nap lengths, and medium and rough surfaces require progressively longer naps. Nap lengths range from as short as 3/16 inch to 1-1/4 inches. As shown below, the type of paint also influences the decision on nap length:

ROLLER NAP LENGTH

Surface	Type of Paint	
	Enamel	*Flat*
Smooth	3/16"–1/4"	1/4"–3/8"
Medium	3/8"	1/2"–3/4"
Rough	3/4"	1"–1-1/4"

504.1c Cover Material. The selection of a cover material depends on the surface to be painted and the type of coating to be used. The most popular roller fabrics are polyester or a modified acrylic and mohair or lambs' wool. Table 504.1c summarizes the selection process for roller fabrics. The various types of roller fabrics have distinguishable characteristics which should be kept in mind when selecting a roller for a specific job. See Fig. 504.1c.

Lambs' wool rollers are available in a full range of nap lengths and are recommended especially for solvent-thinned coatings with rough surfaces. They mat badly with water-thinned paints, and should not be used for this application.

Mohair, primarily the hair of the Angora goat, is solvent-resistant and is available only in small-nap grades for use on smooth surfaces and for applying synthetic enamels. Unlike lambs' wool, it can be used with water paints.

Table 504.1c. *Types of Fiber to Use for Various Surfaces.*[*]

Type of Paint	Type of Surface Smooth	Rough
Interior paints		
Latex	Dynel	Dynel
Solvent-thinned	Dynel	Dynel
	Mohair	Lambs' wool
Exterior paints		
Wood	Dacron	Dynel
Other surfaces	Dynel	Dynel
Paints and lacquers		
containing strong solvents	Mohair	Lambs' wool

[*]From Sidney B. Levinson and Saul Spindel, *Developments in Architecutral and Maintenance Painting: A State of the Art Review* (Federation of Societies for Paint Technology: Philadelphia, 1969).

Acrylic, modified, known as Dynel, can be used on smooth and rough surfaces for water-thinned and solvent-reduced paints, but it should not be used with strong solvents, such as ketones.

Polyester, identified as Dacron in the table above, has a soft nap, which minimizes bubble-formation in latex paints over smooth surfaces. Polyester rollers are also suitable for oil paints, and are mostly used for exterior surfaces.

Miscellaneous, rollers made of carpeting are sometimes used for stippling and for applying viscous mastics. Embossed carpeting fabric is used for applying texture paints.

Fig. 504.1c. Effect of Roller on Leveling.
The type of cover material selected for a roller will determine the leveling effect achieved. Note the difference in effects achieved with mohair (left) and a high-leveling roller (right).

504.2 Roller Classes. A wide variety of rollers are available for different applications.

504.2a Tapered Rollers. Tapered rollers have receding edges. They are designed to aid in painting corners.

504.2b High Leveling Rollers. These provide better leveling with latex gloss paints because they are made of woven fabric. The woven fabric reduces the stippling effect often produced by other types of roller covers.

504.2c Industrial Rollers. Large rollers for industrial maintenance surfaces are available in sizes up to 18 inches in width and 2-1/2 inches in diameter. Their 140 sq.in. of surface hold five times as much paint as the normal 7-inch roller. An industrial roller with extension is shown in Fig. 504.2c.

504.2d Pressure Rollers. Pressure rollers use either compressors or a carbon dioxide cylinder to force paint onto the roller surface. The air compressor device takes paint directly from its regular container and feeds it to the roller. The carbon-dioxide-powered device requires a special holder for loading.

Fig. 504.2c. Industrial Roller with Extension Pole.

504.2e Fence Rollers. These are special rollers with extra long nap. The long nap length lets the paint from the roller surround the wires of the fence so that the painting can be done from one side.

504.2f Pipe Rollers. Pipe rollers are made up of four or five small roller segments placed on a flexible metal holder which enables them to hug the contour of the pipe to be painted. See Fig. 504.2f.

504.2g Calibrated Rollers. Calibrated rollers were designed to help assure adequate film thickness of latex paints, which flow so easily painters tend to spread them too thin. To spread these paints at a rate of about 450 sq.ft. per gallon, calibrated rollers have the word FILL embossed on them. Paint is rolled on the surface to be coated in the form of a large **W** of a definite size. When the word FILL can be seen on the roller, the direction of the roller is reversed and enough paint

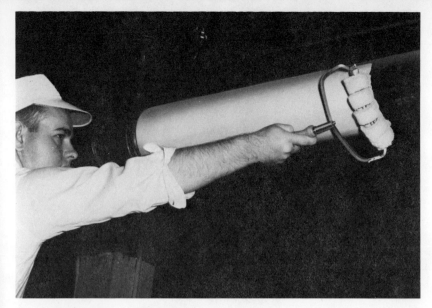

Fig. 504.2f. Pipe Roller in Use.

is still available in the roller to cover the square area bound by the extremes of the W. Paint can be applied rapidly without fear of under-coverage. See Fig. 504.2g.

504.3 Helpful Notes. Various roller cover manufacturers have their own synthetic fiber formulations aimed at getting best performance for the dollar. Since their standards differ, it's up to the buyer to know what to look for to get maximum painting for the lowest hourly cost. Here are some points to watch:

1) *Check paint discharge capacity,* which is the net delivered amount of paint that can be discharged to the surface from one pickup.
2) Note the *ability to yield a level finish,* which depends on fabric composition.
3) Determine the *rate of discharge* by checking the weight of paint delivered by the roller per square foot; or this can be measured in mils of wet film per square foot.
4) Observe the *ability to conform to surface irregularities.*

504.3a New Cover Conditioning. New covers can be conditioned by soaking in a solvent and squeezing out the excess. Alternatively, the conditioning can be done by rolling out the first load of paint on a newspaper. The purpose is to

Fig. 504.2g. Calibrated Roller in Use.

wet the fibers and flush out any lint or dirt that may be present.

504.3b Cleaning the Roller. Thickened paint on the roller cover will result in an uneven surface coating. The roller should be rinsed occasionally in thinner or water to remove such thickened paint. On jobs using fast-drying paint, it is helpful to use two rollers and have one soaking while the other is in use.

505 Flat-Pad Application. These new applicators are probably the fastest way to coat surfaces except for spray guns. Fig. 505 shows a flat-pad applicator in use.

To speed the painting job, several pad manufacturers have designed special paint-trays with varying degrees of success. The most ingenious has a roller in the tray. The paint pad is drawn across the paint-covered roller, much as the inking

Fig. 505. Flat-Pad Applicator in Use.

roller in a printing press passes over a second roller which rotates through an ink fountain. The paint "fountain" in the tray keeps the roller wet, and the pad picks up a load by passing over the wet roller.

505.1 Materials. Flat-pad applicators vary in size from 1" × 2" units for moldings to 4" × 7" units for large surfaces. They consist of a foam base covered with a napped fabric, usually mohair. The foam serves as a reservoir, and the napped face with pile depths of 1/2 or 3/4 inch, and 1-3/4 inches to choose from, can be used on surfaces of varying roughness. Paint application is hastened with these applicators since the 4" × 7" device holds about one-third more paint than a 4-inch brush.

To hold the even pressure of an ever-flat face on the surface being painted a spring-head is attached to the handle.

505.2 Types of Applicators. The types of applicators available include *stain pads, floor-finish applicators,* and *paint mitts.*

505.2a Stain Pads. Stain pads utilize a lambs' wool pad over the foam, instead of mohair, for applying stains to shakes, shingles, or panels. Short-nap wool yields smooth surfaces;

and for rough areas depths up to 1-1/4 inches are used. It has been estimated that as much as seven times more stain than is provided by a 4-inch brush is held by a stain pad, and that application is three times faster.

505.2b Floor-Finish Applicator. Floor-finish applicators have lambs' wool attached to a flat board, with a 4-foot handle. They are about 3 inches across and from 8 inches to 16 inches wide.

505.2c Paint Mitts. Paint mitts can be classified as flat pads because their contact surfaces theoretically are dragged flat across irregular surfaces. They are of lambs' wool, with or without thumbs, and are considered about twice as fast as a brush for painting such odd-shaped surfaces as valve-control wheels and half-round and clover-leaf designs.

505.3 Helpful Notes. Because pressure on the pads will release considerable paint from the reservoir, it is possible to saturate problem surfaces and work in paint with results approaching that of brush application. For primer application and for use over areas requiring careful penetration, pads are superior to rollers.

A considerable number of brush and roller manufacturers are now offering pads. Some foam reservoir materials probably hold more than others, and some nap-face fiber formulations may produce smoother finishes than others. Small pads are inexpensive, and it may be a good idea to purchase specimens from several manufacturers to serve as testing samples and try them out for paint-loading and discharge, in accordance with suggestions under "Helpful Notes" for roller application (Section 504.3), before buying them for big jobs. Similarly, what was said about conditioning rollers also applies to pads.

505.3a Solvents. Strong solvents should be avoided with applicator pads as they may dissolve the urethane foam.

505.3b Extension Poles. Extension poles are available for pad applicators, usually with screw-ends to fit the female screws in the pad handle. These are similar to the poles used for

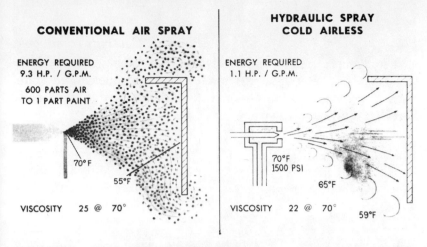

Fig. 506.1a. Conventional Air Spray.
Fig. 506.1b. Hydraulic Spray, Cold Airless.

rollers, which are either of fixed lengths or extendable up to as much as 16 feet.

505.3c Care of Pads. Since these pads have both a foam reservoir and a napped surface that can dry out and harden after use, it is doubly important to remove all paint after each use.

506 Spray Application. Spraying is often favored because of its speed. Specifiers should remember, however, that in many instances time-consuming masking may be required to protect nearby surfaces. The labor cost involved in masking must be balanced against the savings in spray application.

506.1 Types of Spray Systems. The three types of spray systems presently available are *air spray, airless spray,* and *electrostatic spray.*

506.1a Air Spray. With this system, compressed air is used to atomize the paint. The paint is forced through a spray gun orifice onto a surface in a diffused but directed fog. See Fig. 506.1a.

506.1b Airless Spray. In an airless system hydraulic pressure is used to produce a high-speed stream of atomized paint. Because air turbulence is lacking, over-spraying is minimized. Less thinning solution is required because greater pressure is provided. See Fig. 506.1b.

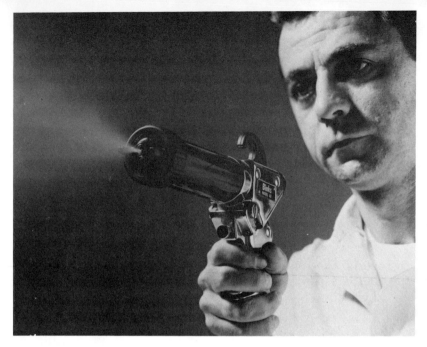

Fig. 506.1c. Electrostatic Airless Spray Gun.

506.1c Electrostatic Spray. In an electrostatic spray system, electrostatic charges as great as 80,000 volts are imparted to paints propelled by systems similar to airless spray. Electrostatic systems are highly desirable for coating irregularly shaped conductive surfaces, particularly where wraparound is desired, but methods are also available for coating nonconductors. See Fig. 506.1c.

Electrostatic spray, because it provides a more concentrated pattern and because it consists of charged particles which are attracted to their intended substrate, requires little if any masking. Most electrostatic systems require specially-formulated paints, but systems are available which utilize a unique system, described as Electro-Gas Dynamics, that will spray many conventional paints as well as highly sophisticated ceramic frits and fluorinated hydrocarbons.

506.2 Heated Paint. Heated paint (120° F to 200° F) can be provided in special systems for feeding any of the three spray systems. Hot spray requires virtually no solvent thinner; and this is becoming very important because of air pollution laws limiting solvent emissions. Application is faster than cold spray methods; and it is cleaner and safer.

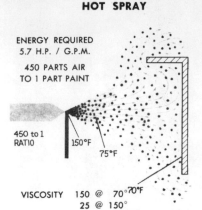 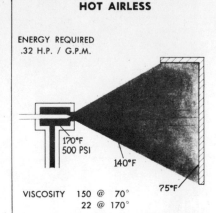

Fig. 506.2. Hot Air Spray Versus Hot Airless Spray.

Hot spray systems will pay benefits in faster production and smoother surfaces. The advantages of a hot airless spray system over a cold airless system are summarized below:

1) Speed
    Paint can often be used directly from the can
    High film build—improved one coat coverage
    Production is increased about 25 percent
    Faster tack free time
2) Cleanliness
    Practically no bounce back from corners
    Very little overspray
3) Safety
    Practically no fog
    Less solvent fumes (no added solvent)
    Lower pressures are required (500 to 750 psi)
4) Better performance
    Little or no orange peel or sag
    Extremely smooth glossy film
    Covers porous surfaces better
    Better adhesion
5) Economy (See Fig. 506.2.)
    Less horsepower is required
    Less air is required—about 40 percent less
    Less paint is required
    No solvent has to be added, then lost when spraying
6) Versatility
    Can spray at lower temperatures and at higher humidity
    Can spray inside tanks

See Fig. 506.2 for a comparison of hot air spray versus hot airless spray.

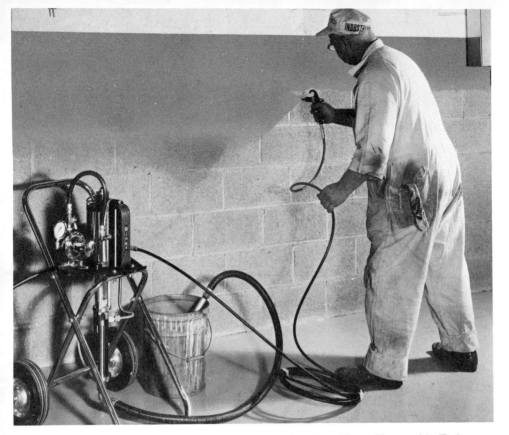

Fig. 506.3. Airless Spray Unit—Suction Hose Type, Air Drive. *Note the evenness of the spray pattern, and the lack of overspray.*

506.3 Comparison of Airless and Air Spray. The advantages of the airless system, as compared with the air spray system, are as follows:

    1) It is faster
        Twice the production rate (fewer passes)
        Less time required for masking and cleanup
    2) It is easier to use
        Smaller compressors are required
        Small direct drive units are available
        Only one hose is required
    3) It is cleaner
        Less overspray; see Fig. 506.3.
        Less bounce-back from corner
    4) It is safer
        Less fog
        Less fire hazard (no air spray)

Application of Coatings

    5) It does a better job
        Higher film build per coat, better coverage
        No entrained moisture or fluids from the compressed air
        No moisture blush
        Better penetration of voids
    6) It is more economical
        Smaller compressors can be used
        Less paint is required (savings of 10 to 30 percent)

The airless system does have certain disadvantages, however, when compared with the air spray system:

    1) Less versatile
        Restricted to large areas
        Nozzles must be changed to change pattern
        Cannot spray—
            Hammer finishes
            Multi-color finishes
            Fibrous materials
            Materials having a short pot life
    2) Clogging is a problem
        Paint must be absolutely clean (most units have built in filters)
    3) Cleaning the gun can be dangerous due to the high pressures used (most units have built-in safety features).

507 Electrostatic Spray, Description. Heavy transformers are usually required to develop the 80,000 volts needed for electrostatic spray systems, but several manufacturers now make complete units mounted on wheels so the cumbersome devices can be transported to a construction site or where maintenance painting is needed. Experienced specifiers, particularly those involved with odd-shaped surfaces, such as channels, angles, piping, and cables, and whose companies have their own crews, prefer electrostatic application because of its reduction of paint loss and its minimal need for masking. However, it has its disadvantages as well as advantages.

507.1 Advantages of Electrostatic Airless Spray:

    1) Complete coverage of odd shapes—practically no touch-up

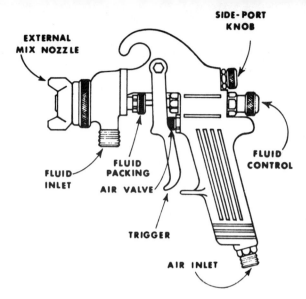

Fig. 507.4A. Cross Section of a Modern Production Spray Gun.

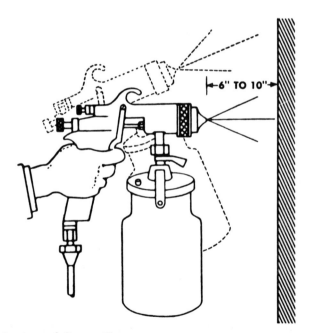

Fig. 507.4B. Positioning of Spray Gun.
The spray gun must be held perpendicular to the work surface to prevent uneven deposit of paint.

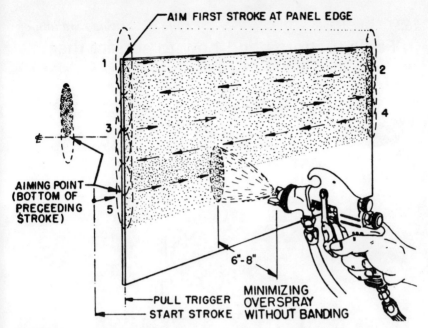

Fig. 507.4C. Spraying Technique for Large Flat Areas.
As illustrated, a definite pattern must be followed by the painter if good results are to be achieved. Note also that there is an optimal distance from the spray gun orifice to the work surface.

2) Most efficient in paint usage—no loss
3) Minimum overspray—no face masks or protection required
4) Most uniform finish even in inaccessible areas and on edges

507.2 Disadvantages. On the other hand, there are a number of disadvantages as shown below:

1) Higher cost of equipment
2) Special formulation required
3) Higher servicing costs
4) Slower operation than cold or hot airless
5) Only one coat can be applied
6) There is a possible shock hazard
7) Can only be used on bare metal.

The fourth advantage (achievement of uniform surfaces in difficult-to-reach areas) is usually cited as a principal reason for using electrostatic spray. Because it is drawn to the surface by electrostatic attraction, the paint spray reaches the recesses and edges, often missed by other spray methods.

Faulty patterns and how to correct them

Pattern	Cause	Correction
	Dried material in sideport "A" restricts passage of air through it. Result: Full pressure of air from clean sideport forces fan pattern in direction of clogged side. 	Dissolve material in sideport with *thinner*. Do not poke in any of the openings with metal instruments.
	Dried material around the outside of the fluid nozzle tip at position "B" restricts the passage of atomizing air at one point through the center opening of air nozzle and results in pattern shown. This pattern can also be caused by loose air nozzle. 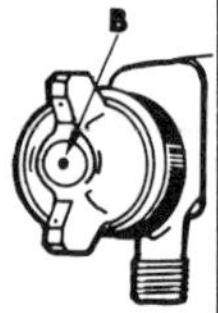	If dried material is causing the trouble, remove air nozzle and wipe off fluid tip, using rag wet with thinner. Tighten air nozzle.
	A split spray or one that is heavy on each end of a fan pattern and weak in the middle is usually caused by (1) too high an atomizing air pressure, or (2) by attempting to get too wide a spray with thin material.	Reducing air pressure will correct cause (1). To correct cause (2), open material control "D" to full position by turning to left. At the same time turn spray width adjustment "C" to right. This will reduce width of spray but will correct split spray pattern. 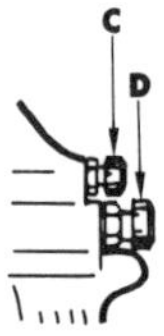
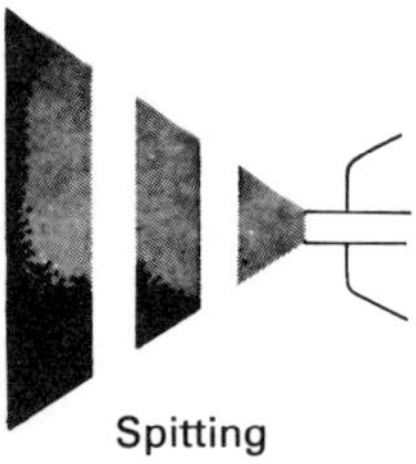 Spitting	(1) Dried out packing around material needle valve permits air to get into fluid passageway. This results in spitting. (2) Dirt between fluid nozzle seat and body of a loosely installed fluid nozzle will make a gun spit. (3) A loose or defective swivel nut on siphon cup or material hose can cause spitting.	To correct cause (1) back up knurled nut (E), place two drops of machine oil on packing, replace nut and tighten with fingers only. In aggravated cases, replace packing.  To correct cause (2), remove fluid nozzle (F), clean back of nozzle and nozzle seat in gun body using rag wet with thinner, replace nozzle and draw up tightly against body. To correct cause (3) tighten or replace swivel nut (G).

Fig. 507.4D. Spray G

Application of Coatings   183

Faulty patterns and how to correct them

Pattern	Cause	Correction
	A fan spray pattern that is heavy in the middle, or a pattern that has an unatomized "salt-and-pepper" effect indicates that the atomizing air pressure is not sufficiently high.	Increase pressure from your air supply. Correct air pressures are discussed elsewhere in this instruction section.

Pointers on cleaning

When the gun is used with a pressure tank or gravity bucket, remove the hose, turn the gun upside down and pour thinner into the fluid opening while moving the trigger constantly. This will flush all passageways.

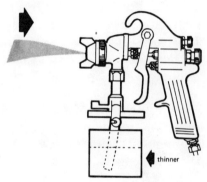

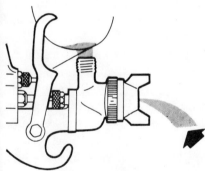

When used with a cup, thinner or suitable solvent should be siphoned through gun by inserting tube in open container of that liquid. Move trigger constantly to thoroughly flush passageway and to clean tip of needle.

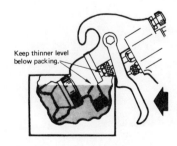

Keep thinner level below packing.

It is extremely poor practice to place an entire gun in thinner. When this is done, the solvent dissolves the oil in the leather packing and causes the gun to spit.
It is good practice to place the nozzle and fluid connection in thinner. Vessel used should be shallow enough to prevent thinner from reaching packing.

justments and Cleaning.

The fifth and sixth disadvantages listed above are only partly justified now, since spray applied by some methods can be used over other paints, particularly if the latter have conductive pigments; and with these pigments, nonconductive surfaces can be primed and then topcoated with an electrostatic spray.

507.3 References. The tables and comparative figures used in this section on Spray Application were developed by Sidney B. Levinson and Saul Spindel of David Litter Laboratories, for the Federation of Societies for Paint Technology's brochure "Recent Developments in Architectural and Maintenance Painting," obtainable from the Federation at 121 South Broad St., Philadelphia, Pa., 19107.

Another helpful book for the specifier charged with the responsiblity for actual physical application, such as painting contractors, heads of maintenance crews for utilities, petroleum firms, or shipbuilding companies, is the *Application Manual for Paint and Protective Coatings,* by W. F. Gross of Binks Manufacturing Co. (New York: McGraw-Hill, 1971).

507.4 Equipment. Since spraying is a mechanical operation, individual manufacturers' equipment is designed to meet user requirements and at the same time, hopefully, to provide performance advantages. A cross section of a typical spray gun is shown in Figure 507.4A.

Specifiers, other than those few involved directly with application personnel, require only generalized knowledge of spray techniques. Helpful hints for maintenance engineers may be derived from Figs. 507.4B, 507.4C, and 507.4D. The variety of requirements for atomizing air pressure, or nozzle openings for paints of various viscosities, is diverse and unnecessary in a guide of this kind. When the specifier lacks confidence in his painting contractor's knowledge of these techniques, he can get specific information for each kind of paint from paint suppliers. Generalizations in this area of interest may do more harm than good.

HELPFUL ADDITIONAL READINGS

1. Taylor, A. J. A., *Paint Technology Manuals,* Vol. 4 (New York: Reinhold, Inc., 1965), pp. 18–49, 158–205.

chapter 6

Deterioration of Coatings

600 General. Most coatings designed for durability shrug off exposure to water. Cheap paints, or flat paints with too much pigment and extender, may sometimes be damaged directly by water, but even these paints are mainly the instruments of their own destruction. The deteriorating effects of moisture on components of these poorly formulated paints—or on microbes in or on them, as this moisture passes through, or around, or rests within them—is due to their poor material rather than the destructive effect of water.

Deterioration of protective coatings and the surfaces they are supposed to protect can be blamed on other things than water. Potential troublemakers include chemicals, sloshed on accidentally or intentionally, or in the form of vapor; or erosion by abrasion, scrubbing, or surface chalking; or poor materials; or even good materials poorly applied; or good materials properly applied to badly prepared surfaces. And, let's be frank, bad architectural design can cause the deterioration of coatings and protectable surfaces.

Since water is an important villain, however, it will engage our attention first.

601 Water Sorption and Release. An odd phenomenon triggered a study of the diffusion of water into and out of films and showed an unsuspected cause of a number of film failures. It seems that it was once noted that a primer applied to metal without a topcoat protected the metal beautifully all by itself during three years of weathering in a frequently moist Florida environment. The same primer with a topcoat of a fine alkyd permitted rust to form on the metal substrate within a few months.

Two curious scientists employed by the Sherwin-Williams Research Center carried out a series of experiments. The experiments showed that the topcoat in the failed system was cramping the style of the primer that had done so well by itself. What was happening? Water that permeated through the primer when it was uncovered was able to leave as soon as the surroundings became dry, and thus no damage was done to the coating or to the substrate. When the topcoat, which probably had a slower rate of diffusion of water, got wet and passed on the moisture to the primer, the topcoat held on to its water longer than the primer would have if it were alone. That blocked the drying out of the primer beneath. The result was too much water trapped for too long a time in the layer of paint closest to the metal, and rust was the result.

Further, experiments have shown that on wood surfaces retention of water in the primer due to the topcoat's slower diffusion can cause other ailments, such as mildew, wood swelling, paint peeling, and blisters.

The Sherwin-Williams scientists, T. Kirk Hay and Garmond G. Schurr, used a concept explained at great length in an earlier section, that of Pigment Volume Concentration and also Critical Pigment Volume Concentration, to explain what happens and why. If you recall, Critical Pigment Volume Concentration (CPVC) describes a relationship of the major parts of a paint formulation wherein just enough resin (or binder if you prefer) is present to bind up or completely surround all the pigments and extender pigments in the paint. In other words, there is no free resin around, and no unbound or incompletely wetted pigment. If the Pigment Volume Concentration is below the CPVC, then free resin is present because all the pigment and extender particles are amply wetted and some resin is left over and can wet more pigment if more is added. If the Pigment Volume Concentration is higher than the CPVC, that means that the pigment particles are incompletely wetted because they are too plentiful to be completely surrounded by the available binder. When the PVC is below the CPVC,

Deterioration of Coatings

meaning that there is an excess of binder, the paint is usually glossy or semiglossy; the paint with a PVC that is higher than the CPVC has unbound, or partially wetted, pigment or extender and is flat.

Hay and Schurr in the course of many experiments found that, sure enough, if the primer and the topcoat have the same low Pigment Volume Concentration, then rust is prevented. However, as the relative amount of pigment and extender in the primer is increased, its ability to diffuse water also goes up. If a topcoat is present with a low PVC, which means that it doesn't have unbound pigment and extender present to speed up the diffusion of water from the primer, then the diffused water present in the primer can't escape fast enough. The slower diffusing topcoat traps the moisture in.

"It becomes apparent," wrote Hay and Schurr in their report in the *Journal of Paint Technology,* the scholarly organ of the Federation of Societies for Paint Technology, "that corrosion inhibition is dependent on the concentration-time history of water in the paint system or at the metal interface. A proposed mechanism for the failure of the primer-topcoat system is that moisture is held in contact with the metal surface or in the paint by the two coat system but [would] move more freely in and out of the primer [by itself]." They point out that the topcoat, where the Pigment Volume Concentration is low and pigment is tied up by resin, delays moisture release.

In other words, in each coat, where pigment and extender are plentiful enough, they help the coating serve somewhat as a two-way wick: letting water enter and leave. If the top wick passes moisture more slowly than the bottom wick, water is trapped in the primer and the undercoat may keep the substrate moist too long; and the ills of the substrate and of the coating, such as mildew, rust, peeling, and blisters, may develop.

602 Permeability. The physical mechanism whereby moisture passes through a coating is known as permeability. Pores of one kind or another exist in many kinds of film, if not all. These can be set up in any number of ways: shrinkage of the resin on setting; solvent or water evaporation; soluble chemicals in the film; or tiny microscopic spaces between pigment particles, particularly those that are not completely wet by binder when the PVC is greater than the CPVC, which is another way of saying that more pigment is present than the binder can completely surround.

Permeability is not necessarily bad, particularly if water or water vapor leaves the film quickly enough. If it stays in too long, as explained above, trouble develops.

An impermeable film can be very troublesome. Witness the walls outside bathroom, kitchens, or laundries, rooms that are often wet and steamy. What happens to the steam if the kitchen or bathroom wall isn't sealed well at the window joints, or if it has small cracks where the bathtub joins it? Steam works through the cracks and gets behind the wall, where it condenses into water. Since moisture wants to take a path to the open space, it builds up pressure in its effort to penetrate the outer wall. If the wall is wood or cement, and is not vented so that moisture can get through the easy way, it wets the wood or cement and tries to push its way through the protective paint. If the paint is permeable, the moisture passes through, without causing immediate harm. If the paint is impermeable, or if the primer coat is and the topcoat isn't, peeling is almost certain to follow.

Architectural design considerations, to be discussed later, can prevent this trouble.

Few comparative studies have been made to show the relative permeability of resins. An interesting table is provided in a "Paint Technology Manual" published by the Oil and Color Chemists' Association of Great Britain. In manual 3, on *Convertible Coatings,* D. M. James reports on studies showing the approximate water permeabilities of common varnish materials. Describing linseed oil as the most permeable with a rating of 100, he listed the following guide to the order of permeabilities:

Varnish constituent	Permeability
Dehydrated castor oil	70
Rosin	30
Long oil, linseed-modified phenolic varnish	25
Long oil linseed alkyd	20
Chlorinated rubber	5

Note: Linseed oil has a permeability rating of 100; all other constituents are ranked in terms of the 100 rating assigned to linseed oil.

Deterioration of Coatings

The high-permeability films are desirable for surfaces that, for one reason or another, have to allow water to pass through them; for example, on wooden exterior walls behind improperly sealed kitchens. Low-permeability coatings are needed to protect corrodible surfaces or those that may deteriorate when repeatedly wetted. The low permeability reported for chlorinated rubber confirms the wisdom of its widespread use for concrete protection, a market it shares with styrene acrylate and butadiene styrene, which are synthetic rubber resins with low permeability. Similarly, the low permeability of chlorinated rubber encourages its growing use for anti-corrosive coatings, particularly in salt environments because of its alkali-resistance. One growing use is in ship paints.

Another study of permeability and the transport of water vapor through various paints was made by a Swedish chemist, G. Gardenas. He found that linseed oil paints (depending on pigmentation), 30 micrometers (or slightly more than 1 mil) thick, transported from 2 to 2-1/2 milligrams of water per square centimeter each day. Alkyd paints transported 1.1 to 1.3 milligrams; and latex paints about 25 milligrams. One reason for low linseed oil permeability may be the use of reactive lead pigments, which form soaps and make tight film.

The James table, cited previously, was obviously limited to solvent-reduced paints, since latex was found to be about ten times more permeable than linseed oil. The Gardenas findings were based on tests with the Payne cup under conditions of uniform humidity.

Another investigation by Thun and Ovregard and reported in the same Swedish journal as Gardenas's findings, confirmed these figures, except that styrene-containing latex paints were found to have roughly the same permeability rating as linseed oil and alkyds. (Styrene-butadiene latex paints are still widely available in the United States from one major manufacturer, but most other producers have ceased offering them.)

Further effects of pigmentation on permeability were reported by various experimenters. Phenolic resin, it was found, has a rate of diffusion of water of 0.718 grams per square centimeter per year. When leafing aluminum powder is used in a phenolic formulation, the rate is cut to 0.191, or considerably less than one-third. The alkyd rate is 0.825, but with the same aluminum powder it is reduced to 0.200, or less than one-fourth.

In many of the causes of film failure to be discussed in this section, permeability or lack of it will play a role.

Permeability should not be confused with absorption, because the latter leads to far more intimate merging of moisture and materials.

603 Absorption. Because research workers often work in a semi-ideal world, they do not give the same importance to absorption that they do to permeation; yet absorption can cause as much, or more, harm as permeability in an imperfectly formulated paint. A permeable coating, it is true, may be harmed when water is diffused through it; but when water is absorbed in a coating because the material has certain water-soluble chemicals remaining in it, the coating is almost always harmed. Swelling, for one thing, often follows water absorption. Another frequent result is the dissolution of soluble materials in the absorbed water and their removal when subsequent moisture enters and leaves. Voids created by this dissolution and removal are among the causes of pores that eventually permit permeability.

Absorption, or the intimate holding of moisture in a film, can also be caused by attraction of molecules A complicated process known as electro-endosmosis causes an increase in this molecular attraction in films covering metals that have started to rust. Some sort of electrostatic force appears to be generated in this case, thereby accelerating deterioration by speeding the rate of absorption.

Pigmentation has an important influence on water absorption as well as permeability. Use of zinc oxide in linseed oil coatings has been found to increase water absorption, but it reduces water permeabiliity. Reactive lead pigments, because they form compounds with linseed oil and other resins and form tight molecular networks, decrease both absorption and permeability of water.

Neither absorption nor permeability are always to be shunned. We've seen that permeability is important for coatings over porous surfaces. In addition, some consumers want anticondensation paints so their humid laundries, kitchens, or baths won't drip condensed water. These can be supplied by buying paints made absorptive or permeable, preferably the latter.

Deterioration of Coatings 191

These soak up the steam and hold it until the room dries out.

Consumers, and specifiers, should be reminded that reference to anti-condensation paints is made here in the section on surface defects, because suitable formulations for preventing condensation may also be "suitable" for compromising film integrity, particularly if the special film depends heavily on absorption as well as permeability.

We have seen in the section on permeability that linseed oil in pigmented paints allows a fairly low rate of water transport. However, in linseed-oil varnishes, which have no pigment, considerable water is absorbed. James, in the same book quoted before, reports that a linseed stand oil film is able to absorb more than 400 percent of its weight of water, but when combined with resins absorption is reduced by factors of 10 to 50.

He also observes that phenolics and epoxy resins, as well as coal-tar resins, absorb very little. He explains this by noting that they are highly crosslinked, which means that the molecules which make up these resins are not merely strung together and joined end to end; they are also joined like Siamese twins: some of the offshoots or limbs of the molecules combine with each other, forming tight networks that fight back intruding water.

Stretching of these tightly knit crosslinks is regarded by some theorists as the reason for swelling of films, a frequent cause of film failure.

604 Coating Defects Due to Water—Any Substrate.

604.1 Blistering. Surface coatings on any substrate can be subject to blistering. Water from the atmosphere meets soluble substances in the film and initially dissolves them. The resulting solution serves as one liquid section of an osmotic cell. Other water brought to the surface of the film by rain or otherwise is drawn into the starting solution by a pressure gradient. As in all osmotic situations, a membrane must separate the first solution from the incoming water. A thin elastic layer in the film can serve this purpose. As water passes to the first solution, the elastic separating layer expands under the pressure and forms a blister. Blisters will not occur if the

Fig. 604.1A. Improper Coating Application Likely to Lead to Blistering.
An open invitation to trouble. Blisters will probably form because an oil paint is being applied over condensed moisture, thereby lessening adhesive forces.

strength of adhesion of the coating to the substrate is greater than the pressure exerted by the growing cell holding the water.

Blisters can be prevented by making certain that the primer-topcoat system keeps out water and that the adhesion of the topcoat to the primer and of the primer to the substrate is strong. Attention is needed to see that specified paints are low in water-soluble salts, either from pigments or from thickeners or emulsifying agents. See Figs. 604.1A and 604.1b.

604.2 Bloom. The thin film that sometimes appears on glossy coatings or varnishes, reducing lustre and retarding the fullness of tones and tints, is called bloom.

Some examples of bloom are caused by condensation of moisture on the film before complete drying has occurred. Sometimes this can lead to crater bloom, which leaves small holes resembling craters.

604.3 Chalking. Moisture joins ultraviolet rays and heat to erode the binder in pigmented films. Presence of anatase, or chalking grades of titanium dioxide, accelerates the process,

Fig. 604.1B. Blistering.
Large blisters formed and broke under this eave. Cause was moisture reaching the crawl space above the eave, below the sloping roof.

but all titanium oxide except the finest nonchalking grades of rutile permit some limited chalking. Some theories claim that titanium dioxide catalyzes the slow destruction of binder on an exterior surface.

Chalking is regarded as helpful in keeping exterior surfaces clean. Unfortunately, chalking is rarely uniform, so after a period of time it can cause unsightly patterns in the coating or on adjacent surfaces. See Fig. 604.3. Chalking can cause poor adhesion of new coatings applied over old painted surfaces. To repaint old exterior surfaces, chalk should be wiped off as thoroughly as possible and paints should contain binders that wet the old surface thoroughly (See Fig. 302.) Since latex paints are poor wetters, they usually are supplemented with additives containing vegetable oils or alkyds.

604.4 Flaking. Paint flakes off most frequently near joints on exterior surfaces. Moisture usually enters at the joints and works under the coating's surface, causing it to peel in sheets. See Fig. 604.4.

604.5 Mildew. Unsightly brown or black stains are often ascribed to dirt when really they result from growth and reproduction of fungus deposited on the surface from the air. Most exterior paints contain antifungal materials. Interior paints

Fig. 604.3. Chalking on Residential Structure.
Chalk from the painted wood shingles has washed down on the brick. Excess pigment in the paint is one of the most likely causes.

intended for use in moist areas, such as kitchens, laundries and bathrooms should be checked to see if they have an adequate mildewcide content. Fig. 604.5 shows what happened in one bathroom when this advice was not followed. Most experts contend that the best defense against interior mildew is the hardest possible gloss paint, one with a low Pigment Volume Concentration, which means it has lots of resin. (See Section 601 for a description of water sorption and release, which points to the importance of having a system whose topcoat and prime coat diffuse and release moisture at about the same speed.)

604.5a Anti-Mildew Additives. Exterior paints containing *zinc oxide* have less trouble with mildew than others. In addition, exterior paints, particularly latex paints, must have mildewcides. Mercury salts have been used effectively for years.

Fig. 604.4. Flaking.
Incompatible topcoat lifted off this wood surface after a few months. Primer and topcoat came from different manufacturers.

These were to be outlawed nationwide, but the threatened ban never materialized. Non-mercurial mildewcides are used in some paints. Most substitute exterior latex formulations have been adequately tested. A considerable number of these substitute formulations have already been proven under such drastic conditions as those prevailing on the Louisiana and Texas Gulf coasts, even on Southern pine. Most national and regional paint producers have effective nonmercurial paints.

604.5b Removal of Mildew. When mildew appears, it can be removed by wiping with a solution consisting of 2/3 cup trisodium phosphate cleaner, 1/3 cup detergent, and one quart of household bleach undiluted.

604.5c Repainting Mildewed Surfaces. If the mildewed surface is to be repainted all traces of the fungal stain must be removed. Particular care should be taken to swab the depressions that the fungi may have eaten out of the surface. Surviving fungi will probably eat their way through the new coating and repeat the trouble.

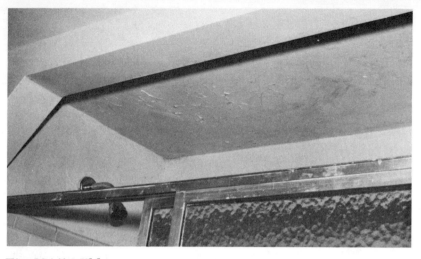

Fig. 604.5. Mildew.
Hot shower steam provided an ideal environment for mildew. Cellulosics in this latex paint provided nutrients, and the coating failed after only ten months.

604.6 Rust on Walls. Unsightly rust spreading down exterior walls can be caused either by overflow of water from rusty gutters, or by rust from ungalvanized nails in the wall. Gutters should be cleaned and painted, or replaced. Nails should be sanded clean, primed, and touched up.

604.7 Poorly Filled Cracks and Holes. Moisture can enter substrates through holes and cracks that were improperly filled prior to priming. Small openings permit moisture to enter hidden areas where deterioration can begin. When pores in the surface coating lead to the substrate, water can eventually work horizontally between coating and substrate and pry loose the coating's bond to the substrate, causing peeling.

605 Defects Due to Water on Wood Only.

605.1 Discoloration from Soluble Dyes. Redwood and cedar, among other woods, contain soluble dyes that must be protected from atmospheric moisture. When moisture penetrates to the surface of the wood the dyes are dissolved, and the solution diffuses through the paint to the surface, forming pink or brown streaks or spots. New lumber should be primed with a good sealant rather than an ordinary porous latex paint. Stains from soluble dyes need not be permanent since they can sometimes be washed off by hand; or rain will eventually remove them. (See Fig. 605.1; see page 253.)

Fig. 605.1. Discoloration from Soluble Dyes.
Soluble dyes in the wood discolor the topcoat because the primer coat did not contain suitable pigments to check them.

605.2 Stresses in Spring and Summer Wood. When wood is cut with the grain, a flat pattern appears, showing fairly wide, irregular waves of alternating hard and soft bands. The lighter softwood represents spring growth of the tree; and the darker hard bands, which are narrower, represent summer growth, which is slower than spring growth. These flat-cut woods absorb applied paint to a greater degree in the softer springwood areas than in the harder summer wood areas. Consequently, the swelling within this flat-cut wood is at two rates: the greater swelling of the more absorptive, softer spring wood and the lesser swelling of the harder summer wood. Stresses are set up in the paint covering the surface that contains these two elements, and these stresses may cause cracking of the paint. Since the hard, denser summer wood will have weaker adhesion to the paint than the softer portion, cracking, when it occurs, will very likely start in the summer wood, and peeling will probably start there.

Flat-grain wood, with its wide bands, often makes attractive furniture or cabinets; but for construction, edge-grain wood is preferable because it is cut directly across the radial growth bands and appears as tight parallel bands, with only small alternating areas of hard and soft wood. Consequently, the swelling that results from entry of moisture is far more evenly distributed than it is on flat-grain wood; and cracking of the coating is less likely to occur since stress is reduced, if swelling occurs.

605.3 Green Lumber. All troublesome water need not come from outside. In green lumber, considerable water is present in the wood. This should be removed by seasoning or by kiln-drying before applying paint. Otherwise, moisture will try to pass through the paint and cause blisters, cracks, and loss of adhesion.

605.4 Plywood Problems. Because they are literally peeled from logs, the plies used for making plywood, when they are flattened, have unresolved stresses that can lead to cracks later on when moisture triggers the stress. If paint has been applied, peeling can occur when the cracks develop.

If plywood is likely to be exposed to moisture, coating it with a water repellent is advisable. Plywood may suffer from entry of moisture between plies. This happens when poor grades of laminating adhesive are used, and the plies part. Moisture enters the plywood and may disrupt adhesion of paint on the surface. To prevent this, apply a sealer and topcoat to the edge of the wood where the layers can be seen.

605.5 Design. A considerable number of coating failures on wood results from faulty design or failure of carpenters or painters to follow architects' instructions. See Section 609 for further discussion of this type of failure.

606 Defects Due to Water—On Cementitious Surfaces.

606.1 Surface Crystals. Small, white crystals sometimes appear on the surface of painted cementitious surfaces. Soluble salts in concrete, stucco, masonry, and plaster may be dissolved

Deterioration of Coatings

in water diffused through it from any source. These salts may pass through the surface's coating and be deposited in a nonuniform manner as tiny white crystals, known as efflorescence. For treatment, see Sections 307.3 and 606.7.

606.2 Loss of Adhesion. The salts referred to above may encounter a nonpermeable coating. The salts are then left at the base of the coating, where adhesion is impaired in spots.

606.3 Popping. Concrete, stucco, and plaster that are improperly formulated or cured cause inconsistent porosities in the dried surface. When moisture enters, the cementitious material softens, swells, and pops up.

606.4 Uneven Gloss. The condition that causes popping can be recognized when coatings have an uneven gloss over the surface. Trouble can be avoided by removing the coating and the obviously soft cement and replacing it. Porous but firm cement can be sealed with a primer-sealer and painted with a coating with low permeability and low absorption, if care is taken that moisture will not enter the cementitious surface from the rear or from junctions with other surfaces. If the area cannot be sealed, a permeable coating should be used, so the moisture can pass through.

606.5 Paint Concrete Crumbling. Wet concrete may be painted. Drying of the paint causes a limited amount of contraction, which eventually lifts the paint and some adherent concrete. See Fig. 606.5. The crumbled concrete should be removed with the paint film. New concrete should replace it, and the entire area should be allowed to dry thoroughly. The surface should be prepared for painting in accordance with the suggestions in Section 307.

606.6 Peeling on Concrete Porches and Steps. If concrete porches and steps rest on concrete foundations that suffer from occasional or frequent wetting by ground water, they may become permeated through a wicking effect as the ground water rises through the concrete. If the coating is impermeable, as so many tough, abrasion-resisting porch floor paints are, peeling follows. The best corrective measure is to lay drains

Fig. 606.5. Paint-Concrete Crumbling.
Concrete in this basement received moisture that seeped from the ground, causing concrete and paint to crumble.

to interrupt the flow of ground water. If this is impractical, use a more permeable coating; but don't expect durability.

606.7 Efflorescence. Water-soluble salts are freqently absorbed by water passing through cementitious surfaces. These salts crystallize on the surface coating causing a chalky deposit—provided that the paint is permeable. If it is not, then the salts seed out beneath the coating, causing it to lose adhesion. Paint must be removed and the surface treated before repainting, using methods described in Section 307.

606.8 Peeling of Paint from Bricks. Bricks are porous and, like concrete, can transport water from ground sources or from wet mortar. Efflorescence can occur on brick. From unpainted bricks it can be removed easily; paint on bricks tends to lift when the salts seed out beneath it.

Ordinary water can cause the paint on bricks to peel, too, if the paint is impermeable; water-pressure builds up behind the stubborn paint and causes lifting. Permeable formulations are necessary to avoid this trouble.

Peeling may also result from a bad practice of some bricklayers. After they have completed a wall, they often use muriatic acid to wipe off flecks and splashes of dried mortar on the bricks. If this acid (dilute hydrochloric acid) isn't washed off thoroughly, chloride salts are left on the surface. These can

Deterioration of Coatings

absorb water from latex paint, causing swelling beneath the coat, which lifts the paint.

607 Defects Due to Water—Metal Surfaces.

607.1 Corrosion. A metal surface is usually uniform in structure and composition; hence, it rarely suffers from variations in expansion and contraction although thin metal may dent, requiring flexibility in a coating. Metal never supplies moisture to its coatings through a wicking action. Metal, therefore, rarely sins but is sinned against by faulty preparation, wrong paint selection, or by atmospheric or individual abuse. Rust on isolated areas should be examined for indications of cause. Chipped paint or dents in the metal show that it has been struck and that nothing fundamental is wrong with the coating, except perhaps poor flexibility and impact resistance. Rust near an edge may mean that the edge was inadequately coated.

Widespread rusting and paint lifting may mean that the metal surface was improperly cleaned prior to painting. (The result of improper removal of mill scale is illustrated in Fig. 703.1.) Wherever a limited amount of rust is present, treatment with a suitable rust remover, followed by paint application, should be satisfactory, provided that surrounding conditions will not cause repeated rusting. This means that the cause of the blow that dented the surface may have to be found and safeguards taken to avoid repetition. If the edge of the surface is subject to wear and rust starts there, some molding or protective barrier may have to be utilized.

If a leaking gutter is letting large amounts of water wear away the paint, the gutter should be repaired. Analysis is required, just as it is when defects on wood or cementitious surfaces are to be corrected.

607.2 Flaking and Peeling on Galvanized Steel. Sometimes when galvanized steel has not been allowed to weather sufficiently prior to painting, the chemicals applied immediately after galvanizing may still be present. These chemicals and the zinc galvanizing deposit when combined with paint materials form compounds that may cause trouble a year or two after application, particularly if the galvanized metal is subjected to pronounced fluctuations in temperature. Adequate weathering

Fig. 607.2. Deterioration on Galvanized Steel Gutter.
This galvanized steel gutter was painted before adequate weathering could remove processing chemicals, and peeling resulted.

and the use of recommended primers prevent this. See Fig. 607.2.

608 Coating Defects—General Causes.

608.1 Bloom. The lustre on gloss paints or varnishes is sometimes dimmed by a thin film that forms on their surface. Some experts ascribe this to condensation of water on the surface at a critical point in the hardening process. Water, while probably an ingredient in the problem, is not the major cause. One explanation is that the use of cobalt or manganese driers in varnish-type binders results in a deposit of a compound called linoxyn on the surface, and this compound, it is contended, attracts moisture and dust, which retard the through-drying of the film.

Another researcher claims that inferior resins are the cause, particularly when exposure to moisture in the air occurs before complete hardening. Whenever oleoresinous binders are used, the possibility of incomplete processing exists, with an attendant quantity of free glycerine available to attract moisture from the air. Two researchers studied the situation exhaustively and concluded that, in addition to free glycerine in the film, another source of trouble could be ammonia in the

drying atmosphere. They made two significant conclusions (see page 243):
1) That even extremely humid conditions or changes in humidity during drying had no effect on blooming.
2) The addition of calcium naphthenate to the formulation prevents or diminishes blooming.

608.2 Chalking by Sunlight. Weather action causes exterior binders to disintegrate slowly, leaving a fine powdery coat on the surface. Often chalking is caused by an anatase-type, titanium dioxide hiding pigment, which serves to catalyze the breakdown of commonly used binders in the presence of ultraviolet rays from the sun. It is hardly noticeable on white areas because the chalk is primarily white. However, tinted surfaces chalk unevenly, as a rule, which can result in an unattractive patchy appearance.

608.2a Removal. Chalk, to some extent, permits a form of self-cleaning since rain often washes it off. Chalk can usually be wiped off.

608.2b Specifications. To prevent accelerated chalking, which is highly undesirable, specifiers should select paints that contain only a small amount of anatase pigment and a large proportion of nonchalking rutile titanium dioxide. Even within the rutile designation, grades vary as to eventual chalking influence.

608.2c Repainting. Repainting chalky surfaces requires careful removal of the dust. Use of paints with good wetting properties is recommended to prevent poor adhesion from occurring. Latex paints, if used for primers over chalk, even if it appears to be removed, should contain additives with alkyds, linseed oil, castor oil, or some adhesion aid.

608.3 Cracking. Several kinds of cracks occur, mainly because coats of paint, upon aging, have lost their ability to expand and contract at the same rate as the surfaces they cover. Several types of cracking have been classified.

608.3a Hair-cracking. Random patterns of cracks are confined to the topcoat.

608.3b Checking. Fine cracks appear in the topcoat, distributed over the surface and resembling small checks.

608.3c Crazing. Deep, broader checks, which penetrate at least one coat, become noticeable.

608.3d Alligatoring. Severe crazing occurs; the patterns resemble an alligator's hide.

608.3e Cracking. This term is specific to a breakdown in which the cracks penetrate at least one topcoat and are likely to lead to complete failure.

608.4 Cratering. Cratering is the result of faulty spraying. Random circular depressions are the evidence. These mainly occur in factory-applied finishes, and when this appears, the supplier of the siding or other part involved should be held accountable.

608.5 Flaking. This term describes the lifting of paint because of loss of adhesion. Flaking may be due to foreign matter on the substrate or the earlier-applied paint. It often occurs on exterior wood surfaces, particularly near joints where moisture can enter, causing swelling and separation of paint.

608.6 Floating. Floating is evidenced by mottled, splotchy appearance in paints that have blended pigments. One of the colors floats to the surface and leaves an inconsistent pattern.

608.7 Flooding. This condition is like floating, except that one of the pigments in the blend rises uniformly and covers the surface. For instance, the blue component in a blend of blue and yellow that was intended to provide a green surface could flood and result in an off-blue. Flooding, as well as floating, can be avoided by selecting paints from reliable manufacturers and refusing to grant permission to painting contractors to switch.

608.8 Orange Peel. Small depressions resembling those of an orange peel result from the failure of sprayed paint to level out before drying. This could be the fault of the paint, because its solvent may be so volatile that the material dries before it can smooth itself, or the painter may have changed the solvent

Deterioration of Coatings

blend by thinning. Avoiding orange peel involves adjusting volatility; care must be taken to avoid over-thinning, which can slow down drying too much, in which case the thin, slower-drying film may flow too much and cause sags and runs, which end up as irregularities.

608.9 Stringiness. Brushed coatings that have not flowed out properly appear stringy. This occurs because of poor formulation, which causes the paint to set up before it can level out.

608.10 Brush Marks. Stroke-prints of brushing result, like stringiness, from poor formulation, or may be the result of water loss in latex paints due to porous substrates or undercoats. Otherwise properly formulated latex paint may set up too fast, and show brush marks, because it is spread on porous materials that draw out water quickly, increasing viscosity and the speed of drying before the coating can flow out and level.

608.11 Sheen Variations. Related to stringiness and brush marks caused by overly rapid drying is the inconsistency of sheen in satin or semigloss films that some times occurs when the film fails to flow out properly because of faulty formulating. Full gloss films rarely exhibit this inconsistency because the irregularities in surface thickness resulting from poor flowout have no influence on gloss films.

608.12 Sulfide Staining. Paints containing lead pigments sometimes react with hydrogen sulfide in the air and end up with a black stain. Where this air contaminant is believed to be present, lead should be avoided.

608.13 Loss of Gloss. Enamels depend on a thin film of free binder for their gloss. When binders are imperfect, or driers precipitate in them, a slight haze forms because light is refracted to a lesser degree than in a clear film. Because the trouble is within the film, it can be corrected only by repainting.

608.14 Wrinkling. The two main causes of wrinkling are excessive use of drier to catalyze the binder and excessive film thickness. Faulty stirring of paint prior to application can lead to an excess of drier coming to the surface; while careless application is the cause of over-thick films. Neither explanation is complete, because the drier may have been oversupplied at the paint factory; and the thick film build may have been due to incorporation of too much thickening agent, which made thin application difficult if not impossible.

608.15 Skin on Surface. A soft, incompletely dried layer near the substrate and a hard skin may mean that the formula had a bad combination of driers, which worked only on the surface and failed to catalyze the area below. Lead or zirconium driers are usually needed for a through-drying of oil and alkyd paints. If they are omitted or are insufficient while cobalt, the top-drier, is present, skinning can occur. Skinning on the surface can also be caused by high winds at the time of painting. Top drying is, thus, accelerated, and the lower portion cannot release its solvent or get its required oxygen for drying. Skinned surfaces are a problem to recoat, because the drying process of new coats can pull off the soft under-portion. The old coat should be removed before repainting.

608.16 Lifting of Coats. When more than one coat is on a surface, they may separate for several reasons. The incompatibility of the successive coats will cause the top coat to draw away from the other and lift. The solvent in the top coat may be so strong that it dissolves the previous coat; xylene in a urethane film, for example, may loosen the previous coat and cause lifting when the topcoat dries. Also, lifting may occur if the undercoat was incompletely dry when the successive coat was applied. The "crawling" of a top coat can occur when the wrong material is applied to a preceding coat. A latex paint applied over a high-gloss enamel would probably fail to adhere unless the enamel had been properly scuffed up to provide a "perch" for the latex. Its failure to set and stick to the enamel would result in a discontinuity of film known as *crawling*.

608.17 Uneven Color and Gloss. When the primer is too thin, some of the topcoat may penetrate it and pass on to the substrate, which, if porous, will absorb enough of it to cause uneven distribution of binder and colorant on the surface. The same may happen to enamels when either the substrate, as in this instance, or the primer itself absorbs some of the topcoat.

608.18 Excess Paint on Surface. Five conditions that can result from applying excess paint: (1) *sagging*, which resembles a frozen cascade; (2) *slow through-dry*, because the portion nearest the substrate is burdened by the heavy over-layer and can't yield its solvent or water; (3) *wrinkling*, because the uneven rate of drying of surface and under-portion creates tensions as the latter finally dries and draws in the previously hardened, but flexible, top; (4) *cracking*, which is due to the stresses created as the different portions dry at varying rates; and (5) *blistering*, caused by the attempt of trapped solvent to escape.

609 Coating Defects—Structural Faults. Builders, knowing that their development houses will be the buyers' headaches, are often not as meticulous as they should be about insisting that their general contractors follow all details of local building codes. The consequence is that architects, concerned about professional ethics, design sound structures only to find that faulty workmanship compounded by indifferent inspection by general contractors leaves residences subject to trouble.

Paint defects resulting from such shirked responsibility are just a part of the entire story. Wiring, plumbing, heating, and floors are subject to the same kind of unsound business practices.

The structural defects depicted in the following drawings are widely understood and easily recognized. They are based on a study made for the three Armed Services of the United States by the David Litter Laboratories as part of a comprehensive study, *Paints and Protective Coatings*, carried out with the cooperation of the National Bureau of Standards.

The study concluded that a poorly built house may have any or all of twenty-six opportunities to suffer from moisture difficulties. Hopefully, the professional architects of the nation may be able to agitate for tighter controls on construction standards. It is also hoped that builders may learn that it costs very little extra to provide a sound structure if they take the trouble to see that their general and subcontractors do what they are paid to do. With this in mind, the following, based on material prepared by Sidney B. Levinson, president of the laboratory, and Saul Spindel, technical director, with the cooperation of the U.S. Armed Forces Tri-Service Painting Manual Committee, is presented.

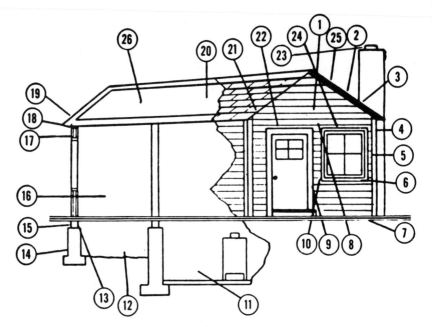

Fig. 609.1A. Moisture from within Structure.
Twenty-six points of potential moisture trouble in a poorly built house: (1) built with green lumber; (2) no cricket where chimney meets roof; (3) no flashing at side of chimney; (4) use of metal corner caps; (5) exposed nail heads not galvanized; (6) no window wash at sill; (7) wood contacts earth; (8) no drip or gutter at eaves; (9) poorly fitted window and door trims; (10) waterproof paper not installed behind trim; (11) damp, wet cellar unventilated at opposite sides; (12) no ventilation of unexcavated space; (13) no blocking between unexcavated space and stud wall space; (14) no waterproofing or drainage tile around cellar walls; (15) lacks foundation water and termite sill; (16) plaster not dry enough to paint; (17) sheathing paper should be waterproof but not vapor proof; (18) vapor barrier omitted—needed for present or future insulation; (19) built during wet, rainy season without taking due precaution or ventilating on dry days; (20) built hurriedly of cheap materials; (21) inadequate flashing at breaks, corners, roof; (22) poorly jointed and matched; (23) no chimney cap; (24) no flashing over openings; (25) full of openings, loosely built; (26) no ventilation of attic space.

609.1 Entrance of Moisture Due to Faulty Structural Conditions. The major cause of abnormal deterioration of coatings, especially those exposed outdoors, is moisture. This moisture may either come from external sources or be developed within the structure. This moisture can produce abnormal deterioration of applied coatings such as wood stain, mildew, blistering, and loss of adhesion resulting in a poor appearance and eventual deterioration by flaking and peeling. A prime reason for this

Deterioration of Coatings 209

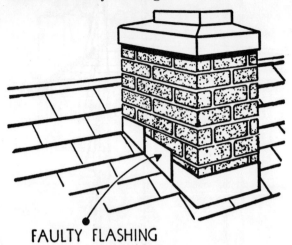

FAULTY FLASHING

Fig. 609.1B. Faulty Flashing.
Faulty flashing can permit the entry of moisture that will lead to coating failure. In the drawing above, exposed ends should be bent 90° and cemented between bricks to prevent the entrance of rain water.

problem is that the major construction materials used, i.e., wood, concrete, stucco, masonry, and plaster, are essentially porous and will allow moisture to pass through. If the walls are wet and the surface is warmed, as by sunlight, the moisture will tend to move to the outside atmosphere. However, if nonpermeable coatings are used (most paints other than latex paints or cement paints) this moisture will be trapped. Increased pressure will eventually cause the coating to either blister or lose adhesion. The multitude of ways moisture can give trouble in a poorly built house are shown in Figs. 609.1A and 609.1B.

609.2 Moisture from Within the Structure. A major cause of excessive moisture is that developed in normal use by the occupants of the structure. There are a number of sources of such moisture.

609.2a Normal Activities. Daily activities by and for the occupants of the structure can account for the following amount of moisture per person each day:

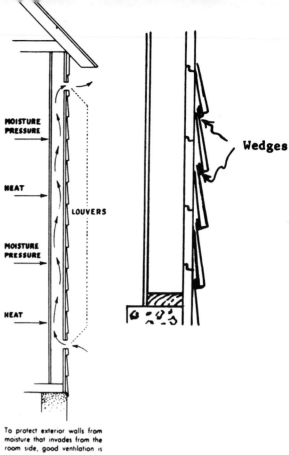

To protect exterior walls from moisture that invades from the room side, good ventilation is necessary.

Fig. 609.2b. Venting of Outside Walls.
Good venting of outside walls is necessary to protect the exterior coating from damage by moisture originating in the inside of the structure.

Breathing and perspiration	2 lbs.
Cooking and dishwashing	1 lb.
Clothes washing and drying	8 lbs.
Showers—daily	1/2 lb.

This adds up to a total of about 1-1/2 gallons of water developed per person per day without including moisture given off by heaters. It is important that venting be used for all equipment and that kitchens and shower rooms have exhaust fans which are kept in operation during use of facilities.

609.2b Humidity. The humidity within a structure should be kept fairly low especially during the cold weather when outside

Deterioration of Coatings

walls are cold. Otherwise moisture will collect and eventually work its way into and through the walls unless the interior paint on the walls is impermeable. This usually is not a problem unless humidifiers are used with heating equipment. The following humidity levels should be the maximum within a structure for indoor air temperatures of 70° F:

Outside Temp. (°F)	Inside Humidity (Max.) (Percent)
Below -20	15
-20 to 0	20
0 to 20	30
Above 20	40

There are two relatively low cost solutions to blistering and peeling problems if the source of moisture is by normal use and no structural defects are involved:

1) Seal the inside surface of exterior walls with aluminum paint or enamel and apply breathing-type paints such as latex paints to the outside surface.
2) Vent the outside walls by the use of vents or wedges. (See Fig. 609.2b.)

HELPFUL ADDITIONAL READINGS

1. Gardenas, G., *Paint Film Defects*, 2nd ed. (New York: Reinhold Publishing Corp., 1965), pp. 5-39.
2. Garlock, Neal, *Diagnosing Paint Problems* (St. Louis: American Paint Journal Co., 1969), pp. 5-39.
3. Hay, J. Kirk, and Garmond G. Schurr, "Moisture Diffusion Phenomena in Practical Paint Systems," *Journal of Paint Technology*, Vol. 43, No. 556, pp. 63-72.
4. Levinson, Sidney B., *Paints and Protective Coatings* (Washington, D.C.: Departments of the Army, Navy, and the Air Force, 1960), Chapters 5-28.
5. *Paint Technology Manuals*, edited by J. A. Taylor, Vol. 3 (London: Chapman and Hall, Ltd., 1962), p. 80.

chapter 7
Selection and Specification of Coatings—Exterior Surfaces

700 General. Surface coatings are often selected for reasons other than best performance. A conscientious architect, true to the tenets of his profession, knows that circumstances demand the best possible coating if durability and good appearance are to be obtained. He is often handicapped in his decision, however. He may be dealing with an owner whose motives are different from his. Perhaps the owner may plan to sell the building immediately, or a few years after completion. The succeeding owner, he knows, will not be able to distinguish an applied coat of high-priced but durable epoxy or urethane paint from a competitively priced latex or alkyd workhorse, which may be long on low-priced clay or calcium carbonate and short on binders and titanium dioxide.

Since later owners will not know if the underlying surface was properly cleaned and prepared prior to application, the original owner may shrug his shoulders and order the architect, or the general contractor—unless the architect stands by his guns—to specify the use of low-grade paints with no more than cursory surface preparation.

Unfortunately, building codes are lax about paint, and unsuspecting future owners are frequently faced with expensive paint-removal operations because an original owner, or a succeeding owner, may have been immorally shrewd about paint buying and application.

When the owner plans to own and operate the building

Selection and Specification—Exterior 213

or expects to live in it, the architect has a practical as well as a moral obligation to convince him that economic considerations dictate selection of the proper quality paint and the use of suitable methods of surface preparation.

The economics of selection and preparation is discussed in Chapter 10. The selection of the proper paint for the surface and the prevailing circumstances will be covered in this chapter and Chapter 8.

700.1 The Importance of Proper Specification. Attention will be given, in addition to ordinary, garden variety paints, to high performance coatings. The ordinary paints, all too often, are regarded by specifiers as routine, and they simply put out paint schedules and call for paints by one of several manufacturers. Doing no more than that, however, ignores the fact that virtually all manufacturers offer a selection of qualities so that the specifier can designate the best coating for either the simplest or most challenging circumstances.

That means that the wall paint for a kitchen or bathroom, or a home laundry—where moisture is often over-abundant—need not be the same used for a bedroom, living room, or dining area. The available selections also mean that an exterior house paint selected for a home by the seashore does not have to be the same as that selected for a mountaintop—even if the substrate is the same. And corridor paints for an old folks' home or a graduate school need not be the same as that for a grammar school or high school where abrasion resistance and ability to resist the impact of flying heels and rocks is important, plus the ability to prevent penetration of marking-pen ink into the surface when graffiti artists become creative.

Selection of coatings, we should know by now, is not simple or routine. It requires homework to understand the circumstances under which the coating will serve; and it requires knowledge of coating materials and costs.

700.2 How To Use This Book for Coating Selection. To simplify the selection of coating systems, coded performance specifications will be presented for individual product-types in a separate section at the end of the book. In addition, where it will be helpful, Performance Comparative Charts will be provided within this chapter and Chapter 8 so the specifier can see at a glance how one material may top the others in the particular performance characteristics that appear desirable for the prevailing circumstances.

The guide to selection of coatings will be organized according to *exterior* (in this chapter) and *interior* surfaces (Chapter 8). Each of these will be further divided into three classes: *cementitious* (covering brick, plaster, masonry, cement-core wallboard, asbestos-cement, and cinder and concrete block); *metal;* and *wood.* Flooring is treated in Chapter 8 and fire retardant paints are discussed in Chapter 9.

Specifiers have two choices in using this section. They may use the Performance Comparative Charts within the chapters to determine what specification comes closest to their needs and may then turn directly to the Detailed Specification Charts at the end of the book. Or they may read the section covering the substrate to be used, which will refer both to the appropriate Performance Comparative Charts and to the Detailed Specification Charts from which the comparisons were made. In almost every instance, the specifier will have more than one or two materials from which to make his choice.

Federal specifications, which in many instances call for excellent products, often, however, only call for marginally satisfactory ones.

We have listed federal specifications on the Performance Comparative Charts and the Detailed Specification Charts. This was done, since even the marginal-performance specifications provide basic standards for comparison. Wherever necessary, and this is frequent, we have provided commercial specifications for products that exceed federal performance requirements.

These improved specifications come in some instances from individual commercial paint manufacturers. Where not obtainable from those sources, specifications were supplied by resin manufacturers who have provided similar formulations to commercial manufacturers. This means that these superior products are available, often from all, or at least many national or regional paint firms.

It may appear that specifications credited to commercial companies here show partiality to a few. That is not the case. Only a few firms provide adequate specification information. Nonetheless, these few firms credited with specifications should

Selection and Specification—Exterior 215

not be regarded as the only ones with outstanding products. Most nationally known manufacturers have lines of comparable quality as do well-established regional manufacturers.

One of the important premises that we are trying to establish in this section, as well as in the book as a whole, is that specification writers have broad possibilities for selecting the best product for a given set of circumstances and that they should fight hard to prevent general contractors and/or painting contractors from persuading building owners to substitute lesser materials in the interest of what is unquestionably false economy.

700.3 General Orientation. To emphasize the importance of circumstances in selecting a coating, broad performance charts, with accompanying use and service ratings are provided in this section. Tables 700.3A, 700.3B, 700.3C, 700.3D, 700.3E, and 700.3F offer rather general evaluations of various types of coatings on a number of characteristics. Table 700.3G then offers an evaluation of paint systems showing the advantages and limitations of each paint as well as the typical applications for each paint.

701 Exterior Surfaces.

701.1. General. Specifiers who, thus far, have not been convinced that casual selection of coatings, or compromises on quality, are costly to their clients may come around when they see results of a highly scientific comparison of paint film performance conducted between 1958 and 1963 by the St. Louis Society for Paint Technology, under the direction of the chairman of the Society's Technical Committee, Dr. Wouter Busch, who heads the department of polymer and paint chemistry at the University of Missouri at Rolla, Mo.

701.2 The St. Louis Study. The Committee tested a large number of exterior paints, all formulated by unquestioned experts with the advice and counsel of the manufacturers of the vehicles tested. The Committee's study is important to the specifier, since, in addition to providing clues on comparative performance, it also shows the importance of the circumstance of exposure.

Table 700.3A. *Economy & Versatility.*

	Polyvinyl acrylic	Acrylic	Oil	Alkyd	Phenol.	Vinyl	Epoxy	Chlorinated Rubber	Styrene Acrylate	Butadiene Rubber
Paint cost			++	++			−			
Total labor cost			++	++						
Use on steel				ALL ARE O.K.			+	+	+	+
Fresh masonry	−	−	−	−		+	+			
On wood			++	++		−	−			
Surface preparation			++	+		−	−			
Solids coat							+			
Average solids—%	56	47	90	65	65	30	65 to 100			

Key:
++ = Outstanding
+ = Above average
Blank = Average
− = Below average
−− = Not recommended

Tables 700.1a through 700.1f are based on information provided in a talk by Sidney B. Levinson and Morris Coffino; the latex information is based upon a talk by Bruin *et al.*, delivered in 1964 at the F.A.T.I.P.E. Congress; information on rubber-based performance has been derived from a number of miscellaneous sources.

Table 700.3B. *Application.*

	Polyvinyl acrylic	Acrylic	Oil	Alkyd	Phenol.	Vinyl	Epoxy
Brushability			+				
Cleanup	+	+	+				−
Odor	+	+	+	+	+		
Speed of dry	++	++	−			++	
Recoating by brush							−

Key:
  ++ = Outstanding
   + = Above average
Blank = Average
   − = Below average
  −− = Not recommended

Table 700.3C. *Film Properties.*

	Polyvinyl acrylic	Acrylic	Oil	Alkyd	Phenol.	Vinyl	Epoxy
Gloss		+		+	+	−	+
Gloss retention		+	−	+		+	−
Color	+	+		+		−	+
Color retention	+	+		+		−	+
Hardness	+		−		+		++
Adhesion to many surfaces	−	−				−	+
Flexibility	+	+	+			++	+
Aluminum leafing		−	−		+	−	

Key:
  ++ = Outstanding
   + = Above average
Blank = Average
   − = Below average
  −− = Not recommended

Table 700.3D. *Film Resistance.*

	Polyvinyl Acrylic	Acrylic	Oil	Alkyd	Phenol.	Vinyl	Epoxy
Resistance to—							
Abrasion	+	+	- -			++	++
Water			-		+	++	
Acid and alkali	-	-	- -			+	+
Strong solvent	-	-	- -	-		-	++
Heat	-	-	-		+	-	

Key:
++ = Outstanding
 + = Above average
Blank = Average
 - = Below average
- - = Not recommended

Table 700.3E. *Comparison for End Uses.*

	Polyvinyl Acrylic	Acrylic	Oil	Alkyd	Phenol.	Vinyl	Epoxy
Interior	+	+	+				
Rural exposure			ALL ARE O.K.				
Marine exposure	- -	-	-			+	+
Corrosive areas	- -	- -	- -	-	+	++	++
Water immersion	-	-	-	-	+	++	
As clears	- -	+			+		
Aluminum paints	- -	- -			+		

Key:
++ = Outstanding
 + = Above average
Blank = Average
 - = Below average
- - = Not recommended

Table 700.3F. *Overall (Summary Tables 700.3B-700.3E)*

	Polyvinyl Acrylic	Acrylic	Oil	Alkyd	Phenol.	Vinyl	Epoxy
Economy and versatility	+2	+2	+4	0	0	-3	+1
Application	+2	+2	+2	+1	+1	+1	-1
Film properties			-1	+4	+1	+2	+4
Film resistance			-8	-4	+2	+3	+6
End uses			-4	-1	+4	+5	+3

Selection and Specification—Exterior

The following vehicles were exposed vertically, both north and south, on the roof of the Chemical Engineering Building on the Rolla, Mo., campus of the University of Missouri:

Linseed oil copolymer
Styrenated oil
Linseed-isophthalic alkyd
Linseed-soybean isophthalic alkyd
Tall oil isophthalic alkyd
Soybean isophthalic alkyd
Raw-bodied linseed oil
Tung-soybean oil
Long-oil alkyd
Vinyl toluene oil copolymer
Polyvinyl acetate
Acrylic emulsion
Epoxy polyamide
Polyester isocyanate (urethane).

Two versions of soybean isophthalic alkyds were tested; three versions of linseed oil, raw-bodied; three versions of long-oil alkyd; two of polyvinyl acetate; and two epoxy polyamides. It must be remembered that these paints were made in 1958 and that most of these materials have been improved, particularly the urethanes, epoxies, acrylics, vinyl toluenes, and the new heavy-bodied linseed oil paints that are regaining some of the market for that type of binder. The same identical material, taken from the same can, performed differently when exposed for 66 months (or 5-1/2 years) while facing the south and while facing north. Paints on only seven of the north-exposed panels and six of the south-exposed panels survived without erosion or severe cracking for the entire 66 months. And even a smaller number in each category was sound enough to be recoated, as is shown in Tables 701.2A and 701.2B.

A number of interesting conclusions can be reached by the specifier who is aware of such considerations as film thickness and Pigment Volume Concentration (PVC), and who realizes the importance of selecting a coating to meet particular circumstances. For example, from Tables 701.2A and 701.2B, note that the acrylic emulsion, which did quite well on the north side, was unable to withstand the heavier influence of the sun on the south exposure.

Table 700.3G. Evaluation of Painting Systems.

	Outstanding Advantages	Chief Limitations	Typical Applications
HIGH CHEMICAL RESISTANCE (Immersion or Splash & Spillage)			
Converted Epoxy	1. Alkali Resistance 2. Abrasion Resistance 3. Recoatability 4. Surface Moisture Tolerance 5. Adhesion 6. Solvent Resistance 7. Water Resistance	1. Two Package 2. Pot Life 3. Yellowing 4. Exterior Chalking 5. Exterior Fading 6. Apply over 50° F.	1. Chemical Plants 2. Oil Refineries 3. Tank Lining 4. Floor Finish 5. Marine Applications 6. Machinery 7. Paper Mills 8. Water & Sewage Plants 9. Cement Trucks
Epoxy-Polyester	1. Abrasion Resistance 2. Stain Resistance 3. Acid Resistance 4. Exterior Gloss Retention 5. Water Resistance 6. High Film Build 7. Exterior Color Retention	1. Two Package 2. Pot Life 3. Alkali Resistance 4. Apply over 50° F. 5. Limited Flexibility	1. Wall Finish 2. Machinery Finish 3. Hospitals 4. Hotels & Motels 5. Schools 6. Kitchens & Bakeries
Converted Polyurethane	1. Acid Resistance 2. Abrasion Resistance 3. Water Resistance	1. Two Package 2. Pot Life 3. Yellowing 4. Exterior Chalking 5. Exterior Fading 6. Recoatability 7. Adhesion to Metal	1. Floor Finish 2. Citrus Industry 3. Tank Lining 4. Machinery 5. Textile Mills
Moisture-Cured Polyurethane	1. Acid Resistance 2. Abrasion Resistance 3. Single Package Converted Coating 4. Water Resistance	1. Package Stability after Opening 2. Yellowing 3. Exterior Chalking 4. Exterior Fading 5. Recoatability	1. Floor Finish 2. Tank Lining 3. Machinery 4. Textile Mills

Coating	Advantages	Disadvantages	Uses
Vinyl	1. Acid Resistance 2. Recoatability 3. Exterior Color Retention 4. Exterior Gloss Retention 5. Water Resistance 6. Adhesion to Metal	1. Spray Application Only 2. Low Solids 3. Low Flash Point 4. Careful Surface Preparation 5. Heat Resistance	1. Tank Linings 2. Chemical Plants 3. Oil Refineries 4. Marine Application 5. Machinery 6. Portable Water Tanks

INTERMEDIATE CHEMICAL RESISTANCE (Chiefly Splash & Spillage)

Coating	Advantages	Disadvantages	Uses
Chemical Resistant Latex	1. Chemical Resistance 2. Exterior Color Retention 3. Exterior Gloss Retention 4. Ease of Application 5. Water Clean-up 6. Non-Flammable 7. Surface Preparation	1. Abrasion Resistance 2. Application Temperature 3. Heat Resistance	1. Storage Tanks 2. Chemical Plant Equipment 3. Oil Refineries 4. Marine Superstructure
Chlorinated Rubber	1. Chemical Resistance 2. Water Resistance (immersion) 3. Abrasion Resistance	1. Solvent Resistance 2. Heat Resistance	1. Chemical Plants 2. Water Immersion 3. Machinery 4. Marine Applications 5. Floor Finish
Epoxy Ester	1. Chemical Resistance 2. Abrasion Resistance 3. Surface Preparation 4. Ease of Application 5. Water Resistance	1. Exterior Gloss Retention 2. Exterior Color Retention 3. Film Build	1. General Interior Plant Use 2. Floor Finish 3. Primers
Urethane Oil	1. Abrasion Resistance 2. Chemical Resistance 3. Ease of Application 4. Water Resistance	1. Film Build 2. Recoatability	1. Floor Finish 2. General Paint Use 3. Wood Finishes

Source: Courtesy of Glidden Coatings and Resins, Division of SCM Corporation.

Table 700.3G. *(Continued)*

	Outstanding Advantages	Chief Limitations	Typical Applications
INTERMEDIATE CHEMICAL RESISTANCE CONT'D. (Chiefly Splash & Spillage)			
Phenolics (Oil Modified)	1. Water Resistance 2. Acid Resistance 3. Abrasion Resistance 4. Hardness 5. Ease of Application	1. Yellowing 2. Recoatability (Intercoat Adhesion) 3. Exterior Fading 4. Exterior Chalking	1. Marine Applications 2. Floor Finishes 3. Water & Sewage Plants 4. Machinery
Tar-Epoxy	1. Water Resistance 2. Chemical Resistance 3. High Film Build	1. Bleeding 2. Exterior Color Retention 3. Exterior Gloss Retention 4. Two Package 5. Pot Life 6. Dark Colors	1. Crude Oil Tank Lining 2. Sewage Disposal Plants 3. Pipe Lining or Exterior Coating
LOW CHEMICAL RESISTANCE (General Purpose Products)			
Oil, Linseed	1. Ease of Application 2. Minimum Surface Preparation 3. Flexibility 4. Excellent Adhesion	1. Slow Dry 2. Soft Film 3. Low Chemical Resistance 4. Poor Solvent Resistance 5. Water Resistance 6. Abrasion Resistance	1. Wooden Buildings (Exterior) 2. Metal Surfaces (Exterior)
Alkyd	1. Ease of Application 2. Surface Preparation 3. Low Cost 4. Good One Coat Hiding 5. Durability 6. Gloss Retention	1. Chemical Resistance 2. Water Resistance	1. Tank Exteriors 2. Structural Metal 3. Machinery 4. Plant Equipment 5. Interior or Exterior Wood or Metal
Exterior Latex Products	1. Water Reducible 2. Non-Flammable	1. Freeze-Thaw Limitations	1. Exterior Wood 2. Exterior Concrete, Stucco

Selection and Specification—Exterior

Product	Properties	Limitations	Uses
Interior Latex Products	3. Ease of Application 4. Blister Resistance 1. Water Reducible 2. Non-Flammable 3. Ease of Application	2. Heat Resistance 1. Freeze-Thaw Limitations 2. Heat Resistance	& Masonry 3. Exterior Metal 1. Interior Wood 2. Plaster, Wallboard, Drywall 3. Interior Brick, Concrete, Masonry
Silicone (Masonry Water Repellent)	1. Colorless 2. Invisible 3. Effective for 10 years 4. Prevents Staining	1. Not For Use on Limestone 2. Use Only on New Masonry	1. New Brick, Mortar, Sandstone & Poured Concrete
HIGH TEMPERATURE COATINGS Silicones	1. Air-Dry Tack Free (some formulations) 2. Service Temperature up to 1000° F. 3. Interior or Exterior Service	1. High Cost 2. Low Film Build 3. Limited Solvent Resistance 4. Limited Chemical Resistance	1. High Temperature Stacks 2. Boilers & Boiler Breeches 3. Exhaust Lines, Manifolds & Mufflers
SPECIAL PRODUCTS Zinc Rich Coatings (Organic)	1. Tolerant of Surface Prep. 2. High Temperature Tolerance 3. High Film Build 4. Abrasion Resistance 5. Primer and Finish Coat	1. Solvent Resistance 2. High Cost 3. Poor Acid or Alkali Resistance	1. Rust Inhibitive Primer 2. Exterior Equipment in Petroleum & Chemical Plants 3. High Temperature Stacks, Boilers, Etc.
Zinc Rich Coatings (Inorganic)	1. Solvent Resistance 2. High Temperature Tolerance 3. High Film Build 4. Abrasion Resistance 5. Primer and Finish Coat	1. Critical of Surface Preparation 2. High Cost 3. Poor Acid and Alkali Resistance	1. Solvent Tank Lining 2. Exterior Equipment in Petroleum & Chemical Plants 3. High Temperature Stacks, Boilers, etc.

Source: Glidden Coatings and Resins, Division of SCM Corporation.

Note also, though, that the acrylic coat that worked satisfactorily on the north exposure was .3 mils thinner than the one that failed on the south exposure, indicating still further the influence of exposure.

The acrylic observation relative to north and south exposures, in order to be completely fair to this highly serviceable binder, should include a notation that the PVC of the acrylic formulation was a very high 58 percent, which means that it had relatively little acrylic and a large quantity of pigments, mainly extender pigments such as barytes and calcium carbonate and had no talc, which helps on exterior exposure.

Its high pigment and extender content (both of which count equally when figuring PVC) explain why the researchers got adequate hiding with only 2.8 mils thickness, while the other paints on the north exposure area ranged from 3.7 to 4.5 mils dry film thickness. But the large pigment content and low binder-resin component affected performance.

Color retention characteristics of the binder can be gauged by what happened to appearance on the north and south exposures. The following table shows the changes, if any:

Binder	North exposure	South exposure	Change
Polyester urethane	9	7	−2
Epoxy polyamide	8	5	−3
Acrylic emulsion	7	1	−6
Long-oil alkyd	7	3	−4
Tung-soy copolymer	6	7	1
PVA emulsion	5	5	—
PVA emulsion II	4	4	—

These results confirm long-accepted contentions that certain binders are harmed by prolonged exposure to the sun and that others are not significantly harmed.

The lowest rating on the sunny side among these survivors of the grueling St. Louis study was the acrylic emulsion formulation, and the probable reason was listed as being an unusually large Pigment Volume Content, which minimized the effect of the acrylic binder. Acrylic emulsions are now generally considered to be among the best available binders when it comes to color retention after prolonged sun exposure.

Table 701.2A. Exposure Data after 66 Months (North).

Binder	PVC	Dry film thickness (mils)	Over-all appearance*	Condition for repainting	Remarks
Polyester urethane	30.4	4.5	9	Excellent	—
Epoxy polyamide	30.3	4.0	8	Excellent	—
Acrylic emulsion	58.0	2.8	7	Poor	Medium mildew
Long-oil alkyd	29.7	4.3	7	Good	—
Tung-soy copolymer	30.0	4.2	6	Excellent	—
Polyvinyl acetate emulsion	34.0	4.9	5	Poor	Excessive cracking
Polyvinyl acetate emulsion II	33.1	3.7	4	Poor	Excessive cracking

Note: 10 is highest rating; 1 is lowest.

Table 701.2B. Exposure Data after 66 Months (South).

Binder	Dry film thickness (mils)	Over-all appearance*	Condition for repainting	Remarks
Polyester urethane	4.8	7	Excellent	Medium mildew
Epoxy polyamide	3.8	5	Good	No mildew
Acrylic emulsion	3.1	1	Eroded	Eroded
Long-oil alkyd	4.2	3	Unsuitable	Badly mildewed
Tung-soy copolymer	3.6	7	Poor	—
Polyvinyl acetate emulsion	4.0	5	Poor	Excessive cracking and mildew for both
Polyvinyl acetate emulsion II	3.0	4	Poor	

Note: 10 is highest rating; 1 is lowest.

The St. Louis study also confirmed that polyester urethanes, because of the structure of the basic polymer, are sensitive to prolonged sun exposure; the epoxy polyamides tested also exhibited their outstanding fault, a tendency to chalk in the sun. Alkyds also, despite their virtues, are short on color retention on the sunny side.

Soybean oil has always been selected for use as the fatty acid in alkyds and polyesters where color retention is important; the wisdom of this choice was confirmed by the study, although tung oil, one of the tough oils, is not noted for its color retention. PVA emulsions were considered fair on exteriors when the tests were made, but new PVA copolymers using acrylate monomers as part of the molecule have substantially improved color retention and performance.

An acrylic gloss paint is available for exterior trim coating. Specifiers should learn carefully the lessons taught by the St. Louis study. Once learned, these lessons may discourage specifiers from being soft in their response when painting contractors ask permission to change required paints to a product "that's just as good, even made by the same company—even made with the same binder."

701.3 Specifying Rules. There are six rules or generalizations the specifier should bear in mind:

1) Just about every paint company has a top quality line, and a competitive line.
2) Just about every paint company offers product lines for each purpose, even containing the same binder but with more binder for the top lines and, significantly, more pigment and low-priced extender for the less expensive lines. (See Section 111.3.)
3) By selecting quality performance specifications from among those set forth in this book, you assure your client top-of-the-line products.
4) By insisting that these specifications be met, you prevent substitutions of lesser competitive materials made by major paint companies or minor firms.
5) Substitutions in exterior coatings are especially unwise because often specified paints are designed to combat troublesome weather conditions leading to deterioration of substrate as well as coating.

Selection and Specification—Exterior

6) Exterior coatings, particularly for metal, are carefully balanced to meet potentially challenging electric-charge factors, microbes, abrading influences, wind-driven rain, atmospheric chemicals, and salt-laden air; and these problems must be respected by use of suitable, and often expensive, products.

702 Cementitious Exterior Surfaces. Under the heading of cementitious surfaces, we include concrete, concrete block, cinder block, brick, stucco, general masonry, and asbestos-cement. These substrates, in general, are characterized as porous to a greater or lesser degree, and because they usually contain portland cement, they are decidedly alkaline, which can be of considerable importance in selecting and applying coatings.

Two other characteristics of these materials must be taken into account: (1) they seem to attract and hold water; and (2) their surfaces are rough. Rough types of cementitious substrates require free-flowing materials that cover the tiny hills and valleys of the surface but still are not so fluid that penetration into pores is excessive. Smooth types, on the other hand, may be so hard and impenetrable that roughening by abraders may be required to get a "perch" for the paint. (See Section 307.)

Primers are necessary for cementitious surfaces, but thinned topcoat materials can often be used for this purpose. For open-pored surfaces, however, such as those found in cinder and cement blocks and aggregate blocks, rather thick fillers are necessary to prevent over-penetration of primers into the material, with resultant waste of paint and poor performance of succeeding coats.

702.1 Exterior Cementitious Surfaces, Ordinary Conditions. Properly aged new masonry (see Section 307), brick, filled-blocks, and stucco can be primed with suitably thinned topcoats selected by the specifier. Several choices are open, depending on drying time required, economics, and required durability.

For ordinary conditions, basic formulations intended to meet Federal Specifications TTP 19 (GPC 23)*, or 24 (GPC 24), or 35 (GPC 25), or 55 (GPC 26) will probably suffice. See

---

*See separate section at end of book for complete listing of all GPC classifications. The GPC classifications are new, prepared especially for this book.

Performance Comparative Chart I, at the end of this chapter. Note that GPC 23 and 26 are set-to-touch in 15 minutes, compared with 4 hours for solvent-thinned GPC 24.

GPC 23, based on acrylic emulsion, calls for a mere 200 oscillations of the abrasion-test device, compared with 1,000 oscillations for the polyvinyl acetate of GPC 26. This illustrates the often marginal quality of the more basic federal specifications, although the heavy-duty specifications based on urethane, epoxies, phenolics, and silicone are usually excellent.

We can compare a typical commercially formulated acrylic house paint with GPC 23. The usual commercial formulation has only 4 percent of anatase TiO_2, which is the grade that chalks, while the federal specification (GPC 23) permits 30 percent, or seven times more of this material, which is inherently eroding. The commercial formula dries through in one-half hour, versus the one hour required by the specification; and, most important, the wet abrasion resistance is 25 times as great as called for by the specification. From this one can see the marginal quality that may be provided if an architect's specification merely states that a federal specification (such as TTP 19, which is the same as our GPC 23), particularly for a simple coating, is all that is required.

Caution must also be exercised in specifying a federal type because a number of companies on the qualified products list for several of the federal specifications make virtually nothing but paints designed to meet these specifications. To meet the stiff competition for high-volume, low-profit government business these firms are obliged to cut raw material costs to the bone while coming as close as feasible to the minimal requirements of the specification. All this is legal but puts into question the desirability of specifying these low-performance materials, which in the end cost the taxpayer and the builder more than better paints. Where the client wants good paint, federal specifications for simple systems, such as exterior latex paints or wall paints, should be avoided.

Table 702.1 illustrates the degree to which a standard acrylic emulsion exterior paint exceeds the federal specification for that particular type of paint.

Similarly, polyvinyl acetate emulsion paints, with copolymers based on maleic anhydride or an acrylate, can exceed Federal Specification TTP 55 (GPC 26), the PVA specification,

Selection and Specification—Exterior

Table 702.1. *Comparison of a Commercial Formulation of Acrylic Emulsion Paint with Federal Specification TTP 19 (GPC 23).*

Physical Properties	Representative Commercial Formulation	Required by Specification TTP 19 Minimum	Maximum
Total solids, percent	56.2	56	—
Pigment, percent by weight of paint	37.9	—	38
Titanium dioxide, pounds per gallon	2.5	2.5	—
Percent anatase TiO_2	4	—	30
Percent rutile TiO_2	96	70	—
Weight per gallon, pounds	11.8	11.0	—
Fineness of grind	6	4	—
Viscosity, Krebs Units	72	65	85
Drying time, set-to-touch, hours	1/4	—	1/4
Dry through, hours	1/2	—	1
Wet abrasion resistance, oscillations	> 5,000	200	—
Daylight directional reflectance	90.0	87	—
Dry opacity contrast ratio	1.0	0.98	—
Freeze thaw stability	Satisfactory	5 cycles	
Heat stability	Satisfactory	140° F.—1 week	
Application properties	Satisfactory	Good brushing properties and appearance	
Accelerated weathering resistance	Satisfactory	300 hrs.—Weather-Ometer	

Source: Information supplied by Rohm & Haas Co.
Note: The underlined data indicates superiority of the commercial formulation. Higher viscosity and weight indicate better coverage. Wet abrasion resistance of 5,000 cycles indicates tougher film. A premium formulation, with talc as an extender and more acrylic resin, offers even higher performance than this.

which calls for 1,000 oscillations (500 cycles) of the abrasion tester. PVA abrasion tests up to 3,000 cycles and beyond are common.

702.2 Exterior Cementitious Surfaces, Problem Conditions. Problems faced in specifying cementitious surfaces include pronounced irregularities; chemical environments; water exposure; graffiti removal; wind-driven rain; immersion; sand abrasion; impact; and repainting chalky surfaces. The main problem-solvers are rubber-based paints (solvent reduced); epoxies; cement-powder paints; oil-modified urethanes; waterproofing

Fig. 702.3. High-Build Coating for Irregular Cementitious Surfaces.
High-built coatings primed with fillers help smooth this rough cinder block.

compounds and clears of silicone, vinyl toluene butadiene, and acrylic. Many of the high-performance, problem-solving coatings require primer-fillers, which may be based on styrene butadiene, acrylic, or cement powder, depending on the recommendation of the manufacturer of the topcoat.

702.3 Wind-Driven Rain and Irregular Surfaces. Texture paints are the answer to these conditions because they provide heavy-build films to cover irregularities that frequently result from the poorly trained cement workers now practicing their trade; and they are able to resist rain driven by winds of hurricane strength. See Fig. 702.3.

Federal Specification TTC 555, Type II (GPC 30),* is an accepted performance-requirement guide. No binder requirements are designated in the federal material, but WeatherOmeter studies indicate a decided edge for vinyl toluene acrylate, a rubber-based, solvent-reduced binder. Even here, attention must be given to specifying PVC. One binder manufacturer's study showed that a PVC of 62 yielded results significantly better than products with PVC's of 70. Low-priced vinyl toluene acrylate texture paints may be found with the higher PVC, and specifiers should insist on the proper level.

Table 702.3 provides the results of a study of vinyl toluene

*Refer to Detailed Specification Charts at end of book, and see also Performance Comparative Chart II at the end of this chapter.

Selection and Specification–Exterior

Table 702.3. *Comparison of Vinyl Toluene Acrylate Texture Paints and Other Competitive Commercial Products.*

Properties	Vinyl Toluene Acrylate Texture Paint	Alkyd Texture Paint	Powder Texture Paint	Latex Texture Paint
Gardner scrub test 1% Tide (2000 strokes pass)	Pass	Pass	Failed (300 Strokes)	Failed (520 Strokes)
Weather Ometer, 500 hours / 240 hours				
White appearance	9/10	3/8	Fail/1	Fail/1
1% Cal-ink 6614 (Phthalo blue)	8F/10	3/8	Fail/0	Fail/1
1% Cal-ink 6610 (Yellow iron oxide)	10/10	3/9	Fail/1	Fail/1
Flexibility, 1-inch mandrel	Pass	Cracks	—	Cracks
Wind-driven rain, 98 mph (50 ± 10 ft. gal)	Excellent	—	—	—

Key: 10 = No change   0 = Worst possible condition   F = Faded
Source: Goodyear Chemical.

acrylate texture paints and high-quality competitive commercial products.

702.4 Chemical Environments. Weak acids and weak alkalis may be present near industrial areas, and salt spray is present near the seashore. Chlorinated rubber and styrene butadiene may be superior to other available materials where weak acids or weak alkalis will reach gloss enamels, but styrene acrylate is effective, otherwise, for resistance to dilute acids or weak alkalis. Oil-modified urethanes have fairly good ratings for alkali-resistance. See Federal Specification TTP 95 (GPC 28), which calls for chlorinated rubber or styrene acrylate, in gloss, semigloss, or flat. (See Performance Comparative Chart II at the end of this chapter.) Considerable attention has been given to acrylic clear coatings for protection of concrete structures and cast-concrete sculpture from the ravages of air-borne chemicals. These seem satisfactory, but attention should be given to inclusion of ultraviolet absorbers to prevent discoloration.

Table 702.5 *Estimated Coverage of Silicone Water Repellent*

Type of Masonry Surface	Porosity	Silicone Concentration	Coverage Sq.Ft./gal.
Terra Cotta Block	1 to 3	3	250 to 400
Terrazzo	1 to 3	3	250 to 400
Dense Brick	2 to 4	3	150 to 200
Concrete	2 to 4	3	150 to 200
Mortar	6 to 10	3 or 5	100 to 125
Stucco	6 to 10	3 or 5	100 to 125
Common Brick	6 to 12	3 or 5	100 to 125
Cement Brick	8 to 11	3 or 5	60 to 100
Concrete Block	8 to 12	3 or 5	60 to 100

Source: Union Carbide Corp.

702.5 Water Exposure. Coatings to withstand extensive exposure to water or high humidity should come as close as possible to preventing absorption. In white or colored coatings, the presence of pigments assures some absorption, and the art in formulating is to hold this to a minimum. Federal Specification TTP 97 (GPC 29),* for white masonry, and TTP 1181 (GPC 35),* for tints and deep tones, call for low water absorption. The PVC of paints used for meeting the problem of water exposure is critical. Research has shown that at 45 percent PVC, the film is virtually impervious, while at 65 percent PVC, the permeability rises to a high of 22 mgs. of water per sq.cm. per millimeter in 24 hours; but at 57 percent PVC, permeability is 10 mgs. This PVC has been found low enough and at the same time it has been found the optimum level for preventing peeling and blistering on new masonry. That level also works out well to prevent mottling caused by varying degrees of surface penetration resulting from inconsistent porosity throughout the surface of the cementitious material, particularly when stucco-like materials are used. A PVC between 52 and 62 percent prevents mottling.

Clear water repellents for cement block, brick, concrete, stucco and other cementitious surfaces are often based on silicone. While they repel water in liquid form from the outside, they permit passage of water vapor from rooms inside, thereby preventing damage due to entrapped moisture.

*Refer to Performance Comparative Charts at end of this chapter, and Detailed Specification Charts at end of book.

Fig. 702.6. Scrub-Resistant Coating for Easy Cleaning.
Hard, scrub-resistant coatings makee removal of "street art" easy.

702.6 Graffiti and Stain Removal. Hard, tough, stain-resistant coatings are needed for prevention of long-term damage from graffiti writers and careless stainmakers. Two-package aliphatic urethanes (see Page 36) have been found more likely to resist removal by the solvents that are strong enough to remove the stubborn dyes and resins used in felt pad markers and aerosols favored by the "folk artists." Successful formulations are based on aircraft coatings and have endured hundreds of solvent washovers without any of the phantom lines characteristically found on most other coatings able to survive use of the required strong solvents.

702.7 Sand Abrasion and Impact Resistance. A test report from Allied Chemical appearing in Unit 15 of the survey course on paint materials issued by the Federation of Societies for Paint Technology* listed a linseed oil-modified urethane as

---

*Paint Technology Manuals (Philadelphia: Federation of Societies for Paint Technology).

Table 702.6. Stain Resistance.

	Polyester Epoxy	Epoxy Polyamide	Alkyd
Stain			
Coffee	10	6	4
Lemon juice	10	9	7
Mustard	10	6	2
Iodine	10	0	0
Merthiolate	10	0	2
Washable ink	10	0	2
Lipstick	10	8	7
Tomato paste	10	9	7
Grape juice	10	9	6
Solvents (24-hours)			
Gasoline	9	10	7
Xylol	4	8	–
Water	10	10	8
Acetone (1 hr.)	3	8	2
Ethyl alcohol (1 hr.) (95%)	7	9	5

Note: 10—No effect; 0—Failure
Source: Ashland Chemical Co.

having a superior rating for withstanding sand abrasion; medium-oil alkyd was second; epoxy was third; and spar varnish was fourth. The urethane rated 32 liters/mil; and the others ranged downward from 25 to 23 liters/mil.

On impact resistance, the urethane, epoxy, and alkyd passed at 28 lbs., while the spar varnish failed.

The entire Allied Chemical chart is reproduced as Table 702.7.

702.8 Recoating Chalking Painted Masonry. Adhesion of a fresh coating is seriously impaired by powder formed on a deteriorating old coat. The powder, or chalk, should be removed by sandblasting, and if that is impractical, by wirebrushing and washing to whatever extent is possible. Then, a coating based on a binder with wetting ability should be applied. Federal Specification TTP 620 (GPC 31) was designed with that in mind. Meanwhile, oil and alkyd additives have been developed for exterior latex paints such as acrylic and polyvinyl acetate, so that they also are capable of adequately wetting chalk. The advantage of using acrylics is that some of these

Selection and Specification—Exterior

Table 702.7. Typical Properties of Four Resins.

Properties	Linseed Oil-Based Urethane	Spar Varnish	Epoxy	Medium-Oil Alkyd
Solvent	Min. Sp.	Min. Sp.	Xylene	Min. Sp.
Dry time, tackfree (hrs.)	2-1/2-3-1/2	3-4	2	3-3-1/2
Hardness, Sward				
1 day	14	27	8	12
1 week	30	29	36	22
1 month	43	35	—	41
Flexibility, 1/8-in. mandrel	Pass	Pass	Pass	Pass
Impact—28 in.lbs.	Pass	Fail	Pass	Pass
Sand abrasion, liters/mil	32	23	24	25
Weather-Ometer	Very good	Very good	—	Poor
Florida exposure	Very good	Good	Good	Poor
Alkali resistance	Good	Fair	Good	Poor

Source: Allied Chemical Corp.

binders require only 11–15 percent alkyd replacement rather than approximately 20 percent for other latex binders. Better durability has resulted when less alkyds are added.

Caution must be exercised in repainting masonry surfaces because the combination of porosity of chalk and porosity of substrate can lead to wicking of the paint, a situation where the binder migrates through the porous chalk and substrate, causing uneven coloration of the coating and excessive pigment buildup on the surface. This results because binder has migrated, leaving a large amount of pigment on the surface where absorption occurs.

702.9 Waterproofing Against Seepage. We have seen that wind-driven rain poses problems that can be overcome by using a heavy-textured coating. Moderate water pressure below-grade, or above, can be overcome by a more fluid coating. Federal Specification TTP 1411 (GPC 36) has been devised to provide a paint able to withstand hydrostatic pressure where concrete or cinder blocks are used. Vinyl toluene butadiene binders are required in the specification, and extensive testing shows the wisdom of the requirement. Tests performed at Moore Research Laboratories, outside of Washington, D.C., revealed that coatings based on these binders plus added portland cement withstood water pressure equivalent to that exerted by 100 mph winds and 70 gals. of water per hour. Tested at the same time were latex formulations and alkyd. The alkyd coatings were as good as the vinyl toluene butadiene when applied to dry masonry, but when the substrate was wet, the latter was clearly superior. The rating of excellent earned by the butadiene copolymer coating signified that no leakage of water took place in eight hours and that not more than 25 percent of the test area was dampened in that period.

Because the butadiene material here uses portland cement, it is sometimes incorrectly used as a filler to end porosity of stucco, concrete, and cinder blocks. The formulation was not designed for that purpose, however; and an actual filler-grout of the type required in Federal Specification TTF 1098 (GPC 33) is recommended.

702.10 Filler for Porous Surfaces. In addition to the rubber-based, or butadiene styrene, filler in the federal specification, materials are available with polyvinyl acetate and acrylic binders, but little performance data appear to be available.

A test conducted by Moore Laboratories in Silver Springs, Md., indicated that a styrene butadiene filler coat topped by an exterior masonry paint of the same material repelled 70 gals. of water driven by the equivalent of 100 mph winds. Caution must be taken on new construction to remove dirt, loose particles, and flaking material. On previously painted surfaces wire brushing, sandblasting, or scraping—depending on the seriousness of the deterioration—must precede filling. (See Section 307.)

703 Metal Exterior Surfaces. Under metal exterior surfaces we include iron, steel, galvanized metals, aluminum, and bronze and copper. Since iron and steel are the major metal surfaces in construction, they will be given far more attention than the others.

In discussing cementitious surfaces in the preceding section, a distinction was made between *ordinary* conditions and *problem* conditions. Ordinary conditions are nonexistent where metal surfaces are concerned.

All primers used in metal protection are formulated to meet the basic problem of nullifying the effect of moisture, which is always—except in the desert—engaged in a constant, seem-

Selection and Specification—Exterior 237

ingly unrelenting attack on the metal molecule.

Topcoats for metal, if we want to risk oversimplification, can be described as armor for the primers. Most of the primers outlined in Performance Comparative Chart III can also be used as topcoats. The exceptions are those with red lead (GPCs 49-52); and even these paints can be used as topcoats under some circumstances if the owner doesn't mind the color.

703.1 Primer Pigments. Older paint types feature red lead, an effective anticorrosive where moisture conditions are mild and where the surface is not likely to be scratched. Federal Specification TTP 86, Types I-IV (GPCs 49-52), call for red lead. Only one of these types has a water resistance provision and that only calls for 14 days of cold water immersion and 7 hours boiling water. That contrasts with 9 years for a linseed alkyd with basic silico lead chromate under actual weather conditions and 1-1/2 years of tidal water exposure for an epoxy ester with basic silico lead chromate and red iron oxide. An epoxy polyamide (GPC 57) listed in Performance Comparative Chart III was tested for three months of fresh water immersion and the same in salt water with no indication of deterioration.

Thus expectations of protection increase with use of basic silico lead chromate or zinc dust in organic coating systems, particularly those using epoxy, urethane, or chlorinated rubber as binders. Combinations of zinc dust with inorganic silicate binders offer outstanding protection against corrosion. See Fig. 703.1.

703.2 Topcoats. As mentioned earlier, for hidden and many kinds of exposed structural steel, most of the primers in Performance Comparative Chart III are suitable for topcoats. Various manufacturers offer a selection of colors in these coatings so that attractive topcoats are available if the surface will be exposed to view.

Topcoats characterized by gloss as well as performance are listed in Performance Comparative Chart IV. In specifying semigloss or high gloss enamels, it is important to select a primer similar to Federal Specification TTP 659 (GPC 43), which is designed to provide enamel holdout. This means that the solvent and resins in the enamel will not draw pigment

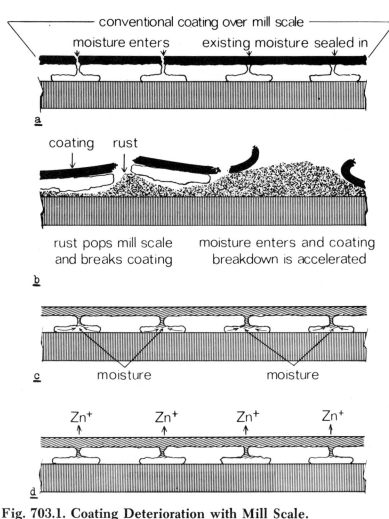

Fig. 703.1. Coating Deterioration with Mill Scale.
The four accompanying panels illustrate why conventional coatings are ineffective over metal surfaces when mill scale is permitted to remain. First, the coating can seal in moisture which has already penetrated the mill scale. Then, as shown in Panel a, additional moisture enters through pinholes and breaks in the coating. In the next stage (Panel b), the moisture rusts the steel surface, popping the mill scale and breaking the coating. Additional moisture enters through the broken coating, accelerating the deterioration process.

Panels c and d illustrate the protective effect of a coating containing zinc dust. The moisture that succeeds in reaching ferrous metal reacts with the zinc dust. Zinc dust serves to prevent the electric charge, set up when moisture reaches the metal, from corroding the ferrous surface. It continues this sacrificial role until it is consumed.

Selection and Specification—Exterior 239

from the primer and cause irregular areas of low gloss to appear in the topcoat. All manufacturers provide suitable primers for this purpose, which will meet the anticorrosive pigment requirements of metal primers.

Weather resistance, salt spray performance, and water immersion resistance are key considerations. In some instances abrasion and impact resistance may be factors influencing choice. Compromises may be necessary. For instance, where water resistance and impact resistance are needed, the proprietary vinyl chloride formula (GPC 53) in Performance Comparative Chart IV may be selected although its impact resistance is only 80 in.lbs., as against 160 in.lbs. for other topcoats. Its 3,000 hours of salt and fresh water immersion resistance outrank most others in that characteristic. If its impact resistance is inadequate, then a compromise will have to be made, which may result in selecting another coating whose water immersion resistance may not be as good as the vinyl chloride.

On the other hand, the situation may also call for excellent acid resistance and ability to withstand water immersion. If the compromise to get high impact resistance inclined the specifier to a urethane, where we find a proprietary formulation (GPC 22) that resists 160 in.lbs. of impact and has excellent water resistance, he would have to pass this up unless he is satisfied with limited ability to withstand acids. See Tables 703.2A and 703.2B.

He may then decide that an epoxy polyamide with impact resistance of 160 in.lbs. offers the best compromise because it has outstanding water resistance and good acid resistance. See Table 703.2C for water resistance.

Other factors may enter in, of course, but the specifier can study the Performance Comparative Charts and select the topcoat most likely to survive in the envionment in which it is to function and then decide the proper primer to serve with it. The selection of a primer may come first, and then the topcoat to protect it from the elements. The system, or combination of primer and topcoat or topcoats, is what counts.

703.3 Basic System Recommendations for Exterior Metal Surfaces.

703.3a Rural Environment (rough substrate). *Surface preparation*—requires simple cleaning with hand or power tools (See

Section 305.). *Primer*—linseed oil (TTP 86 Type I; same as GPC 49). *Topcoat*—linseed oil, or linseed oil with alkyd modification. Oil-modified alkyd primer may be used with an alkyd topcoat for medium or deep colors or a phenolic topcoat may be used with aluminum pigments.

703.3b Rural Environment (smooth substrate). *Surface preparation*—requires commercial blast cleaning. (See Section 309.) *Primer*—alkyd, plus alkyd topcoat. For a high-gloss topcoat, a topcoat of the TTE 489 (GPC 44) type is suitable and for semigloss with high durability, TTE 490 (GPC 45), a silicone alkyd, will give outstanding wear.

703.3c Industrial Environment with Sulfide Fumes. *Surface preparation*—commercial blast cleaning. (See Section 309.) *Primer*—TTP 86, Type I (GPC 49). *Topcoat*—lead-free linseed oil or alkyd.

703.3d High Humidity, Heavy Rainfall, or Fresh Water Immersion. *Surface preparation*—commercial blast cleaning. (See Section 309.) *Primer*—phenolic (TTP 86, Type IV; same as GPC 52). *Topcoat*—chlorinated rubber (TTP 95; same as GPC 28) (for bad conditions—use TTP 641 (GPC 54) with zinc dust for primer).

703.3e Marine or Mildly Corrosive (rough structural). *Surface preparation*—flame-cleaning or brush-off blasting. *Primer*—same as rural, except two coats of primer. *Topcoat*—same as rural.

703.3f Marine or Mildly Corrosive (smooth). *Surface preparation*—commercial blast cleaning. *Primer*—same as rural, except two coats. *Topcoat*—same as rural.

703.3g Marine or Moderately Corrosive. *Surface preparation*—commercial blast cleaning. *Primer*—zinc dust with chlorinated rubber. *Topcoat*—alkyd for high gloss (TTE 489; same as GPC 44), or silicone alkyd (TTE 490; same as GPC 45) for semigloss.

Table 703.2A. Properties of Five Classes of Urethanes.

Class Urethane	Package Type	Method of Cure	Type of Cure	Corrosion Resistance					Ease of Pigmentation	Solvents Used
				Chemical	Solvent	Water	Wear			
1a	1	Air-dry	Oxidation	Fair	Fair	Fair	Good	Excellent	Aliphatics
1b	1	Room temperature	Atmospheric moisture	Good	Good	Good	Good	Very limited	Aromatics and esters
1c	1	Bake	Polymerization	Excellent	Excellent	Excellent	Excellent	Good to excellent	Ketones aromatics and esters
2a	2	Room temperature	Atmospheric moisture and polymerization	Good	Good	Good	Good	Very limited	Aromatics and esters
2b	2	Room temperature	Polymerization	Very good	Very good	Excellent	Excellent	Very good	Aromatics and esters

Table 703.2B. *Resistance of Two-Component Air Drying Urethanes to Corrosives.*

Acids, Mineral		Gases	
Hydrochloric, 10%	LR	Chlorine, dry	LR
37%	LR	wet	LR
Sulfuric, 10%	LR	Ammonia	LR
70%	NR	Hydrogen sulfide	R
Nitric, 10%	LR	Solvents	
70%	NR	Ethyl acetate	LR
Phosphoric, 10%	R	Butyl acetate	LR
85%	LR	Acetone	NR
Chromic, 10%	LR	Methyl ethyl ketone	LR
50%	NR	Methyl isobutyl ketone	LR
Hydrofluoric, 48%	LR	Cyclohexanone	NR
Hypochlorous	LR	Isophorone	NR
Acids, Organic		Methyl alcohol	NR
Acetic, 10%	LR	Ethyl alcohol	NR
Glacial	NR	Fatty alcohols	R
Anhydride	R	Glycols	R
Formic	NR	Trichloroethylene	LR
Lactic	R	Perchloroethylene	LR
Cresylic	NR	Carbon tetrachloride	R
Oleic	R	Methylene chloride	NR
Oxalic	R	Ethylene dibromide	LR
Maleic	R	Toluene	R
Stearic	R	Benzene	R
Benzene sulfonic	R	Aromatic hydrocarbons	R
Fatty acids	R	Ethyl ether	R
2-Ethyl butyric acid	R	Aliphatic hydrocarbons	R
Alkalis		Gasoline	R
Ammonium hydroxide dilute	LR	Jet Fuel	R
Ammonium hydroxide conc.	LR	Orthodichlorobenzene	R
Calcium hydroxide	LR	Carbon disulfide	R
Potassium hydroxide	LR	Dimethyl formamide	NR
Sodium hydroxide, 15%	LR	Turpentine	R
Sodium hydroxide, 50%	LR	Miscellaneous	
Acid Salts		Cutting oils	R
Aluminum sulfate	R	Vegetable oils	R
Ammonium chloride, nitrate, sulfate	R	Lubricating oils	R
		Diester lubricants	R
Calcium chloride, nitrate, sulfate	R	Styrene monomer	R
Zinc chloride, nitrate, sulfate	R	Glycerine	R
Alkaline Salts		Pyridine	NR
Sodium bicarbonate	R	Detergents	R
Sodium carbonate	R	Formaldehyde, 37%	R
Sodium sulfide	R	Distilled water	R
Sodium sulfite	R	Tap water	R
Trisodium phosphate	R	Salt water	R
Sodium nitrate	R	Condensate water	R
Oxidizing Agent		Fruit juices	R
Sodium hypochlorite	LR	Milk products	R
Potassium permanganate	LR	Whiskey	R
Sodium chlorate	R	Carbonated beverages	R
Hydrogen peroxide	LR	Beer	R

Selection and Specification—Exterior 243

Table 703.2B. *Continued*

Miscellaneous *continued*			
Phenol in alcohol	NR	Anionic wetting agents	R
Sewage waste	R	Acetonitrile	LR
Hydrazine	LR	Butyraldehyde	R
Glyoxal	R	Monoethanolamine	LR

(R = Recommended; LR = Limited Recommendation; NR = Not Recommended)
Source: *Materials Protection,* official publication of National Association of Corrosion Engineers, Houston, Tex.

703.3h Marine, Immersion, or Highly Corrosive. *Surface preparation*—white metal or near-white blasting. *Primer*—vinyl chloride. *Topcoat*—vinyl chloride (*or alternatively: Primer*—chlorinated rubber; *Topcoat*—vinyl; or *Primer*—inorganic zinc-rich; *Topcoat*—urethane, or epoxy or vinyl).

703.3i Galvanized Steel. *Surface treatment*—none. *Primer*—linseed oil (TTP 641, Type I, II, or III; same as GPC 54) or (TTP 1046; same as GPC 58). *Topcoat*—same as primer, or for bad conditions use the same as for steel and iron.

703.3j Aluminum, Tin, Copper, or Brass. *Surface preparation*—solvent cleaning, if new, or hand cleaning, with polyvinyl butyral-phosphoric acid wash as pretreatment. *Primer*—linseed or phenolic (TTP 645; same as GPC 55). *Topcoat*—same as for iron and steel, if opaque coating is desired. Usually, these are coated clear, in which case an acrylic, epoxy, urethane, or silicone clear may be used.

703.4 Selection Guide for Heavy-Duty Systems for Metal Surfaces. For surfaces expected to be subject to unusual environmental factors, in corrosive industrial areas, for example, heavy-duty systems may be needed for metal surfaces. Table 703.4 provides a general guide for selecting heavy-duty painting systems for bare ferrous metals. Space does not permit listing individual GPC numbers for easy reference to the Detailed Specification Charts. However, at the end of the section on selections a listing of specifications is provided, identified as to resin systems.

Once a decision is made as to the type of paint system that will give best results for a particular circumstance, refer

Table 703.2C. Effect of PVC of Primer on the Properties of Three-Coat, Epoxy Coating Systems Subjected to One and One Half Years' Tide Range Exposure.

Identification of resin system	PVC of Primer	Film thickness (mils)	Blisters (ASTM D-714-56)	Adhesion to Substrate* (Reverse Impact)	Intercoat Adhesion* (Direct Impact)	Underfilm Corrosion
Epoxy/Polyamide	25%	7.1	10	6	10	None
Epoxy/Polyamide	35%	7.3	10	8	10	None
Epoxy/Polyamide	40%	7.1	10	8	10	None
Epoxy Ester	25%	6.6	10	6	10	Yes
Epoxy Ester	35%	6.8	10	6	10	Yes
Epoxy Ester	40%	7.9	10	10	10	None

*Determined by Falling Ball Impact Test. For both tests, a steel ball weighing 6.3 lbs. (2860 gms.) was dropped from a height of 36 inches (91.4 cm.) onto the test panel supported on a wood block. The panels were rated as follows: 10—no failure; 8—coating fractured but no loss in adhesion to substrate; 6—coating fractured and definite loss in adhesion to substrate. Pigment Volume Concentration of the primer played a role in improving substrate adhesion of the epoxy ester and the epoxy polyamide. Higher PVC, from 35% to 40%, improved adhesion of the epoxy ester, while raising the epoxy polyamide from 25% to 35% was helpful.

Source: Ciba-Geigy.

Selection and Specification—Exterior

Table 703.4. Selection Guide—Heavy-Duty Systems for Metal Surfaces.

Generic Description of Coatings	Total DFT (dry film thickness)	Flexibility and Impact Resistance	Abrasion Resistance	Water Resistance @ 77° F.	Solvent Resistance Aliphatic	Solvent Resistance Aromatic	Solvent Resistance Ketone	Corrosion Resistance	Alkali Resistance	Acid Resistance	Maximum Temperature Resistance Dry	Gloss and Color Retention	Surface Preparation° Steel	Specification No.	Page‡	Typical Coverage Sq.ft./gal. at DFT Recommended	Recommended Application Method(s)	Thinner and Cleanup Solvent
CHLORINATED RUBBER																		
conventional build top coat over chlorinated rubber primer	6 mils	Limited	Good	Immersion up to 120° F.	Limited	Not recommended	Not recommended	Good	Excellent	Good	150° F.	Slight chalk	SP-6	71S1	24	280 @ 2 mils 140 @ 4 mils	Brush or spray Brush or spray	Xylol Xylol
conventional build top coat over organic zinc rich primer (one package)	6 mils	Limited	Good	Intermittent immersion	Limited	Not recommended	Not recommended	Excellent	Excellent	Good	150° F.	Slight chalk	SP-10	71S2	24	280 @ 2 mils 195 @ 4 mils	Brush or spray Spray	Xylol Xylol
conventional build top coat over inorganic zinc rich primer (two component)	6 mils	Limited	Good	Intermittent immersion	Limited	Not recommended	Not recommended	Excellent	Excellent	Good	150° F.	Slight chalk	SP-5	71S4	24	280 @ 2 mils 240 @ 4 mils	Brush or spray Brush or spray	Xylol Water
hi-bild top coat over chlorinated rubber primer	8 mils	Limited	Good	Immersion up to 120° F.	Limited	Not recommended	Not recommended	Good	Excellent	Good	150° F.	Slight chalk	SP-6	71S5	24	130 @ 4 mils 140 @ 4 mils	Brush or spray Brush or spray	Xylol Xylol
hi-bild top coat over organic zinc rich primer (one package)	8 mils	Limited	Good	Intermittent immersion	Limited	Not recommended	Not recommended	Excellent	Excellent	Good	150° F.	Slight chalk	SP-10	71S6	24	130 @ 4 mils 195 @ 4 mils	Brush or spray Spray	Xylol Xylol
hi-build top coat over epoxy zinc rich primer (three component)	8 mils	Limited	Good	Intermittent immersion	Limited	Not recommended	Not recommended	Excellent	Excellent	Good	150° F.	Slight chalk	SP-10	71S7	24	130 @ 4 mils 230 @ 4 mils	Brush or spray Spray	Xylol Xylol

Table 703.4. Selection Guide (Continued).

Generic Description of Coatings	Total DFT (dry film thickness)	Flexibility and Impact Resistance	Abrasion Resistance	Water Resistance @ 77° F.	Solvent Resistance Aliphatic	Solvent Resistance Aromatic	Solvent Resistance Ketone	Corrosion Resistance	Alkali Resistance	Acid Resistance	Maximum Temperature Resistance Dry	Gloss and Color Retention	Surface Preparation° Steel	Specification No.	Page‡	Typical Coverage Sq.ft./gal. at DFT Recommended	Recommended Application Method(s)	Thinner and Cleanup Solvent
hi-bild top coat over inorganic zinc rich primer (two component)	8 mils	Limited	Good	Intermittent immersion	Limited	Not recommended	Not recommended	Excellent	Excellent	Good	150° F.	Slight chalk	SP-5	71S8	24	130 @ 4 mils 240 @ 4 mils	Brush or spray Spray	Xylol Water
EPOXY																		
modified epoxy top coat (one package) over modified epoxy primer (one package)	5 mils	Good	Good	Good	Limited	Not recommended	Not recommended	Good	Limited resistance	Not recommended	250° F.	Slight chalk	SP-7	71S9	24	195 @ 3 mils 290 @ 2 mils	Brush, roll or spray Brush or spray	
modified epoxy top coat (one package) over long oil alkyd primer	6 mils	Good	Good	Good	Limited	Not recommended	Not recommended	Good	Limited resistance	Not recommended	200° F.	Slight chalk	SP-3	71S10	24	195 @ 3 mils 310 @ 3 mils	Brush, roll or spray Brush or spray	
conventional build epoxy top coat (two component) over epoxy primer (two component)	6 mils	Good	Excellent	Intermittent immersion	Excellent	Good	Limited	Excellent	Excellent	Good	275° F.	Slight chalk	SP-6	71S11	24	360 @ 2 mils 180 @ 4 mils	Brush, roll or spray Brush or spray	
conventional build epoxy top coat (two component) over organic zinc rich primer (one package)	6 mils	Limited	Excellent	Intermittent immersion	Excellent	Good	Limited	Excellent	Excellent	Good	275° F.	Slight chalk	SP-10	71S12	24	360 @ 2 mils 195 @ 4 mils	Brush, roll or spray Spray	
conventional build epoxy top coat (two component)				Intermittent												360 @ 2 mils	Brush, roll or	

Selection and Specification—Exterior

System	DFT		Immersion							Temp	Chalking	Surface Prep	Spec No.		Coverage	Application
over epoxy zinc rich primer (three component)	6 mils Limited	Excel-lent	immersion	Excel-lent	Good	Limited	Excel-lent	Excel-lent	Good	275° F.	Slight chalk	SP-10	71S13	24	230 @ 4 mils	spray Spray
conventional build epoxy top coat (two component) over inorganic zinc rich primer (two component)	6 mils Limited	Excel-lent	Intermittent	Excel-lent	Good	Limited	Excel-lent	Excel-lent	Good	275° F.	Slight chalk	SP-5	71S14	24	360 @ 2 mils / 240 @ 4 mils	Brush, roll or spray / Spray
hi-bild epoxy top coat (two component) over epoxy primer (two component)	8.5 mils Good		Intermittent immersion	Excel-lent	Good	Limited	Excel-lent	Excel-lent	Good	275° F.	Slight chalk	SP-6	71S15	24	165 @ 4.5 mils / 180 @ 4 mils	Brush or spray / Brush or spray
hi-bild epoxy top coat (two component) over organic zinc rich primer (one package)	8.5 mils Limited	Excel-lent	Intermittent immersion	Excel-lent	Good	Limited	Excel-lent	Excel-lent	Good	275° F.	Slight chalk	SP-10	71S16	24	165 @ 4.5 mils / 195 @ 4 mils	Brush or spray / Spray
hi-bild epoxy top coat (two component) over epoxy zinc rich primer (three component)	8.5 mils Limited	Excel-lent	Intermittent immersion	Excel-lent	Good	Limited	Excel-lent	Excel-lent	Good	275° F.	Slight chalk	SP-10	71S17	24	165 @ 4.5 mils / 230 @ 4 mils	Brush or spray / Spray
hi-build epoxy top coat (two component) over epoxy zinc rich primer (three component)	8.5 mils Limited	Excel-lent	Intermittent immersion	Excel-lent	Good	Limited	Excel-lent	Excel-lent	Good	275° F.	Slight chalk	SP-5	71S18	25	165 @ 4.5 mils / 240 @ 4 mils	Brush or spray / Spray
two coats of coal tar epoxy (two component)	16 mils Good	Excel-lent	Immersion up to 150° F.	Excel-lent	Limited	Limited	Excel-lent	Excel-lent	Good	250° F.	Slight chalk	SP-10	71S19	25	145 @ 8 mils	Brush, roll or spray

URETHANE

moisture-cured urethane top coat (one package) over epoxy primer (two component)	6 mils	Excel-lent	Intermittent immersion	Excel-lent	Good	Limited	Excel-lent	Excel-lent	Excel-lent	250° F.	Slight chalk	SP-6	71S20	25	305 @ 2 mils / 180 @ 4 mils	Brush or spray / Brush or spray
moisture-cured urethane top coat (one package) over epoxy zinc rich primer (three component)	6 mils Limited	Excel-lent	Intermittent immersion	Excel-lent	Good	Limited	Excel-lent	Good	Excel-lent	250° F.	Slight chalk	SP-10	71S21	25	305 @ 2 mils / 230 @ 4 mils	Brush or spray / Spray

Table 703.4. Selection Guide (Continued).

Generic Description of Coatings	Total DFT (dry film thickness)	Flexibility and Impact Resistance	Abrasion Resistance	Water Resistance @ 77° F.	Solvent Resistance Aliphatic	Solvent Resistance Aromatic	Solvent Resistance Ketone	Corrosion Resistance	Alkali Resistance	Acid Resistance	Maximum Temperature Resistance Dry	Gloss and Color Retention	Surface Preparation* Steel	Specification No.	Page‡	Typical Coverage Sq.ft./gal. at DFT Recommended	Recommended Application Method(s)	Thinner and Clean-up Solvent
VINYL																		
hi-bild vinyl top coat over hi-bild vinyl primer	8 mils	Excellent	Good	Intermittent immersion	Good	Limited	Not recommended	Good	Good	Good	150° F.	Excellent	SP-10	71S22	25	65 @ 6 mils 170 @ 2 mils	Spray	Spray
hi-bild vinyl top coat over vinyl wash primer	6.5 mils	Excellent	Good	Intermittent immersion	Good	Limited	Not recommended	Good	Good	Good	150° F.	Excellent	SP-3	71S23	25	65 @ 6 mils 220 @ .5 mil	Spray	Spray
hi-bild vinyl top coat over organic zinc rich primer (one package)	10 mils	Limited	Good	Intermittent immersion	Good	Limited	Not recommended	Excellent	Good	Good	150° F.	Excellent	SP-10	71S24	25	65 @ 6 mils 195 @ 4 mils	Spray	Spray
hi-bild vinyl top coat over epoxy zinc rich primer (three component)	10 mils	Limited	Good	Intermittent immersion	Good	Limited	Not recommended	Excellent	Good	Good	150° F.	Excellent	SP-10	71S25	25	65 @ 6 mils 230 @ 4 mils	Spray	Spray
hi-bild vinyl top coat over inorganic zinc rich primer (two component)	10 mils	Limited	Good	Intermittent immersion	Good	Limited	Not recommended	Excellent	Good	Good	150° F.	Excellent	SP-5	71S26	25	65 @ 6 mils 240 @ 4 mils	Spray	Spray
ZINC RICH																		
one coat of organic zinc rich (one package)	4 mils	Limited	Excellent	Immersion up to 85° F.	Excellent	Limited	Limited	Excellent	Not recommended	Not recommended	250° F.	Not applicable	SP-10	71S27	25	195 @ 4 mils	Spray	Spray

Selection and Specification—Exterior

Coating	Thickness											Temp		Surface Prep	Spec No.		Coverage	Application
one coat of epoxy zinc rich (three component)	4 mils	Limited	Excellent	Immersion up to 120° F.	Excellent	Good	Limited	Excellent	Not recommended	Not recommended	500° F.	Not applicable	SP-10	71S28	25	230 @ 4 mils	Spray
one coat of inorganic zinc rich (two component)	4 mils	Limited	Excellent	Immersion up to 120° F.	Excellent	Excellent	Excellent	Excellent	Not recommended	Not recommended	700° F.	Not applicable	SP-5	71S29	25	240 @ 3 mils	Spray
ALUMINUM																	
two coats of silicone alkyd aluminum	1.5 mils	Good	Limited	Good	Excellent	Limited	Not recommended	Limited	Not recommended	Not recommended	500°–1000° F.	Not applicable	SP-5	71S30	25	500 @ .75 mils	Brush or spray
two coats of alkyd aluminum	2 mils	Good	Limited	Good	Good	Limited	Not recommended	Limited	Not recommended	Not recommended	200°–400° F.	Not applicable	SP-3	71S31	25	500 @ 1 mil	Brush or spray
moisture cured urethane aluminum over epoxy primer (two component)	6 mils	Excellent	Excellent	Intermittent immersion	Excellent	Limited	Good	Excellent	Good	Excellent	250° F.	Not applicable	SP-6	71S32	25	180 @ 4 mils 335 @ 2 mils	Brush or spray

Source: Sherwin-Williams, Inc.

*Numbers in this column refer to surface preparation specifications established by Steel Structures Painting Council.

‡Page nos. refer to Sherwin-Williams Specification Book.

to the list, arranged by resin types, and choose the most promising specification, or design your own specification from the information provided by several of the available specifications.

703.5 Shop Coating. For industrial installations, where conditions require high-performance coatings, specifiers should investigate the economies of shop application. Steel fabricators in many instances can ship material to specialized firms that can sandblast or shotblast the surface and apply required coatings for less money than field applicators; and since they are specialists working under optimum conditions with expert inspectors, builders and engineers are assured of top quality. Field application by reliable firms is reasonably safe, but the subjective judgment of field inspectors checking work under sometimes adverse atmospheric conditions is likely to be more lenient in interpreting tolerances than inspectors in specialized plants working under cover.

Labor costs of application in the shop are about 25 percent less than in the field. Sandblasting in the shop may be as much as 50 percent less than in the field, depending on the cost of the field job, which is based on difficulty of reaching the surface, number of seams in the finished shape, and difficulty in removing sand.

Shop work has limitations. Each shop has a maximum size that it can handle. Some can accept pieces with a maximum of six feet in one of its dimensions. The other dimensions have no limit. Another limitation arises from economics and could be regarded as a negative limit: the minimum quantity of structural steel that is worth surface treating and coating in the shop.

Roughly 125 tons has been suggested as a reasonable minimum. At approximately 250 sq.ft. per ton (structural steel runs between 200–300 sq.ft./ton), this comes to about 31,000 sq.ft. of surface.

A comparison of field preparation costs versus shop costs is provided below:

	White metal blast (sq.ft.)	Commercial blast (sq.ft.)	Wire brush (sq.ft.)
Shop	$0.25 max.	$0.20 max.	$0.10
Field	$0.40–55	$0.30–.45	$0.10

Selection and Specification—Exterior

703.6 Special Alloys. Considerable interest has been shown in special steel alloys that form their own protective coatings upon contact with water. Protection comes when components of the alloy oxidize, or form a pseudo rust, which has the right electric potential to impede the destructive effect of moisture.

One such material, Cor-Ten, described as high-strength, low alloy steel, has been used as supporting beams for bridges—without protective paints. Aside from high cost, a decided disadvantage of this steel is its tendency to stain adjacent structural areas, as the pseudo rust is absorbed by rain and runs down whatever surface it can reach. The makers of Cor-Ten, evidently aware of this, now advocate the use of protective coatings on this alloy as a means of getting extremely durable structural steel.

Exposures of 14 years in a semi-rural atmosphere on painted carbon steel and on painted Cor-Ten showed that coatings with red lead, iron oxide, or zinc chromate gave blemish-free protection to Cor-Ten, while carbon steel panels with the same coatings showed signs of deterioration. A 12-year study of railroad hopper cars with adjacent panels of Cor-Ten and copper steel showed the improved paint life on a Cor-Ten substrate. Two observations are worth making:

1) Cor-Ten is subject to limited corrosion.
2) If Cor-Ten is too costly for a particular application, it may pay to use copper steel, which outperformed low-carbon steel, both when coated with paint alone or when painted after surface treatment with Bonderite.

The improved performance of copper steel points up an often ignored probability: the alloy used likely plays a role in the ability of a coating to protect it.

703.7 Galvanized Steel. Steel sheets to be exposed to the elements are often dipped in molten zinc to provide the extra protection of that metal. The name *galvanize* is derived from the galvanic protection which zinc affords by diverting to itself the destructive galvanic electric charge that water otherwise sets up with steel.

Paint is applied to the zinc surface to delay the sacrificial destruction of the thin zinc coating and thus prolong the life of the substrate. A.S.T.M. Specification A361, which applies to galvanized steel sheets, only calls for 1.25 ounces of zinc coating per square foot, while A.S.T.M. Specification A93 calls for 2.75 ounces per square foot. That's not a lot of armor either against rain or blows that can chip away the zinc and lead to rapid rusting.

703.7a Special Galvanizing Specifications. Specifiers may want to require certain factory-applied treatments to facilitate painting. Some steel producers provide either one or both of the following treatments:

1) *Galvannealed*—Galvanized sheet is heat treated to change the surface characteristics of the zinc film to make it more receptive to paint.
2) *Phosphatized*—The galvanized sheet is treated with either a hot or cold phosphate solution to convert a small portion of the surface zinc to an inert zinc phosphate film, which is more receptive to paint than ordinary zinc.

Either type is ready to paint unless storage or handling has caused contamination, in which case cleaning, with care that the zinc phosphate surface is not ruptured, is needed.

703.7b Inhibitive Coating Removal. To prevent moisture from staining the uncoated zinc surface, most galvanized steel producers coat the sheets with an inhibitor. Since these may not be suitable for painting, it is recommended by the Committee on Galvanized Steel Sheet Research of the American Iron and Steel Institute that these sheets be weathered for six months or more so the inhibitors will be oxidized away before painting.

If immediate painting is necessary, it is necessary to consult the manufacturer for treatments. This should not be left to chance nor to a decision by painting contractors selected to apply the coating.

Primers with zinc dust and topcoats based on TTP 641-type paints (GPC 54), using phenolic resins, are among paints recommended for this surface. Do not be casual about selecting primers for galvanized surfaces because some binders and pigments react adversely with zinc. Don't forget that zinc as a pigment depends for effectiveness on its reactivity with some binders.

Selection and Specification—Exterior 253

704 Wood Surfaces. Wood poses problems as much because of its own characteristics as because of outside influences. A review of Section 605 on deterioration of wood coatings will be helpful at this point for an understanding of the factors that influence selection of wood coatings, particularly for exterior surfaces exposed to the vagaries of temperature.

Wood consists of alternating portions of soft and hard material caused by the quick growth and soft texture of the fiber produced in spring and the relatively hard, tight formation of the slower-growing summer wood; further, wood often is inadequately dried and contains moisture that must be sealed off; and, some woods have dyes that must be held in check.

704.1 Primers. Wood primers have to be able to withstand the stresses caused by the differing degrees of swelling and contracting of the spring and summer bands of the wood; they must be able to check moisture (note the reference to Whatman Paper tests in Performance Comparative Chart V under the heading "Water Resistance"); and finally, they have either to neutralize the dyes in such materials as redwood and cedar or they must check them.

704.1a Lead-Based Primers and Substitutes. Before 1973, exterior wood primers contained basic white carbonate of lead (white lead) which effectively combined with fatty acids in linseed oil or alkyds to keep moisture and dyes out of the topcoat. Lead in house paints has been banned by federal regulations. Fortunately, several companies and the U.S. Housing and Urban Development Department (HUD) have been able to bridge the gap. The HUD product (GPC 72) is outlined on Performance Comparative Chart V. A new and effective acrylic latex paint to counteract staining has a proprietary lead-free additive. The new lead-free commercial primer contains about one pound of the additive per gallon and has a PVC of 21.8 percent. For some reason, the additive, which is an emulsion, is considered part of the Pigment Volume Concentration (PVC). About one-half pound of zinc oxide per gallon is used, and about 1.4 lbs. of rutile titanium dioxide.

Extensive tests by the manufacturer of the additive and the acrylic emulsion, Rohm & Haas Co., show that in quelling wood stains the product as formulated works as well as or

Table 704.1a. *Resistance to Topcoat Staining over Cedar and Redwood Substrates in Acrylic Formulations with PR 26 (GPC 72).*

	Recommended formulation with PR 26	XPR-35-3 without PR 26	Self-primed topcoat	Lead silicate primer
Designation	XPR-35-3	—	W-35-1	PR-35-2
Substrate: Cedar				
Initial stain resistance	Good	Poor	Poor	Good
Long-term stain resistance				
Fog box	Good	Fair	Poor	Good
Blister box	Good	Fair	Poor	Good
Humidity cabinet	Good	Fair	Poor	Good
Substrate: Redwood				
Initial stain resistance	Good	Poor	Poor	Fair
Long-term stain resistance				
Fog box	Good	Fair	Poor	Good
Blister box	Good	Fair	Poor	Good
Humidity cabinet	Good	Fair	Poor	Good

better than lead-silicate primers. The company stresses the importance of proper formulation. The company's test results are shown in Table 704.1a.

704.1b Primers for Nonstaining Wood. One of the most difficult woods to coat is Southern yellow pine, which is widely used for siding. A wood with great structural strength and low cost, Southern yellow pine is an example of wood with varied texture and ability to swell, shrink, twist, and warp.

This wood and the paint industry have been having a running battle for generations. In the Gulf of Mexico region, two coats of oil-base paints have been known to need repair within 18 months of application, and three coats commonly fail within two or three years. One candidate for meeting such problem conditions was an all-acrylic latex system. After six years, it stood the swelling and contraction of the wood; and now paint companies are offering it.

Selection and Specification—Exterior

704.2 Topcoats. Aside from the need to expand and contract with the wood substrate, the requirements for topcoats for wood surfaces are fairly similar to other coatings. Topcoats for wood are usually general purpose, in flat and gloss; trim paint, usually gloss; and shake and shingle paint, which is semigloss; new plywood paint; and heavy-duty paints for industrial areas.

For a dependable flat finish, the best bet, now that lead is to be banned, is the companion acrylic latex that was tested on Southern yellow pine along with the acrylic primer described above. A linseed oil flat for general use is TTP 105 (GPC 41), which also serves as a primer.

704.3 Trim Paint. TTP 37 (GPC 64) indicates a lead-free alkyd with high gloss which offers weather resistance and adequate flexibility for most requirements in exterior industrial environments.

704.4 Shake and Shingles. An alkyd with good water resistance is TTP 52 (GPC 67), designed to withstand water from the clouds and from overflowing gutters. Because this specification requires relatively little flexibility, usage should be only on rather stable substrates.

704.5 Plywood. For most new plywoods, a seal coat, TTS 176 (GPC 68), should be applied. An acrylic system should be used for at least two coats. A topcoat of TTP 19 (GPC 23), an acrylic emulsion, with a suitable acrylic primer should provide adequate protection. Plywood with its end-grain exposed should be sealed to prevent moisture migration. A top quality exterior primer, TTP 105 (GPC 41), should be used before painting. An aluminum paint for wood also holds off moisture. Sanded plywoods should always be finished off with high quality paint. Unsanded plywood may be painted or stained. The American Plywood Association recommends that plywood primers be free of zinc oxide.

704.6 Textured Plywood. Textured plywood brings the problem of end-grains, where moisture may enter. These end grains often appear in the groove of the texture. Stain finishes are recommended here, because it is felt that the minor amount

Table 704.8. *Adhesion of Urethanes to Wood.*

Primer	Northern Maple		Type of Removal
	Average Force to Remove Coating		
	Lb./in.-width	Thickness, mils	
Vinyl urethane	20.0	3.4	Peel
Vinyl	5.3	3.4	Peel
Alkyd	6.5(4.5)	5.0(3.4)	Chip
Oil modified urethane	7.7(5.9)	4.6(3.4)	Chip
Oil primer A	3.2(2.6)	4.3(3.4)	Chip
Oil primer B	3.5(2.9)	4.3(3.4)	Peel

Primer	Douglas Fir Plywood—Springwood		Type of Removal
	Average Force to Remove Coating		
	Lb./in.-width	Thickness, mils	
Vinyl urethane	23.9	3.3	Peel
Vinyl	7.7	3.0	Peel
Alkyd	9.6(6.6)	4.8(3.3)	Chip
Oil modified urethane	12.7(6.5)	6.4(3.3)	Chip
Oil primer A	3.7(2.5)	4.8(3.3)	Peel
Oil primer B	3.7(2.0)	6.2(3.3)	Peel

Primer	Douglas Fir Plywood—Summerwood		Type of Removal
	Average Force to Remove Coating		
	Lb./in.-width	Thickness, mils	
Vinyl urethane	17.9	2.3	Peel
Vinyl	6.1(3.1)	4.5(2.3)	Peel
Alkyd	7.7(4.9)	3.6(2.3)	Chip
Oil modified urethane	12.7(4.5)	6.4(2.3)	Chip
Oil primer A	3.7(1.3)	6.4(2.3)	Peel
Oil primer B	3.5(1.1)	7.0(2.3)	Peel

Primer	Coatings on Overlaid Plywood		Type of Removal
	Average Force to Remove Coating		
	Lb./in.-width	Thickness, mils	
Vinyl urethane	(°)	—	—
Vinyl	5.8	3.7	Peel
Alkyd	6.5	3.3	Chip
Oil modified urethane	(°)	—	—
Oil primer A	3.3	5.2	Peel
Oil primer B	5.9	4.2	Peel

(°) Primer stronger than substrate; substrate failed.
Source: The Baker Castor Oil Co.

Selection and Specification—Exterior 257

of checking that is likely to occur because of the end-grain exposure looks best when the product is stained to provide a rustic effect.

704.7 Gloss on Plywood. These have been found to perform poorly on plywood and are not recommended as a general rule. Clear coatings are not recommended on plywood because they do not provide ultra-violet protection and the substrate changes color.

704.8 Industrial Areas. For heavy-duty exterior wood surfaces, a primer-topcoat system of linseed oil, TTP 105 (GPC 41), provides a representative product. If the circumstances demand an excellent product, Proprietary EC 167 (GPC 72), a vinyl urethane, has given outstanding test results on wood and plywood (see Performance Comparative Chart V and Table 704.8).

PERFORMANCE COMPARATIVE CHARTS

The following Performance Comparative Charts have been prepared. They are based on the extensive Guide to Paints and Coatings (GPC's) in the marked pages at the end of the book and enable the specifier to compare at a glance the characteristics of various available specifications. These Performance Comparative Charts are mentioned in Chapter 7, and are also referenced in Chapter 8.

For quick reference in comparing the available specifications for a particular job, scan the following Performance Comparative Charts, and then turn to the detailed Guide to Paints and Coatings (GPC's) at the end of the book to choose the best specification for your application. More general information, such as test results, is provided in the text of Chapters 7 and 8 (and Chapter 9 for fire-retardant paint).

When writing specifications for the job, you can use the federal specification number, since these are generally understood and readily available, or you can use the GPC classification by extracting the relevant information from the appropriate Guide to Paints and Coatings (GPC's) and putting this information in the job specification. THE WIDEST POSSIBLE USE OF *GPC* NUMBERS WILL AID IN GAINING ACCEPTANCE FOR THIS NEW SYSTEM AND WILL MAKE YOUR JOB EASIER IN THE FUTURE.

Performance Comparative Chart I *Exterior Cementitious Surfaces—General Use.*

GPC Specification Specification Surce:	GPC 23 TTP 19	GPC 24 TTP 24	GPC 25 TTP 35	GPC 26 TTP 55	GPC 27 TTP 96
Binder type	Acrylic emulsion	Oil or varnish	Cement powder	Polyvinyl acetate emulsion	Vinyl chloride acrylate or PVA
Suitable surfaces	Concrete, masonry, primer, topcoat	Concrete masonry topcoat	Cement; primer, topcoat	Masonry; primer, topcoat	Masonry, etc. Topcoat
Coverage, sq.ft./gal.			40-60		
Gloss	Flat	Eggshell	Flat	Flat	Flat
Set-to-touch, hrs.	1/4	4		1/4	1/12
Hard dry, hrs.	1	18		1	1

Selection and Specification—Exterior

Abrasion resistance	200 oscillations		1,000 oscillations	750 oscillations
Impact resistance				
Weather resistance		Good		Fair
Water resistance		18 hrs. immersion		
Chalk resistance	Excellent	Fair		Good
Flexibility		1/4" mandrel		1/8"
Primer required	Self	Depends on conditions	Self	Self
Surface preparation references	See Par. 306, 307, 308	See Par. 306, 307, 308	See Par. 306, 307, 308	See Par. 306, 307, 308
Deterioration of coatings references	See Par. 606	Par. 606	Par. 606	Par. 606

Performance Comparative Chart IIA Exterior Cementitious Surfaces—Problem Areas.

GPC Specification Specification Source	GPC 28 TTP 95* Types I & II	GPC 29 TTP 97	GPC 30 TTC 555 Type II, for wind-driven rain	GPC 31 TTP 620
Binder type	Chlorinated rubber or styrene acrylate	Styrene butadiene	Not specified	Modified alkyd
Suitable surfaces	Moist areas	Moist areas	Heavy rain and imperfect surfaces	Chalking, painted masonry
Coverage, sq.ft./gal.	250	400	40-60	
Pigment content	45.6 PVC°	57.1 PVC	62.1 for vinyl toluene	
Permeability°°	Adequate	Adequate	Exceed 0.4 perms°°°	
Gloss	Type I, gloss; II, flat			
Set-to-touch, hrs.	3/4	1/4-3/4		4
Recoat time, hrs.				

Hard dry, hrs.	24	2	24	24
Impact resistance			Fair	
Weather resistance		Excellent	98 mph wind-rain	
Alkali resistance	Excellent			Good
Water resistance**	Excellent	Excellent	Outstanding	Good
Stain resistance			Good	
Flexibility	1/4" mandrel	1/8"	1"	1/8"
Scrubbability	500 strokes in alkali			
Primer required	Self	TTF 1098 filler	TTF 1098 filler	
Surface preparation references	Par. 307	Par. 307	Par. 307	Par. 303
Deterioration of coatings references	Par. 606	Par. 606	Par. 606	Par. 604.5; 606.6; 606.8

*TTP 95 II & IV, gloss & semigloss, same, except Gloss PVC is 24.5 for styrene acrylate; and semigloss 27.

**See Par. 209 for specifier's role in setting needs.

***See Federal specification for procedure using vapor permeable chart (Form HK) from Leneta Co., Ho-Ho-Kus, N.J., based on procedure A of ASTM E 96.

Performance Comparative Chart IIB *Exterior Cementitious Surfaces—Problem Areas*

GPC Specification Specification Source	GPC 22 Proprietary* Gloss	GPC 20 TTC 542	GPC 18 TTC 540	GPC 13 TTC 535	GPC 38 Proprietary (Moisture & chemical resistant paint)†
Binder type	Urethane (moisture)	Urethane (moisture cure)	Urethane (air-dry)	Epoxy polyamide	Acrylic latex
Suitable surfaces	All cementitious	All cementitious	All cementitious	All cementitious	All cementitious
Coverage, sq.ft./gal.		480–700 (at 1 mil)			270 (2 mil)
Permeability	Adequate				
Gloss	High or semigloss	Gloss & clear		High	Flat
Set-to-touch, hrs.	1-1/2–2	1-1/2–4, depending on type		1	1/5–1/3
Hard dry, hrs.	5–14			6	2
Abrasion resistance		Excellent	50 mgs. loss 2,000 strokes	1,000 strokes	

Impact resistance	160 in.lbs.		160 in.lbs.	
Weather resistance	Excellent			Good
Salt spray resistance, hrs.	1,000			
Alkali resistance	Excellent	Excellent	Excellent	Good
Acid resistance	Good			Good
Solvent resistance	Excellent			Fair
Water resistance	Excellent			Fair
Chalk resistance	Excellent			Good
Stain resistance	Excellent		Excellent	Fair
Flexibility	1/8"			
Scrubbability	Excellent			
Washability	Excellent			
Surface preparation references	Par. 307	Par. 307	Par. 307	Par. 307
Deterioration of coatings references	Par. 606	Par. 606	Par. 606	Par. 606

*Hughson Chemical Corporation formula.
†Glidden Coatings & Resins.

Performance Comparative Chart IIIA Exterior Metal Primers.

GPC Specification / Specification Source	GPC 49 TTP 86 Type I	GPC 50 TTP 86 Type II	GPC 51 TTP 86 Type III	GPC 52 TTP 86 Type IV	GPC 41 TTP 105	GPC 55 TTP 645 Types I, II, & IV	GPC 43 TTP 659	GPC 77 TTP 636
Binder Type	Linseed oil	Alkyd & Linseed oil	Alkyd	Tung oil phenolic	Linseed oil	I & II Linseed; IV phenol.	Alkyd	Soya alkyd
Suitable surfaces	Structural steel	Structural steel	Structural steel	Structural steel where exposed to water or chems.	Metal, wood; primer, topcoat	Metal	Metal, wood; primer & topcoat	Primer
Coverage, sq.ft./gal.	500	500	500	500				
Pigment content	77% by wt.	66% by wt.	67% by wt.	65% by wt.	59-61% by wt.	57-64% by wt.		
Special Pigments	Red lead	Red lead & red iron oxide	Red lead	Red lead, talc & silica	Zinc oxide	Basic silica lead chromate	None	None

Set to touch, hrs.	6	4	1/4-1	1/2-1				
Recoat time, hrs.	36	16	6	6	18			
Weather resistance					200 hrs. accel.			
Salt spray resistance								200 hrs.
Solvent resistance						4 hrs.		
Water resistance				7 hrs. boiling, 14 cold	1 in.* absorption	14 days distilled	18 hrs. immersion	96 hrs., 100% humidity
Flexibility, mandrel	1/4"	1/4"	1/4"	1/4"			1/8"	
Surface preparation references	Par. 304-306; 309			Meticulously clean surface		Same as TTP 86, Type I	Same as TTP 86, Type I	Same as TTP 86, Type I
Deterioration of coatings references	Par. 601-604.5c; 607	Same as Type I	Same as Type I	Same as Type I	Same as TTP 86 Type I			

*See Par. 209 for specifier's role in setting needs.

Performance Comparative Chart IIIB *Exterior Metal Primers.*

GPC Specification	GPC 58	GPC 54	GPC 54	GPC 59	GPC 56	GPC 57	GPC 60
Specification Source	TTP 1046	TTP 641 Type I	TTP 641 Type II & III	Proprietary Formula*	Proprietary Formula**	Proprietary Formula†	Proprietary Formula‡
Binder type	Chlorinated rubber	Linseed oil	Phenolic	Epoxy polyamide	Linseed alkyd	Epoxy ester	Inorganic silicates
Suitable surfaces	Metal; primer & topcoat	Galvanized surfaces; primer & topcoat	Galvanized surfaces; primer & topcoat	Metal; primer & topcoat	Metal primer & topcoat	Metal primer & topcoat	See Sect. 110 & 110.1
Pigment content	90% by wt.	78–81% by wt.		92.5% by wt.		40% PVC	
Special pigments	Zinc dust	Zinc dust & ZnO	Zinc dust & ZnO	Zinc dust	Basic silico lead chromate	Basic silico lead chromate & red iron oxide	
Gloss		Egg & semigloss	Egg & semigloss	Flat	Flat	Flat	
Set-to-touch, hrs.	1/4		1/2–4	1/6			
Recoat time, hrs.		18	18				

Selection and Specification—Exterior

Property						
Abrasion resistance, Taber, 1000 cycles			78 mgs. loss			
Impact resistance						
Weather resistance				5 yrs. New Jersey	3-1/2 yrs. South expos.	
Salt spray resistance	Very good					
Solvent resistance				Very good	Excellent	
Water resistance		24 hrs. @ 20-30° C	24 hrs. @ 20-30° C	3 months, fresh & salt	9 yrs. equiv.	1-1/2 yrs. tidal
Flexibility		1/8"	1/8"			
Surface preparation references	Same as TTP 86, Type I	Par. 304-306, 309	Same as TTP 641 Type I	Par. 304.3; 309.2c	Par. 304.3; 309.2c	Par. 307.4
Deterioration of coatings references		Par. 607.2	Par. 607.2	Par. 601, 603, 607		Par. 607
Primer required					Self	Self

*Matthiesen & Hegeler
**N L Industries
†Ciba-Geigy Corp.
‡New Jersey Zinc Co.

Performance Comparative Chart IV *Exterior Metal Topcoats.*

GPC Specification Specification Source	GPC 62 TTE 485	GPC 44 TTE 489	GPC 45 TTE 490	GPC 46 TTE 522	GPC 53 Proprietary Formula*	GPC 22 Proprietary Formula†
Binder type	Alkyd	Alkyd	Silicone alkyd	Phenolic	Vinyl chloride	Moisture-cure urethane
Suitable surfaces	Sheet metal; primer & topcoat	Metal	Metal topcoat	Topcoat	Metal topcoat	Metal, wood, cementitious; topcoat
Coverage, sq.ft./gal.				500		
Pigment content					23.5% by wt.	
Gloss	Semigloss	High gloss	Low gloss (eggshell)	Low gloss (eggshell)	High & semigloss	High & semigloss
Set to touch, hrs.	3	2	2	2	1/3	1/2-2
Dry through, hrs.	72	8	8	2	1	5-14
Hardness, pencil					2B	
Abrasion resistance, Taber					19.2 mgm. loss	12 mils loss, 2000 cycles
Impact resistance					Exceeds 80 in.lbs.	160 in.lbs.

Weather resistance	168 hrs. accel.	168 hrs. accel.	300 hrs. accel.	300 hrs. accel.	400 hrs. accel.	400 hrs. accel.
Salt spray resistance, hrs.	192			300	700	1,000
Alkali resistance					Excellent	Excellent
Acid resistance					Excellent	Fair
Solvent resistance, hrs.	4	4				Excellent
Water resistance, hrs.	18	18	18	21	Salt & fresh 3,000	
Permeability					Low	
Stain resistance						Excellent
Flexibility	1/8"		1/4"		1/8"	
Washability		Depends on service	Depends on service	Depends on service		
Primer required	Self				Par. 309.4	Self
Surface preparation references	Par. 304–306; 309	Par. 304–306; 309	Same as TTE 489		Par. 309.2c	Par. 309.2c
Deterioration of coatings references	Par. 607.2	Par. 601–604; 604.5c; 607	Same as TTE 489	Same as TTE 489	Par. 607	Par. 607

*Glidden Coatings & Resins.
†Hughson Chemical Co.

Performance Comparative Chart V-A Exterior Wood Surfaces.

GPC Specification Specification Source	GPC 63 TTP 25 Wood primer	GPC 68 TTS 176 Sealer	GPC 67 TTP 52	GPC 41 TTP 105 Primer-Topcoat	GPC 64 TTP 37 Wood trim (lead free)
Binder type	Linseed oil & coumarone	Varnishes or oils	Alkyd	Linseed oil	Alkyd
Suitable surfaces		Wood sealer	Shakes, shingles & siding		Trim
Pigment content	Lead carbonate & Lead sulfate*	Lead-free for home use	42.5% by wt. TiO$_2$	59-61% by wt.	
Special pigments			Zinc oxide 10%	Zinc oxide 34% of pigment	
Gloss		Even sheen	Semigloss		High
Set-to-touch, hrs.	1/2-4	1-2	3		3

Selection and Specification—Exterior

Recoat time, hrs.	18			18	18
Adhesion	Tape test—1/8" pickup	Knife test			Knife test
Hard dry, hrs.			3		
Weather resistance		Within 18		200 hrs. accel.	168 hrs. accel.
Water resistance	Whatman** 2 paper— 3/8" migration	24 hrs. immersion	18 hrs. distilled	1 in. Whatman** paper	
Stain resistance		Resist blue-black ink			
Flexibility	1/8"		1/2"		1/8"
Surface preparation references	Par. 308	Par. 308	Par. 308	Par. 308	Par. 308
Deterioration of coatings references	Par. 605 & 608	Par. 605 & 608	Par. 605 & 608	Par. 605 & 608	Par. 605 & 608

*Lead pigments will be banned from household paints after Jan. 1, 1973. Substitute formulations will be available.

**A given amount of paint is applied on Whatman filter paper and a specified amount of water is poured on the paint. The water reaching the filter paper and spreading is measured according to the specification.

Performance Comparative Chart V-B *Exterior Wood Surfaces.*

GPC Specification	GPC 71	GPC 72	GPC 23	GPC 70
Specification Source	HUD-HM HMG 7482.1§ Lead-free wood primer	Stain Inhibiting Formula PR-26* Lead-free primer	TTP 19	Proprietary EC 167†
Binder type	Alkyd	Acrylic emulsion	Acrylic emulsion	Vinyl urethane
Suitable surfaces				
Coverage, sq.ft./gal.		450		
Pigment content	PVC 39-41	PVC 21.8	70% rutile TiO$_2$	PVC 15-45
Special pigments	Magnesium silicate & Montmorillonite	Additive PR-26 & Zinc oxide		Antimony oxide‡
Gloss	50 max.	2		
Set-to-touch, hrs.			1/4	7 minutes
Recoat time, hrs.				

Selection and Specification—Exterior

Hard dry, hrs.	18	1	10
Abrasion resistance		200 oscillations	
Weather resistance	Good (simulated)	Good	10 yrs. Weather-Ometer equivalent
Water resistance	1/8" absorption in[w] Whatman paper	Good	Tough alternate wet, dry, hot & cold tests
Stain resistance	Good		
Surface preparation references	Par. 308	Par. 308	Par. 308
Deterioration of coatings references	Par. 605 & 608	Par. 605 & 608	Par. 605 & 608

[o] Rohm & Haas Co.
[†] Baker Castor Oil Co.
[‡] Antimony oxide will be banned by Federal regulations after Jan. 1, 1973, substitutes are available.
[§] Housing and Urban Development Dept, Washington, D.C.
[w] See note p. 271.

chapter 8
Selection and Specification of Coatings—Interior Surfaces

800 General. Coatings formulated to protect interior surfaces are less specialized, in general, than those for exteriors. Some exceptions are anticorrosion primers for metal; wood sealers; and certain primer-sealers for porous substrates. However, most general-purpose interior topcoats and high performance materials can be used on cementitious, metal, or wood surfaces. (Floors will be covered separately in Section 808.)

801 Basic Purposes of Interior Painting. Coatings used on interior surfaces are not subject to the rigors of sun, wind, rain, sand, dust, dirt, and fungus against which exterior paints must contend. Interior paints are intended to:

1) Improve appearance
2) Enhance sanitation efficiency
3) Aid illumination and visibility
4) Protect substrates
5) Boost safety.

The listing above cannot be set in order of importance, because every property owner would assign a different ranking to each purpose. The housewife usually puts appearance first in living room, dining room, and bedrooms. If she thinks about it, she would probably put sanitation efficiency first in selecting a coating for her kitchen, while putting that and protection of substrate first for a bathroom coating, since humidity and

Fig. 801A. High Visibility in an Interior Industrial Setting.
The white ceiling reflects enough light to aid illumination and yet does not cause glare. A semigloss finish was specified for the open supporting members and a high gloss for the high ceiling. A low ceiling would have required a semigloss to avoid eyestrain of workers in this bakery.

mildew often cause problems. For her den, where she may sew, embroider, or read, illumination and visibility may be uppermost in her rating.

Similarly, a specification writer will have difficulty ranking the order of importance, whether he is specifying for a home, an institution, or an industrial plant. Dairies, food packing plants, atomic energy installations, or biological laboratories or hospitals would require considerations of illumination, impact resistance, and easy cleanup. Machine shops require illumination, visibility, and protection of substrate; and the specifier has to bear in mind the potential glare of some coatings if he wants his client to provide a headache-free environment in which his workers can function efficiently and without fatigue-inducing eye strain. See Fig. 801A. In most industrial

Fig. 801B. High Impact Resistance Requirements.
Tough coatings able to withstand the roughhouse tactics of youngsters is required for school stairwells (left); *hospital corridors* (right) *likewise must be able to withstand impact and rapid, frequent, and thorough cleaning.*

plants or institutions, lower sections of walls likely to be bumped by carts, barrels, pallet-trucks, or boxes should be protected by impact-resistant coatings. See Fig. 801B.

Thus, interior coatings should be selected to fill a need. In this chapter, as in Chapter 7, Performance Comparative Charts will be used to list coatings providing various standards of performance. These will enable the specifier to know what is available from paint producers. Information available from these charts and the Detailed Specification Charts on which they are based will free the specifier from dependence on paint companies for recommendations of products their men believe are suitable. In writing specifications, if you use the charts within this chapter and at the end of the book properly,

you will have a ready fund of information on which to base pertinent requirements.

802 Cementitious, Metal, and Wood Coatings. So many different interior coatings can be used on all kinds of substrates that selection will be simplified by placing materials for general use all together. Later in the chapter specialized coatings for cementitious, metal, and wood substrates will be covered, particularly those primers and sealers required for specific substrates.

Interior coatings will be divided into two categories:

1) General purpose
2) High performance.

Each category includes coatings characterized as to sheen. These in order of gloss, or sheen, are *flat, eggshell,* and *semigloss* and *gloss,* which are known as enamels. See Tables 802A, and 802B, 802C, and 802D.

Flat and eggshell, in general, are used for walls and ceilings. Gloss and semigloss are mainly used for trim, molding, and doors. But there is nothing which limits the specification of

Table 802A. *Properties of Egg Shell Enamel Compared to a Premium Quality Flat Paint.*

	Acrylic Latex Egg Shell	Premium Quality Acrylic Latex
Equilibrated viscosity (hand stir)	82	86
Sheen (85° Gloss)	30	5
60° Gloss	11	5
Flow and leveling		
Brushed	Good	Good
Hiding properties		
Opacity	Equal	Equal
Brushed	Good+	Good
Film integrity (cycles)		
Machine	500	500
Manual-water	100+	100+
Ajax	65	30
Burnish resistance (manual)		
Dry	Fair	Good
Wet	Good−	Good
Stain removal	Excellent	Excellent
Ease of stain removal	Very good	Good

Source: Rohm & Haas Co.

Note: The sheen and gloss of the eggshell is significantly higher than the flat. The film integrity with abrasive cleaning (Ajax) was more than twice as good. Ease of stain removal was better, but dry and wet burnish resistance was not quite as good. This is characteristic of latex paints whose Pigment Volume Concentration is just below the critical PVC. The lower viscosity indicates more binder resin, less pigment.

gloss. Semigloss enamels have been selected for walls, and flats have been used for trim. It is partly a matter of taste. High gloss enamels, however, should be used sparingly for walls and ceilings, and only when circumstances demand their use, because resulting glare may lead to eye strain. When they must be used, if circumstances require their ultimate performance in sanitation, then light sources should be adjusted.

Eggshell, sometimes called satin or low-gloss, has the modest sheen of a clean egg. Its appearance is scarcely distinguishable from a flat, but it has the advantage of containing more resin than flat paints and less pigment and extender (lower PVC). Hence, eggshell paints generally outlast flats and should survive more washings. When specifying these, however, one should be careful to specify that it be able to take a reasonable number of rubs, such as those used to remove grease or stains, without

Table 802B. Properties of Premium Quality Flat Wall Paints.*

Acrylic Flat Emulsion Formulations vs.
Commercially Available Latex Flat Paints

	Recommended Formulations			Commercial Latex Flat Paints		
	Acrylic A	Acrylic B	Acrylic C	A	B	C
Pigment volume concentration	42.7	43.0	42.6	—	—	—
Flow and leveling:*						
Brushed	3	3.5	3	3	3	3
Rolled	3	4	3	3	3	3.5
Hiding properties:						
Opacity (contrast ratio)	0.967	0.955	0.963	0.945	0.922	0.964
Brushed hiding*	3	3	3	3	4	3
Rolled hiding*	3	3	3	4	4	3
Film integrity: (Nylon bristle brush and abrasive medium)						
Machine (cycles to failure)	920	990	1025	735	625	350
Manual (150 cycles) (Cheesecloth and Ajax)	Pass	Pass	Pass	Pass	Fail (70)	Fail (90)
Burnish resistance: (200 cycles)						
Machine (increase in 85° gloss)	2.2	1.9	2.0	0.7	7.0	12.4
Manual Wet*	3	3.5	2	2.5	2	5
Dry*	3	3.5	2	3	2	5
Stain removal:*						
Machine	3	2.5	2	3	1	1
Manual	3	3	3	3.5	3	3

*For purposes of this evaluation, Acrylic A was rated 3 for all subjective evaluations and used as a control. Rating is 1—best, 5—poorest.
Source: Rohm & Haas Co.

Table 802C. *Properties of Typical White Semigloss Paints.*

	New 100% Acrylic Semigloss Paint	*Solvent Alkyd Semigloss Paint*	*Early Latex* Semigloss Paint*
Water cleanup	Yes	No	Yes
Open-time (length of time a coat of paint remains amenable to brushing)	15 minutes	15–20 minutes	5
Flow and leveling (0 = poor, 10 = perfect)	4	5	0
Sagging (running)	Very slight to none	Fair amount	None
Gloss (60°)	53 to 63	45	45
Hiding	Fair	Fair plus	Poor
Dry adhesion to old enamels	Good	Fair	Fair plus
Wet adhesion to old enamels (after 1 week air dry)	Good	Good	Poor

Selection and Specification—Interior

Odor	Very slight	Strong solvent	Fairly strong
Resistance to yellowing and embrittling	Good	Fair	Good
Tack-free time (77° F., 50% RH, low air circulation)	4 to 5 hrs.	3 to 5 hrs.	20–30 minutes
Recoat time (77° F., 50% RH, low air circulation)	4 to 5 hrs.	7 to 16 hrs.	1 hr.
Fire hazard	No	Yes	No
Print resistance (after 1 week air dry)	Very slight	Trace to none	None
Stain removal	Good	Good	Fair

Source: Rohm & Haas Co.

*Recently developed PVA latex semigloss paints are also regarded highly. However, acrylic versions have had such a head start that they are used by most companies. Acrylic semigloss enamels are widely used in hospitals and hotels because of their ease-of-cleaning and because unskilled maintenance staffs can apply them. Superior high-performance epoxies and urethanes usually require skilled painters.

Table 80 D. *Semigloss Latex Comparison, Acrylic vs. PVacrylic.*

		Acrylic	PVAc
Viscosity	KU	93	89
pH		9.0	7.7
Ease of brushing	Score	7	9+
Ease of rolling	Score	10	10
Foam & cratering	Score	8	6
Leveling	Score		
Brushed		9	8
Rolled		8	7
Lapping	Score	7	9
Speed of dry			
Set-to-touch	min.	12	12
Tack free	hr.	3.0	4.5
Dry hard	hr.	4.5	4.5
Dry thru	hr.	5.3	O/N
Reflectance	%	92.5	91.0
Opacity	%	96.5	96.0
Gloss			
Drawdown—60°		60	54
Drawdown—20°		8	12
Brushed—Sealed		22	32
Brushed—Porous		15	20
Rolled—Sealed		17	20
Rolled—Porous		12	13
Adhesion to Enamel—# Hatch			
Dry	%	99	95
Blocking	Score	9+	9
Scrubbability	Cycles	1400	800

10 = Excellent;   0 = Poor

Source: Independent Testing Laboratory; appeared in *American Paint Journal* (Nov. 16, 1970).

Note: Comparisons were made to determine if lower-priced PVA Copolymers recommended for semigloss enamels performed as well as more costly acrylics.

significant increase in gloss. Some eggshell paints can take 200 rubs with little increase in gloss. This is particularly important for latex paints. A properly formulated eggshell paint will have sufficient binder-resin to minimize sheen, or burnishing, as it is also called. Burnish-resistance should be specified in an eggshell paint.

High-gloss enamels offer advantages in ease of cleanup because their hard, smooth surface discourages dirt and bacteria retention. Since glare may be objectionable, compromises may be necessary, and semigloss enamels may have to be used.

For the hard, resistant, easy-to-clean characteristics of an

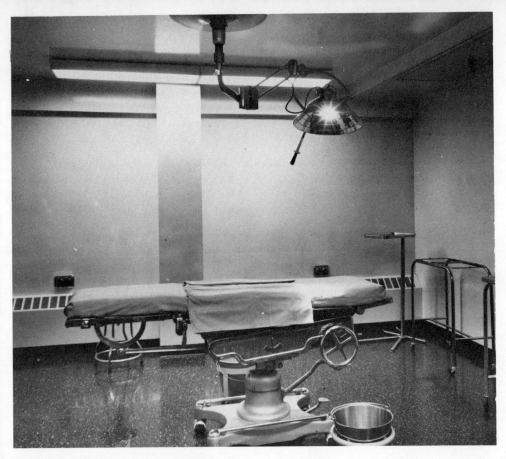

Fig. 802. Flat Enamel for Interior Use.
A flat enamel was used for this operating room where glare must be avoided and frequent scrubbing requires a finish that permits complete cleanup.

enamel, coupled, however, with a dull finish, a new coating has been perfected. It is described as a *flat enamel*, a seeming contradiction in terms. Actually, it contains the low Pigment Volume Concentration of an enamel (which means high binder-resin content) and all the advantages of an enamel, but it has a so-called flatting agent, usually a silica, which eliminates the gloss. The Housing and Urban Development Departmtment (HUD) has a paint based on this principle in one of its wall paint specifications. See Fig. 802.

803 General Purpose Coatings for Interior. For ordinary interior conditions, latex paints and alkyds divide honors. Flat wall paints are rapidly being monopolized by latex products because of their low odor and easy cleanup. Semigloss enamels, too, are going the latex route, mainly acrylics. Alkyd high-gloss trim paints, though, are likely to continue in favor unless air pollution restrictions tighten and all mineral spirits are banned.

Alkyds, in general, are moderately priced, dry well, and will adhere directly to most surfaces, usually with less surface preparation than is required for latex. They should not be used, however, on masonry or other alkaline materials unless an alkali-resisting primer or sealer has been applied first.

Latex flat wall paints are based mainly on polyvinyl acetate emulsions or acrylics. One major paint company still offers styrene butadiene emulsion paints. Latex paints offer ease of application, quick-dry, negligible odor, low fire hazard in open cans, and easy cleanup. Acrylic semigloss paints have become popular, but they have been challenged by PVA's. PVA, as well as acrylic eggshell (or satin) paints are available.

Many latex flat paints are self-priming on new surfaces unless they are porous. Manufacturers' instructions concerning priming should be followed, and this is equally true of alkyds. One important specification for primers for semigloss and high-gloss enamels is that they have enamel holdout, which means they will not rob the enamel of its solvent and/or binder upon application, and thus, cause dull spots to appear on the surface.

Washability of most of the coatings in the accompanying chart of performance characteristics is adequate. (See Performance Comparative Chart VI at end of chapter.) Scrubbability, though, varies among them from 3,000 strokes of a testing device for a flat latex, which is good, to as much as 10,000 for gloss paints. While few walls will ever get that kind of treatment, scrubbability is considered an index of durability. Stain resistance and water resistance of some of the coatings are moderately good for general conditions, but they trail *high performance* coatings, which will be discussed later, by a goodly margin in these and other performance characteristics.

Specifiers who trust blindly in big name paint manufacturers will have their eyes opened to the need to set up very definite specifications for ordinary garden-variety paints when they read of the performance tests made on leading U.S. paints by a major United Kingdom pigment manufacturer. Research

Selection and Specification—Interior

chemists in the company's laboratory were free of bias because they were unaware of the identities of the paint companies manufacturing the products. The wall paints, identified by number, tested out as follows:

	PVC	TiO_2 as % of PVC	Resin Type	Scrub* Resistance
US 3	53	26	Polyvinyl acrylate	Total removal 2,000 scrubs
US 4	50	21	Styrene acrylic	Slight removal 2,000 scrubs
US 5	45	20	Styrene acrylic	Total removal after 150 scrubs
US 8	47	24	Polyvinyl acrylate	Total removal after 850 scrubs
US 7	Eggshell 58	(Satin) 21	Polyvinyl acrylate	Total removal after 500 scrubs
US 10	30	30	Acrylic	Slight removal after 2,000 scrubs
US 2	Semigloss 25	25	Styrene acrylic	Film unharmed after 2,000 scrubs

*2,000 scrubs on REL machine; used nylon bristle under 200 gms. pressure.

Source: Sales manager, United Kingdom pigment supplier, by letter communication.

Of the flat wall paints, only US 3 and US 4 reached a full 2,000 scrubs. US 4 showed up somewhat better than US 3, but not by a wide margin. US 5, although it was made by one of the biggest firms in the business, failed after a mere 150 scrubs; but it cost more than US 3 and US 8. US 4 cost about the same as US 5 but it survived 2,000 scrubs as against 150 for US 5.

From this we should learn the importance of specifying scrub resistance as an index of durability although few flat wall paints will ever be scrubbed 2,000 times.

The loose terminology used also emphasizes the importance of careful specification. US 7, described on its label as a *Wall Satin*—a term synonymous with eggshell—had a higher Pigment Volume Concentration (58) than the PVC of any of the flat wall paints tested by the British company (and satins, or eggshells should have substantially lower PVC's than flats). US 10, on the other hand, also described as a Satin, had a PVC of 30, which indicates that it has adequate binder-resin to impart durability, as one has a right to expect from an eggshell-satin.

US 10 survived 2,000 scrubs with slight removal of film. US 7 failed after 500 scrubs. Both are famous, high-rated brands in the public's mind; both cost about the same.

Notice that US 2, a semigloss, had the lowest PVC of all of the interior paints tested. It survived 2,000 scrubs unharmed—the only one of those tested that did so. It is interesting to note that the same manufacturer produced US 8, which survived 850 scrubs, and US 2, which lasted through 2,000. The difference in price is not significant in view of the life expectancy; however, not everyone will want a semigloss on his walls. An eggshell with specifications embracing the scrub resistance of US 10, an eggshell based on an acrylic latex, and the general characteristics described in Table 802a should assure a livable level of gloss and outstanding durability—at little, if any, higher cost.

The important lesson here is: Don't specify by brand alone. Be specific! Then demand that general contractors hold to your specifications.

Getting adherence to your specifications may help overcome some complications. Painting contractors who successfully bid for your job may not be able to supply products from the particular paint companies if you limit company selection in your painting schedules. Dealers supplying products of these manufacturers may not offer credit to the chosen painting contractor who prefers to go to the dealers who know him and provide credit; and they may not sell the specified brands.

By giving performance and component specifications, you oblige the painting contractor to find out which of his paint dealers' brands can meet these specifications; and he makes his bid accordingly.

Otherwise, what often happens is that he will make every effort to use brands supplied by the dealer who gives him credit; and he has to give the general contractor and the owner a big song and dance about the substitute brand's equivalency, while ignoring the important details about its performance characteristics.

If the specification writer spells out the performance characteristics, the painting firm will be obliged to meet them, at

Selection and Specification—Interior 287

pain of not getting paid if its men apply material of lesser quality.

804 High-Performance Interior Coatings. When a specifier is working on a project for professionals who understand the economies of quality coatings, if circumstances require them, several families of excellent high performance materials provide a choice of characteristics to fit just about every need. See Table 804.

A satisfactory service-life of three to six years may please the average user of the alkyds and latex paints described in the previous pages; but when we come to high performance coatings, we talk in terms of 10 to 25 years. On the other hand, we talk about two-tenths of a cent for a 1-mil coat per square foot when we talk about alkyds; but for high performance coatings—epoxies, urethanes, epoxy-polyesters, and polyesters—we have to spend from one-half to seven-tenths of a cent.

For some high-performance applications, notably on metal and some hard, smooth concrete surfaces, we have to be extra careful about surface preparation and may even have to use sandblasting or chemical pretreatment—depending on conditions.

However, when we realize that a properly prepared surface and a high performance coating may last four or five times as long as a general-purpose coating, with or without proper surface preparation, it's easy to see why the more costly materials are to be preferred. See Fig. 804.

First of all, the extra cost of the top-grade coating for one application is only a fraction of the cost of that for four or five less expensive ones, which would be applied at five- or six-year intervals.

Second, the cost of surface preparation and application is met only once over a period of 10 to 25 years, instead of once every five or six years.

Third, the interruption of business or production that is necessary every five or six years with general purpose materials is delayed.

Performance Comparative Chart VII, for high performance interior coatings, is filled with excellent ratings for many characteristics of the ten materials analyzed. See end of chapter.

Table 804. Test Data for High-Performance Coatings.

For test purposes the films were spray applied to steel "Q" panels with the exception of the polyester/peroxide system, which was brush coated to its limited working time of less than one hour. All films were cured at room temperature for seven days at 25° C (76° F) before testing. Dry film thickness is 1.0–1.5 mils except for the polyester/peroxide system which had a dry film thickness of 6.0 mils.

Film Properties	Epoxy Polyamide	Epoxy Ester	Urethane Polyester	Urethane Castor	Polyester Peroxide	Polyester Oxirane
Rate of dry, hours:minutes						
Set-to-touch, minutes	0:10	0:10	0:20	0:85	0:20	0:50
Cotton free, minutes	0:20	0:30	0:50	1:50	0:30	3:00
Tack free, hours	2:40	3:50	3:10	10–15	1:00	10–15
Dry hard, hours	4:30	5:10	5:50	10–15	1:00	10–15
Dry through, hours	6:00	6:40	6:50	10–15	1:00	10–15
Hardness						
Pencil—1 day	F-H	3B-2B	H-2H	F-H	B-HB	HB-F
Sward—1 week	38%	26%	54%	26%	6%	28%
Sward—1 month	46%	38%	58%	30%	14%	32%
Sward—3 months	56%	48%	76%	52%	18%	44%
Gloss, 60°						
Initial	100	94	100	99	61	93
Adhesion—cross hatch percent of film retained	100	100	0	0	0	0
Distensibility, percent elongation	30	30	0	0	30	0
Direct impact resistance, in.lbs.	160	160	32	32	108	36

Reverse impact resistance, in.lbs.	60	100	0	0	16	0
Abrasion resistance, falling sand. liters/mil	54	53	49	43	47	36
Water resistance at 25° C (77° F) immersion time			3 weeks			
Effect	No. 8 few blisters; slight softening	No. 8 medium blisters; slight loss of gloss; softening	No. 6 dense blisters; softening	No. 6 medium dense blisters; slight discoloration; softening; loss of gloss	No. 6 dense blisters; softening; discoloration; loss of gloss	No. 8 dense blisters; softening;
Recovery—24 hours	Complete	Partial	Slight	Slight	Partial	Partial
Boiling water resistance immersion time, hours	4.0	0.5	0.3	2.0	0.3	1.0
Effect	No. 8 medium blisters; softening; slight discoloration	No. 8 dense blisters; softening	No. 8 dense blisters; slight loss of gloss; softening; slight discoloration	No. 8 medium dense blisters; slight discoloration; slight loss of gloss; softening	Film lifted	No. 8 dense blisters softening
Recovery—24 hours	Partial	Partial	Partial	Partial	Slight	Partial
Alkali resistance (NaOH, 10%) at 25° C (77° F) Immersion time, days	56	9	28	19	6	1

Table 804. *Continued*

Film Properties	Epoxy Polyamide	Epoxy Ester	Urethane Polyester	Urethane Castor	Polyester Peroxide	Polyester Oxirane
Recovery—24 hours	Excellent	Fair	Poor	Poor	Failed	Failed
Acid resistance (hydrochloric acid, conc.) at 77° F						
Immersion time, hours	24	48	5	24	5	5
Effect	No. 6 dense blisters; slight loss of gloss; moderate discoloration; softening	No. 8 dense blisters; moderate discoloration; slight loss of gloss; softening	Film lifted	Film lifted	Film lifted	Film lifted
Recovery—24 hours	Partial	Partial				
Gasoline resistance 3 months immersion	Very good	Good	Very good	Very good	Very good	Very good
Ethyl alcohol immersion						
Days	11	11	11			
Results	Poor	Poor	Poor	Failed	Failed	Failed

Source: Ciba-Geigy.

Fig. 804. General Purpose Enamel Versus High-Performance Urethane Enamel.
In a large bakery, one-half of an area above the oven was coated with general purpose enamel, and the other (right) had a high-performance urethane enamel. The left side shows the general purpose product one year after application.

For quick, immaculate cleanup, most of the coatings on the chart meet the needs of hospitals, food plants, dairies, clean rooms of biological laboratories, and chemical and atomic energy plants where biological contaminants or other dangerous materials must be removed quickly.

Where abuse is common, such as in corridors and lobbies of public and industrial buildings and schools, epoxies and urethanes with incredible impact resistance will withstand the most brutal treatment.

Table 804 compares epoxy polyamide, epoxy ester, and urethane and polyester coatings using a wide variety of performance characteristics.

From either, or both charts, deciding which family of coatings is best for a particular set of circumstances should be simplified. The Performance Comparative Chart also provides a guide to the Detailed Specifications Charts at the end of the book, with indications as to the resin family involved. Thus, specifications can be identified from the first chart and then studied in greater depth in the Detailed Specification Charts.

804.1 Rate of Dry. With the exception of the urethane-castor oil coating and the polyester oxirane, dry-to-touch times are fairly rapid. Tack-free time, which is important because that's when contaminants in the air will no longer adhere, is best for polyester peroxides, but the urethane polyester and the epoxies rate reasonably well.

804.2 Hardness. Urethane polyester is far and away the hardest of the coatings, with epoxy polyamide a reasonably good second.

804.3 Adhesion. The high rating of both epoxy families enables them to be used as primers over a wide variety of substrates and is one of the big advantages of these materials.

804.4 Impact Resistance. While federal specifications for epoxy polyamides call for a mere 24 in.lbs., members of this family have little trouble meeting 160 in.lbs. on direct impact and 60 in.lbs. on reverse impact (which means the blow is struck from the rear) with the coating tested on a thin steel panel. Polyester peroxides also perform well on direct impact, but the particular urethanes on this test do not come out well here. However, as shown in Performance Comparative Chart VII, several reached 160 in.lbs. (See Fig. 804.4.)

804.5 Abrasion Resistance. All families do well in abrasion resistance, with the edge going to epoxy polyamides. Polyester epoxies, although not on this test chart, also have excellent abrasion resistance. Several proprietary moisture-cure urethanes rate best of all, but these were not included in the tests.

804.6 Water Resistance. After 3 weeks' immersion in water at 77° F, epoxy polyamide was the only family that recovered completely from the few blisters and slight softening that occurred. The same family withstood boiling water for four hours with modest damage, while the others lasted considerably less time.

804.7 Alkali Resistance. Immersion for 56 days in sodium hydroxide had no effect on the epoxy polyamide. The others performed poorly in the test, although it must be observed that tests of more diversified alkalis found the urethanes performing well.

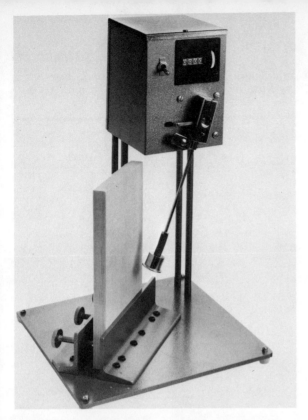

Fig. 804.4. Behm-Litter Flail Impact Tester.
Used for testing very high quality furniture finishes, this tester can be utilized for any wood finish. The pendulum flails the surface with a 100-gram rimmed weight until observable damage is done. The number of blows until damage occurs indictes the impact-resistance rating.

804.8 Acid Resistance. The epoxies were clearly superior with hydrochloric acid.

804.9. Gasoline Resistance. Three months' immersion in gasoline resulted in very good ratings for most of the high-performance coatings tested.

804.10 Ethyl Alcohol. Immersion for eleven days in ethyl alcohol found all materials doing poorly.

To extend test reports to polyester epoxies, see Tables 804.10A and 804.10B which show the performance of these materials as compared with epoxies and alkyds. They tend to confirm the test data cited above for epoxies and to prove the outstanding qualities of polyester epoxies.

Table 804.10A. *Chemical and Stain Resistance of Polyester Epoxy Systems vs. Epoxy Polyamide and Alkyd Coatings.*

Chemical or Solvent	Exposure Time (hours)	Polyester Epoxy	Epoxy	Alkyd
Chemicals				
5% NaOH	4	10	10	0
	24	7	10	0
10% NH_4OH	24	10	10	3
5% Acetic acid	1	10	7	8
	4	10	2	6
50% Acetic acid	4	10	0	2
	24	10	0	0
Formic acid	1	10	0	–
10% HCl	4	10	7	9
	24	10	6	8
10% H_2SO_4	24	10	7	–
1% Tide	24	10	10	8
Solvents				
Gasoline	24	9	10	7
Xylol	4	8	10	8
	24	4	8	–
Water	24	10	10	8
Acetone	1	3	8	2
Ethyl alcohol 95%	1	7	9	5
50%	20	10	—	7

10 = Best

0 = Worst

Source: Ashland Chemical Co.

805 Cementitious Interior Surfaces. Protecting interior cementitious surfaces is relatively simple compared with exterior surfaces where rain, wind, sun, and sand are vicious antagonists. Indoors, the aim is serviceability within the demands of circumstances.

All of the recommended coatings for interior cementitious substrates have been met before in the discussion of the corresponding exterior surfaces, or in the discussion of coatings that can be used on all interior surfaces. See Performance Comparative Charts I and II at the end of Chapter 7. The three types are general purpose (where problems, if any, are minor), moisture resistant, and heavy duty.

Selection and Specification—Interior

Table 804.10B.

	Polyester Epoxy	Catalyzed Epoxy	Two-Component Urethane	Two-Component Polyester	Air Dry Alkyd
Resin cost for a dry film 1 mil thick ($ per sq.ft.)	.0050	.0050	.0070	.0030	.0020
Applied cost of 8-mil coating ($ per sq.ft.)	.25–.40	.30–.50	.40–.55	.20–.35	.15–.25
Estimated years of wear	10–25	10–25	10–25	10–25	5–8
Number of coats to obtain 8-mil film	2	3	2–3	1–2	3
Gloss retention	Excellent	Poor	Poor	Poor	Good
Pot life	16–48 hrs.	4–24 hrs	2–24 hrs.	1/4–2 hrs.	Unlimited
Nonyellowing properties	Excellent	Poor	Poor	Fair	Fair

805.1 General Purpose. Latex primers are usually recommended for cementitious surfaces because they are less sensitive to alkali and water than alkyds. Topcoats for flat finishes may be the same latex selected for the primer, which can range from a marginally acceptable federal specification paint, TTP 29 (GPC 1), to an acrylic with a relatively low Pigment Volume Concentration. Or the topcoat may be an ordinary alkyd, TTP 47 (GPC 3), or an odorless alkyd, TTP 30 (GPC 2).

805.2 General Purpose Gloss Coats. With a latex TTP 29-type latex primer, alkyd high gloss coats are available, TTE 506 (GPC 7), and odorless types TTE 505 (GPC 6) and 509 (GPC 9), may be selected. For semigloss, soy alkyds, TTE 509 (GPC 9), or TTE 529 (GPC 47), latex, or an acrylic semigloss can be used as the basis for specifications.

805.3 Moisture Resistance. The excellent federal specification TTP 95 (GPC 28) is recommended by many experts as an all-purpose specification for moisture problems on cementitious surfaces. Class C is suitable for flat finishes, Class B for semigloss, and Class C for high gloss. TTP 97 (GPC 29), which dries hard faster than TTP 95, may be preferred. Both are self-priming, and both require fillers on rough masonry.

805.4 Heavy-Duty Conditions. Where heavy buffeting is likely by trucks, trays, or kicking heels of exuberant youth, the specifier may turn to gloss epoxy polyamides TTC 535 (GPC 13), or urethanes, TTC 542 (GPC 20), for primer and topcoat. All of the high-performance coatings listed in the general interior section will function on cementitious surfaces if circumstances such as chemical exposure demand any of the virtues that they have. Where a glaze coating is desired, TTC 550-type (GPC 21) materials are available. When textured, flat finishes are preferred, TTC 555 (GPC 30) may be used. (TTC 555 is also listed in the exterior cementitious performance charts because of its ability to withstand strong winds, rain, and water pressure.) See Performance Comparative Charts I and II at the end of Chapter 7.

Interior cementitious surfaces, like exteriors, require fil-

Selection and Specification–Interior 297

lers where surfaces are rough and porous, such as cinder block or concrete block. Federal Specification TTF 1098 (GPC 33) covers filler-sealers for this purpose. They are based on styrene-butadiene. Heavy-duty fillers based on epoxy esters are also available. Low-priced latex fillers are offered at about two-thirds the price listed for the TTF 1098 and epoxy ester types. These are preferred by some painting contractors, but the success of just about all jobs requiring fillers depends on superior materials.

805.5 Sealers. Several proprietary sealers are on the market, which serve to protect surfaces from water and at the same time provide durability. One low-priced product is a vinyl acrylic whose producer claims that it has survived 50,000 scrubs. The material seals and finishes cement block, drywall, plaster, and precast concrete. In addition it is described as hiding taped joints and nail patches on drywall.

Another masonry sealer with high solids is a clear epoxy polyamide with excellent adhesion to masonry and good water and alkali resistance.

For successful coating of plaster and cinder block and any porous cementitious material, it is important that moisture working from behind the surface is fended off. Sealers help, but primary protection from this potential troublemaker is to correct the condition that lets water get behind.

805.6 Sanitation, Extreme Conditions. The interior walls of pharmaceutical laboratories, dairies, food plants, and chemical firms, in many instances, must be cleaned meticulously. Formerly, tile surfaces were used, but the grout used to cement the tiles has now been recognized as offering hiding places for germs. At present, tilelike finishes made of epoxy, urethane, or epoxy-polyester are used to provide unbroken spreads of cleanable surface where nothing can resist the cleansing action of detergents and water.

805.7 Atomic Energy Plants. While most of the high performance coatings listed in Performance Comparative Chart II (see end of Chapter 7) are suitable for surfaces needing meticulous cleanup, specifiers working on atomic energy plants should be aware that coatings may have been formulated

Fig. 806A. Metal Interior Surfaces in School Kitchen.
Metal coatings for this school kitchen required heat resistance and easy cleanability.

especially for these installations. Many of them are epoxy polyesters, which may cost more than $30 per gallon and may only cover 130 sq.ft. per gallon. Adding about $0.12-.15 per sq.ft. cost of epoxy ester filler, we come up with a material cost in excess of $0.50 per sq.ft. However, they are very durable. Although no incidents of escaping atomic materials are normally expected, these are regarded as coatings likely to succeed in permitting washup without damage if radioactive leakage should occur. High quality, two-component polyurethanes have also been used in atomic energy plants.

806 Metal Interior Surfaces. Topcoats for interior metal have much in common with other interior topcoats. However, three important differences bear watching. First, and most obvious, is the need to keep moisture from the substrate. Second, metal surfaces are often flexible and require coatings that can take flexing. Third, metal surfaces are generally smooth, so some

Fig. 806B. Metal Interior Surfaces in Brewery.
Shutdowns in this brewery are costly. High performance metal coatings are able to take impact and withstand frequent washing.

sort of roughing is required to obtain satisfactory adhesion. See Figs. 806A, 806B, and 806C.

Specifiers should acquaint themselves with the sections on surface preparation of metal (especially Section 309), because this is as important in metal specifying as selection of material.

In studying coatings on Performance Comparative Chart VIII for high performance materials, specifiers will note that all except one calls for a 1/8-inch mandrel bend as a flexibility test. Unless the surface is thick and inflexible, this requirement is important. Epoxies and urethanes are usually recommended for high-performance service with interior metals, but the more expensive polyester-epoxy coatings may be needed where extreme hardness, washability, and abrasion resistance are indicated.

806.1 Primers. Since protection of metal is to a great degree determined by pigments used in the primer, four basic metal

Fig. 806C. Metal Interior Surfaces in Dairy.
In this dairy, high performance metal coatings are used for rapid cleanup. (The brick floor provides a refuge for germs. A tough, seamless coating would have aided sanitation.)

primers for interior use are listed in Performance Comparative Chart IX. For extreme conditions, most of the primers listed for exterior metal use are also suitable.

806.1a Primers, Lead Content, Caution. In specifying primers for homes, it is important to remember that lead content of paints used in or near households is limited by federal law and you will not be able to select primers or topcoats with chrome yellow (lead chromate), red lead, or basic silico lead chromate. Household useage of metal surfaces is not very extensive and usually not extremely demanding, so a primer such as TTP 659 (GPC 43), which uses titanium dioxide and barytes for protection is probably adequate for many interior metal priming jobs. Numerous paint companies have formulations with zinc dust that can be used for conditions that demand high quality protection. Since problems are mainly involved

Selection and Specification—Interior 301

with moisture, systems using TTP 1046 (GPC 58) as primer and two or three coats of TTP 95 (GPC 28) as topcoats will meet the safety requirements of federal anti-lead regulations and provide excellent protection, but at prices higher than those cited above.

806.1b Lead Safe in Industrial Areas. Metal protection in factories, utility plants, or in large spacious production areas with exposed steel beams can be obtained with coatings containing lead, since these structures are not covered by federal hazardous substance restrictions.

806.2 Topcoats. A wide variety of flat topcoats are available to fulfill the requirements of Performance Comparative Chart VI, covering general interior coatings. Two soya alkyds TTE 505 (GPC 6) for high gloss and TTE 509 (GPC 9) for semigloss are listed on Performance Comparative Chart VIII. An acrylic latex, with the added feature of quick-drying is also listed as a semigloss. A fairly quick drying alkyd, TTE 506 (GPC 7) is also listed.

Where moisture-resistance is important, TTP 95 (GPC 28) is recommended as a topcoat for TTP 1046 (GPC 58). These may be made of chlorinated rubber or styrene acrylate, both classified as rubber. Data on chlorinated rubber appears in Table 806.2.

Specifiers interested in the improved performance and economies of shop coating should refer to Section 703.4, which deals with the cost of surface preparation for exterior metal surfaces.

807 Interior Wood Surfaces. Surface preparation is of prime importance in obtaining satisfactory results with any of the several types of coatings for interior wood. Careful specifications for sanding are needed whether the coating is clear or pigmented (see Section 308), and fillers are necessary for certain kinds of wood, mainly open-grained hardwood to be coated with clears or stains.

807.1 Transparent Finishes. The natural beauty of wood has long been a favorite of homeowners with either simple or sophisticated tastes. To preserve these decorative natural sur-

Table 806.2. *Properties of Chlorinated Rubber.*

Thermal	
Burning rate	Nonflammable
Effect of dry heat on film (continuous exposure)	Stable up to and at 125° C.
Softening point of film, ° C	140
Physical-Chemical (clear, unplasticized film)	
Sunlight, effect of	Discolors and embrittles
Cold water, effect of	None
Hot water, effect of	Slight blush
Moisture absorption at 80% relative humidity in 24 hours, %	0.14
Water-vapor transfer of free film, grams/100 sq.in./mil in 24 hours at 95° F. and 100% R.H.	1.0
General Resistance to:	
Acids, weak	Excellent
Acids, strong	Excellent
Alkalies, weak	Excellent
Alkalies, strong	Excellent
Salt spray	Excellent
Alcohols	Excellent
Ketones	Soluble
Esters	Soluble
Hydrocarbons, aromatic	Soluble
Hydrocarbons, aliphatic	Softened slightly

Oils, mineral	Good	
Oils, animal	Poor	
Oils, vegetable	Poor	
Mechanical (ASTM methods used for all tests)	Unmodified Parlon S125	50% Parlon S125 50% Koppers Rezyl 869
Tensile strength, lb./sq.in.	5200	
Elongation, %	1.6	
Flexibility, MIT double folds	139	
Hardness, Sward index, % of glass	70–80	
Electrical (ASTM methods used for all tests) (All tests on clear film)	Unmodified Parlon S125	50% Parlon S125 50% Koppers Rezyl 869
Volume resistivity, ohm-cm.	3.4×10^{15}	0.85×10^{15}
Dielectric strength, volts/mil (7-mil films)	2200	2200
Dielectric constant at 25° C.		
60 cycles	2.1	2.5
1,000 cycles	2.1	2.5
1 megacycle	2.4	2.9
Power factor at 25° C.		
60 cycles	0.0055	0.016
1,000 cycles	0.0025	0.011
1 megacycle	0.008	0.021
1,000 cycles, after immersion in water for 140 hrs. and surface wiped dry	0.0027	—

Source: Polymers Department, Hercules, Inc.

faces and at the same time reveal the beauty of their grain, clear varnishes and stains, which are varnishes with small amounts of dyes or pigments, are used.

807.1a Clear Transparent Finishes. Two specifications meeting the requirements for clear transparent wood finishes are TTC 540 (GPC 18), TTV 109 (GPC 71), and TTV 85 (GPC 70), outlined on Performance Comparative Chart X. TTC 540, a polyurethane modified with linseed oil is more costly than the others and should only be specified when service conditions justify such qualifications as high impact and abrasion resistance, and ability to withstand certain acids and alkalis. Formulations covered in the four specifications can serve as primers and topcoats.

807.1b Stains. A stain similar to TTS 711 (GPC 69) is generally available from all paint manufacturers. It's composed of a drying oil with added earth colors combined to supply the desired tone. Some stains use organic dyes, but these are not as sunfast as the metallic oxides.

807.1c Sanding. Prior to application of clear transparent finishes or stains the surface must be as smooth as possible. Unless the specifier knows that wood surfaces have been thoroughly smoothed at the factory, he should require careful sanding. (See Section 308.) If the surface requires rough sanding, No. 80 Grit, Grade 1/0, should be specified. Preparatory sanding on hardwoods requires No. 120 Grit, Grade 3/0, or fine. Preparatory sanding on softwoods requires either 100 or 120, Grade 2/0 or 3/0, fine. Finish sanding on hardwood requires 220 to 280 Grit, Grades 6/0 to 8/0, very fine, and extra fine; and finish sanding on softwoods needs 220 Grit, Grade 6/0, very fine.

807.1d Wood Fillers. Open-grained hardwoods, such as ash, chestnut, elm, hickory, mahogany, oak, teak, and walnut, require wood fillers when they are to be coated with clears or stains. Stain is applied before the filler and must be allowed to dry 24 hours before filling is done.

The filler may be either natural, if no staining is to be done, or may be colored with some of the stain to accentuate

Selection and Specification—Interior

the grain pattern of the wood. A word here about applying fillers. The material should be thinned with mineral spirits until it becomes somewhat creamy. It should be brushed liberally *across* the grain and then lightly brushed off with the grain. When most of the thinner has evaporated, after about five or 10 minutes, it will lose its gloss and become mealy. At this time, before it has set and hardened, use burlap or other coarse cloth to wipe it off *across* the grain while rubbing filler into the pores.

Finally, the operation should be finished with stroking along the grain, using clean rags. All excess filler must be removed.

Specifiers, just in case the painting contractor doesn't assign an experienced journeyman to the job, should stipulate that wiping begin at the right time, since this is important. He should specify a simple test for determining when to start wiping. The journeyman rubs his finger over the surface. If a ball of filler forms under his pressure, the time to wipe has come. If the filler slips as the finger rubs, it is too soon. Allow fillers to dry 24 hours before finish coats are applied.

807.2 Pigmented Finishes. Before applying an opaque finish to wood, a smooth surface is required, just as it is with a transparent finish. In addition, the first coat, or primer, when it dries completely, should be lightly sanded with 5/0 garnet or 3/0 flint paper. The purpose is to remove dust and debris that may have settled on the primer while it dried.

Knots and sap-moistened areas in the wood should be treated with sealers, represented by TTS 176 (GPC 68) to prevent bleed-through by natural dyes in the wood.

For general purpose alkyd semigloss enamels, TTE 543 (GPC 10) may serve as primer for topcoats of TTE 508 (GPC 8), or TTE 529 (GPC 47); and for high gloss, it serves as primer for TTE 506 (GPC 7), for whites and tints, and for TTE 489 (GPC 44) for colors.

Odorless semigloss alkyds are available based on TTE 545 (GPC 12) as primer and TTE 509 (GPC 9). The same primer can be used with TTE 505 (GPC 6) for odorless high gloss enamels.

807.2a Heavy Duty. A heavy-duty pigmented gloss for wood can be based on TTC 535 (GPC 13) serving as primer and topcoat, or TTC 542 (GPC 20) pigmented. For a glaze, a topcoat based on TTC 550 (GPC 21) can be applied over a compatible primer.

808 Floor Finishes. The grinding of leather and rubber soles and heels is harsh punishment for floor finishes in homes and institutions, but even that is mild compared to the abuse to which most industrial floors are subjected. Tires on motorized pallet trucks, the scraping of loaded pallets, and the plodding tread of vigorous men with heavy industrial shoes make it necessary to select coatings with maximum abrasion resistance for the floors of industrial structures. See Fig. 803.

For ordinary household and institutional conditions, we have alkyds and phenolics, and for heavy duty we have urethanes, epoxy esters, coal-tar epoxies, coal-tar urethanes, and epoxy polyamides. Concrete, metal, and wood floors use many of the same topcoats, but primers are different.

808.1 Concrete Floors. For general purpose concrete floors, two specifications can be used as guides. One is TTE 487 (GPC 73), and TTP 91 (GPC 74). The first will also be found useful for metal and wood floors. It provides a moderately demanding performance specification for indoor and outdoor use. It is not based on any particular resin, although in practice selection is made most often from various phenolics or alkyds containing a fair amount of tung oil or dehydrated castor oil, which impart toughness.

808.1a Intermediate Problems. Where wear is not excessive but water resistance and ability to withstand detergents, mild acid, alkali, and abrasion are required, a tough styrene-butadiene formulation is available in TTP 91 (GPC 74).

808.1b Heavy Duty. For heavy duty service on concrete floors, TTC 542 Type II (GPC 20), a urethane, can serve as both primer and topcoat finishes.

808.1c Damp Environments. A coal-tar epoxy similar to the proprietary product in GPC 75 is recommended for damp environments.

Other coal-tar coatings have been studied for damp environments by the National Association of Corrosion Engineers and described in *Materials Protection,* Vol. 1, No. 6, PP 97-100 (1962). The Society's researchers compared coal-tar epoxies

Fig. 808. Industrial Interior Floors.
Floors in this brewery would have been improved if the tile were replaced with concrete and a tough coating for easy cleanup.

and coal-tar urethanes. Their findings can be summarized as follows:

Coats: Single-coats, each with dry film thickness of 5 to 15 mils should be topcoated to eliminate pinholes.

Hardness and flexibility: These characteristics can be varied by adjusting components.

Abrasion and impact resistance: Rated from good to excellent for both. Coal-tar epoxies are harder and are particularly resistant to penetration by rocks and marine organisms.

Resistance to solvents: Both are good, but urethanes are better in splash resistance. Both have inconsistent records in resistance to aromatic, chlorinated, ester, or ketone solvents.

Resistance to neutral salts: Both are excellent.

Resistance to acids: Both are poor against oxidizing acids such as nitric and to lower aliphatics such as acetic; but the consumer is advised to try various formulations because some are good.

Weathering: Both show excellent weathering in all environments for more than five years. Both chalk badly. Both get harder with age, but the coal-tar urethanes are less susceptible to brittle-hardening.

Temperature resistance: Both resist dry heat from $-40°$ to $200°$ F, but continuous exposure to $200°$ F or higher is not recommended because of embrittlement. Complete immersion in hot water can be withstood better than steam. An upper limit of $150°$ F in complete immersion is recommended as a safe limit.

Flame resistance: Both will char, but neither will support combustion.

808.1d Epoxy Esters. One-package epoxy esters are also available for concrete floor finishing. Table 808.1d compares two epoxy esters based on high-rosin tall-oil fatty acids, and two phenolics, a urethane alkyd, and a soy alkyd. The two epoxy esters in this table offer the specifier an opportunity to select excellent abrasion resistance (T-8 HR) or excellent impact resistance (T-8 LR) combined with better water, gasoline, and alkali resistance and shorter drying time.

On the other hand, if quick drying and early hardness are important, note that the urethane alkyd dries hard and dries through more quickly than either epoxy ester, but that T-8 HR ends up with greater Sward hardness after one month; and selection of the urethane alkyd will sacrifice greater flexibility, which usually doesn't matter on floors.

It should be remembered that the epoxy ester floor finishes described in these charts are designed for moderately tough duty and are not considered expensive or very heavy duty in their wear characteristics.

808.1e Epoxy Polyamide. An epoxy polyamide floor finish is reported to have an improved abrasion resistance, rated by the same falling sand method used for the figures in Table 808.1d, 66.3 liters/mil, versus 33 for T-8 HR. Chemical resistance and gloss are also better. However, the raw material costs, and hence, the selling price, are about half again as much as for the epoxy ester formulation.

808.1f Non-Skid Finishes. To cut down the likelihood of slipping on hard floor finishes, formulators have added abra-

Selection and Specification–Interior

sive chips to enamels. They may be added to any of the floor coatings listed, but best results naturally are obtained with high performance, heavy-duty finishes.

808.2 Metal Floors. The epoxy esters and epoxy polyamides described in Section 808.1d and 808.1e are suitable for metal, as are the coal-tar epoxies and the general purpose coatings. The significant difference between metal floor finishes and others is the primer. For interior or exterior semigloss, TTE 485 (GPC 62) provides a tough primer and topcoat for general use, and TTP 86, Type III (GPC 51) provides a primer for a sizeable number of metal deck paints, which are not germane to this book, since they are mainly used on naval vessels. (For those interested in naval design, Military Specifications MIL-P-18210; MIL-E-18214; MIL-C81346; MIL-E-698 and 699; and MIL-W-5044 are references for metal-deck paints.)

808.2a Heavy Duty. For heavy duty use, TTC 542 (GPC 20) is recommended.

808.2b Corrosive Environments. A floor finish based on coal-tar epoxy or coal-tar urethane is recommended.

808.2c Surface Preparation. Section 309 should be reviewed; proper preparation is vital to durability of metal floors.

808.3 Wood Floors. In describing coatings for wood floors, we add an additional class. Clears must also be considered. For these interior gloss coatings we have a primer-sealer, TTS 176 (GPC 68) and clear varnish TTV 71, which provides a low-cost system for general purpose clears. When a low-gloss finish is desired, TTS 176 may be used as the topcoat as well.

For interior or exterior wood floors, a clear gloss consisting of a primer and topcoat using TTE 487 (GPC 73) is recommended.

808.3a Heavy-Duty Clears. Heavy-duty clears may be used on materials of the same class as TTC 542, Type I (GP 20) or any of the heavy duty materials described earlier in the other sections on floors.

808.3b Topcoats. At least two topcoats are recommended for finishing wood floors.

Table 808.1d. *Test Data on Paint for Concrete Floors.*
For test purposes the films were applied by spray application to steel "Q" panels and the dry film thickness is 1.3 to 1.5 mils. All films were cured seven days at room temperature before testing was initiated.

Film Properties	Epoxy Ester T-8 LR	Epoxy Ester T-8 HR	Phenol Modified Alkyd	Phenolic Varnish	Urethane Alkyd	Alkyd Medium-Oil Soya
Ease of brushing	Good	Good	Very good	Very good	Very good	Good
Leveling	Good					
Rate of dry, hrs.:min.						
Set-to-touch	0:45	0:45	0:30	0:55	0:20	0:45
Cotton free	3:00	1:30	3:10	2:40	0:45	3:00
Dry hard	6:30	6:00	8:00	6:30	2:30	6:00
Hardness						
Pencil—1 day	2B	2B	HB	HB	HB	3B
Sward—1 month	32%	46%	38%	39%	36%	28%
Conical mandrel flexibility, % elongation						
1 month	30	30	30	30	7	30
Direct Impact Resistance, in.lbs.						
1 week	160	36	28	96	68	160
1 month	160	24	20	56	44	160
Abrasion resistance						
Falling sand, liters/mil	31	33	25	35	28	23
Water resistance						
Immersion time, days	7	14	10	2	2	2
Effect	No. 9 dense blisters; very slight discoloration	No blistering; slight discoloration; slight	No. 8 dense blisters; slight discoloration	No. 8 dense blisters; softening	No. 8 medium blisters; softening	No. 8 dense blisters; very slight discoloration

Selection and Specification—Interior

moderate softening	softening	softening			moderate softening

Alkali resistance, 5% NaOH						
Immersion time, hours	2	2	2	2	2	2
Effect	No. 8 dense blisters; slight discoloration; softening	No. 6 dense blisters; very slight discoloration; softening	No. 8 dense blisters; slight discoloration; softening	Film lifted	No. 8 few blisters; slight discoloration; softening	No. 4 dense blisters; discoloration; loss of gloss; softening
Gasoline resistance						
Immersion time, days	14	14	14	7	14	7
Effect	Slight discoloration; slight loss of gloss; moderate softening	———	Discoloration; slight loss of gloss; softening	Film lifted	Discoloration; moderate loss of gloss; softening	Discoloration; slight loss of gloss; moderate softening
Detergent resistance, 1% "Tide" Solution						
Immersion time, days	2	8	1	2	2	1
Effect	No. 8 dense blisters; very slight discoloration; slight loss of gloss; moderate softening	No. 8 medium dense blisters; very slight discoloration; very slight loss of gloss; moderate softening	No. 8 dense blisters; slight discoloration; slight loss of gloss; softening	No. 8 dense blisters; very slight discoloration; very slight loss of gloss; moderate softening	No. 8 dense blisters; very slight discoloration; very slight loss of gloss; softening	No. 8 dense blisters; slight discoloration; slight loss of gloss; softening
Alcohol resistance, 50% Ethanol						
Time for film to lift, hrs.	48	48	1	1	1	48

Source: Ciba-Geigy Corp.
Note: For description of tests, see Chapter 4.

Performance Comparative Chart VI-A *Interior Primer, Sealers and Topcoats—General Use.*

GPC Specification Specification Source	GPC 1 TTP 29	GPC 2 TTP 30	GPC 3 TTP 47	GPC 4 *Proprietary Acrylic Flat**
Binder type	Latex (odorless)	Alkyd (odorless)	Oil alkyd	Acrylic latex
Suitable surfaces	Most walls	Most walls	Porous masonry, wood, etc.	Most walls
Coverage, sq.ft./gal.	630	450	630	
Gloss	Flat	Flat	Flat	
Set-to-touch, hrs.	1	1/2	1/2-2	1
Hard dry, hrs.		7	7	
Alkali resistance	Good			Good
Water resistance	Fair		Fair	Fair
Stain resistance				Good
Flexibility	1/4" rod			
Scrubbability, strokes	3,000			
Washability	Good	Good	Good	
Primer required	Self	TTP 47		Self with oil additive

*Rohm & Haas Co.

Selection and Specification—Interior

Performance Comparative Chart VIB *Interior General Use.*

GPC Specification Specification Source	GPC 9 TTE 509	GPC 11 TTP 1511	GPC 6 TTE 505	GPC 7 TTE 506
Binder type	Soya alkyd	Any latex	Soya alkyd	Alkyd
Suitable surfaces	Most walls and trim	Most walls and trim	All	All
Coverage, sq.ft./gal.		450		
Gloss	Semigloss	Semigloss	High	
Set-to-touch, hrs.	2–6	1/5	5–6	1/2–2
Hard dry, hrs.	12	6	8–12	16
Solvent resistance				Good
Water resistance				Good
Stain resistance	Good	Good	Very good	
Flexibility	1/8″ rod	1/8″ rod	1/8″ rod	1/8″
Scrubbability		10,000 strokes		
Washability		Good		Very good
Primer required	TTE 543	TTS 179		

Performance Comparative Chart VIIA *Interiors, High Performance, All Surfaces.*

GPC Specification Specification Source	GPC 4 TTC 545	GPC 14 Proprietary Formula*	GPC 19 Proprietary Formula†	GPC 13 TTC 535	GPC 16 Proprietary Formula‡
Binder type	Polyester epoxy	Polyester epoxy	Epoxy polyamide	Epoxy polyamide	Epoxy polyamide
Suitable surfaces	All	All	All	All	All
Coverage, sq.ft./gal.		265, 4-mil	200, 3-mil		
Gloss	Gloss and semigloss	Semigloss		High gloss	High gloss
Durability		High	Low	High	High
Set-to-touch, hrs.	1–6	1–6	3	1/2	
Hard dry, hrs	16	5	7 days	6	10 days
Hardness, Sward	28	28–36	2-H pencil	Very good	20
Weather resistance			551 hrs.		Excellent
Salt spray resistance			3,000 hrs.		Excellent

Selection and Specification—Interior

Property					
Alkali resistance	Very good	Very good	Excellent	Good	Very good
Acid resistance	Very good	Very good	Very good	Good	Good
Solvent resistance			Very good		
Water resistance	Excellent	Excellent	1,872 hrs.	Excellent	Outstanding
Chalk resistance	Fair	Fair	Fair		
Stain resistance	Excellent	Excellent	Excellent	Excellent	Excellent
Abrasion resistance	Excellent	Excellent		Excellent	Excellent
Flexibility		1/8"	1/8"		1/8"
Scrubability	Excellent	Excellent	Excellent	Excellent	Excellent
Washability	Excellent	Excellent	Excellent	Excellent	Excellent
Impact resistance	24 in.lbs.	24 in.lbs.	24 in.lbs.		160 in.lbs.
Primer required recommendation	Manufacturer's recommendation				

*PPG Industries.
†Glidden Durkee, Chem-Gard
‡Ciba-Geigy 7071.

Performance Comparative Chart VIIB *Interiors, High Performance, All Surfaces.*

GPC Specification Specification Source	GPC 18 TTC 540	GPC 17 Proprietary Formula*	GPC 22 Proprietary Formula†	GPC 20 TTC 542	GPC 21 TTC 550
Binder type	Polyurethane	Epoxy polyamide	Urethane	Urethane (moisture cured)	Not specified
Suitable surfaces	All	All	All	All	All
Coverage, sq.ft./gal.			480		
Gloss	Clear gloss	High	High	High	High
Durability	High				
Set-to-touch, hrs.	1		1/2-1	1-2-1/2	
Hard-dry, hrs.	8	10 days	5-14		7-15 days
Hardness, Sward	20			Very good	40-85

Selection and Specification—Interior

Property				
Alkali resistance	Excellent	Excellent	Excellent	
Acid resistance	Good	Very good	Excellent	
Solvent resistance	Outstanding	Good	Very good	
Water resistance		Outstanding		Excellent
Chalk resistance				Excellent
Stain resistance				
Abrasion resistance		Excellent		Excellent
Flexibility	1/8"	1/8"	1/8"	1/8"
Scrubbability	Excellent	Excellent	Outstanding	Excellent
Washability	Excellent			
Impact resistance	160 in.lbs.	160 in.lbs.	160 in.lbs.	

*Ciba Geigy 7072.
†Hughson Chemical Co.

Performance Comparative Chart VIII *Metal Interior Topcoats.*

GPC Specification	GPC 28	GPC 6	GPC 7	GPC 9	
Specification Source	TTP 95 (Moisture resistant)	TTE 505	TTE 506	TTE 509	Formula* IG-90-2
Binder type	Rubber	Soya alkyd	Alkyd	Soya alkyd	Acrylic latex
Gloss		High	High	Semigloss	Semigloss
Set-to-touch, hrs.	3/4	2-6	1/2-2	2-6	1/4
Recoat time, hrs.				12	
Hard dry, hrs.	24	12	16		4-5
Impact resistance			Very good		
Alkali resistance	Excellent				
Solvent resistance			Good		
Water resistance	Excellent		Good		
Stain resistance		Very good		Good	Good
Flexibility	1/4"	1/8"	1/8"	1/8"	
Scrubbability	500 strokes in alkali				Adhesion after 10,000 strokes
Primer required				TTE 543	Modified latex or alkyd

*Rohm & Haas Co.

Performance Comparative Chart IX Metal Interior Primers.

GPC Specification Specification Source	GPC 62 TTE 485	GPC 55 TTP 645 Type II	GPC 58 TTP 1046	GPC 43 TTP 659
Binder type	Alkyd	Alkyd	Rubber	Alkyd
Suitable surfaces	Metal; primer & topcoat	Metal; primer & topcoat	Metal; primer & topcoat	Metal; primer & topcoat
Pigment content	58% by wt.	57–64% by wt.		
Special pigments.	Chr. yellow, red lead, zinc oxide	Basic silico lead chromate	Zinc dust	
Gloss	Semigloss			
Set-to-touch, hrs.	3		1/4	1/6 to 2
Recoat time, hrs.			1	
Hard dry, hrs	72		1	
Water resistance	18 hrs., distilled	14 days, distilled		

Performance Comparative Chart X *Wood, Clear and Pigmented.*

GPC Specification Specification Source	GPC 18 TTC 540	GPC 71 TTV 109	GPC 69 TTS 711	GPC 70 TTV 85
Binder type	Polyurethane, linseed-modified, air-dried	Alkyd spar varnish	Drying oil	Alkyd spar varnish
Suitable surfaces	Wood, brick, metal	Wood	Wood; primer for stains	Wood
Pigment content	Clear unless iron oxide added for stain	Clear unless iron oxide added for stain	Earth colors	Clear
Gloss	High	High	Primer	Flat
Set to touch, hrs.	1	1–3		1–3
Recoat time, hrs.				7
Hard dry, hrs.	8	18	8	18
Hardness, pencil	20 Sward			

Selection and Specification—Interior

Abrasion resistance	50 mgms.			
Impact resistance	160 in.lbs.			
Alkali resistance	Ammon. & sod. hydrox.			
Acid resistance	Acetic			
Solvent resistance	Mineral spirits, xylol	4 hrs. immersion hydrocarbons	Mineral spirits, ethyl alcohol, turpentine	Gasoline 4 hrs.
Water resistance	72 hrs, distilled; 15 mins. boiling	19 mins., boiling		Boiling, 19 mins.
Flexibility		1/8"		1/8"
Surface preparation references	Par. 308	Par. 308	Par. 308	Par. 308
Deterioration of coatings references	Par. 605	Par. 605	Par. 605	Par. 605

chapter 9
Fire Retardant Coatings

900 General. Virtually all coatings, organic as well as inorganic, retard the spread of fire at least to a limited degree. Cellulosic lacquers are one major exception. Some coatings, such as epoxy polyamides, are difficult to burn, and others are actually nonflammable.

The term "fire retardant," as used here, will apply to those coatings that are specially formulated to slow the advance of fire and, just as important, to retard the effects of flame or intense heat by insulating them as long as possible from substrates coated with these special paints.

900.1 Fire Resistant Coatings. The term "fire resistant" will be used to describe those coatings that do not burn at all or do so only after prolonged exposure to intense flame and then contribute nothing to the volume of the fire.

900.2 Intumescence. One big feature of a *fire retardant* coating—one that retards the effects of intense heat—is intumescence. Fire resistant coatings *do not* intumesce.

Intumescence is the formation of a charred, swollen, foamlike material upon a coated surface that has been subjected to intense heat. Paints able to sacrifice themselves and form this char are called *intumescent paints*. Intumescent paints are formulated to protect underlying flammable surfaces, such as wood, by insulating them from the intensity of the heat and thus delaying ignition. See Fig. 900.2. They protect nonflamma-

Fig. 900.2. Fire Retardant Coating.
An inspector checks the surface of the protected structure. When the char is removed, the substrate is almost intact and may be repainted.

ble materials such as steel by preventing the buckling that comes with heat; and they protect cementitious materials by insulating them against the drying effect of fire, which eventually can cause collapse, particularly if steel girders behind the concrete buckle. Intumescent chars delay ignition of wood substrates. Fire resistant coatings may not do so to a significant degree.

To simplify terminology we will refer to fire retardant paint and intumescent paint interchangeably.

900.3 Temporary Benefit. The spreading of fire is only delayed by intumescent paints. Some firms claim evidence that these materials can delay the spread of fire by one-half hour, but even five to ten minutes is a big help in aiding evacuation of schools or hospitals and in obtaining the assistance of fire fighters. See Fig. 900.3.

900.4 Examples of Effective Service. One of the pioneer companies in the manufacture of intumescent paints lists about ten

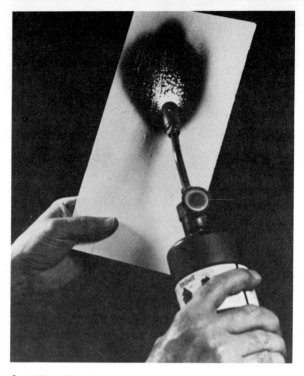

Fig. 900.3. Testing for Fire Resistance.
A blow torch can be put against a protected surface, and the fire retardant coating will form an insulating layer of spongy char that will protect the substrate for as much as 40 minutes.

examples of schools and other public buildings where flames were contained because of fire retardant coatings. The superintendent of schools at Bainbridge Island, Wash., reported: "Fire retardant paints played an important part in preventing our recent high school fire from spreading from the manual training area to other parts of the building."

The safety engineer at Illinois State University claimed that, "fire retardant paint, used on the walls, prevented the fire from igniting the structure."

The Buffalo Evening News proclaimed in a large headline: "Paint Credited with Saving Lockport Hotel."

These are only a few examples, but the lives that are lost every day and the vast property losses that could have been reduced with the use of fire retardant coatings make one wonder why all municipalities do not follow the lead of Hartford, Conn., Chicago, Pittsburgh, and other large cities where fire catastrophes have led to statutes requiring the use of these coatings in public buildings.

Fire Retardant Coatings

901 Methods of Rating Fire Retardancy. The National Fire Protection Association has adopted five classifications for fire retardancy of coatings. They are based on flame spread as determined in tests established by ASTM. They are as follows:

Class A—flame-spread between 0 and 25
Class B—flame-spread between 25 and 75
Class C—flame-spread between 75 and 200
Class D—flame-spread between 200 and 500
Class E—flame-spread over 500.

Certified laboratories, usually Underwriters' Laboratories, conduct tests in a special 25-foot tunnel. The test known as the Steiner Test is used for other building materials, and the paint can be applied to substrates at what would be desirable film thicknesses and checked in the tunnel for fire retardance against the burning rate of the uncoated material.

The method uses two standards. The first is uncoated and untreated asbestos-cement board, which rates as 0. The second is uncoated and untreated red oak flooring, readily flammable, which provides the standard of 100 for flame-spread, 100 for fuel contributed, and 100 for smoke development.

A material with a flame-spread rating of 25 has 25 percent as much flame-spread potential as the standard, red oak flooring. It then is classified as Class A in flame-spread. If it develops 50 percent of the smoke developed by red oak flooring in the 25-foot tunnel, it then has a smoke development rating of 50.

Both those ratings must be met by a coating to pass Federal Specification TTP 26 (GPC 76).

902 Kinds of Fire Retardant Coatings. Either water reducible paints or solvent-reducible may serve as intumescent coatings. Both use the same principle. Both have a carbonific material to provide the char, a phosphate to serve as a catalyst to cause the char to form, a gas producer to cause the char to foam, and a resinous material to hold it all together. The carbonific is usually pentaerythritol, serving as a nonresinous source, and some resinous material such as melamine formaldehyde, which also lets off a gas for foam-forming and provides a nonburning resinous film to contain the foam.

The catalyst is almost always a diammonium phosphate,

selected because it is leached to a lesser extent than most phosphates. This leaching, for many years, was a cause of trouble, because intumescent coatings lost their effectiveness when older forms of phosphate were leached away in moist environments, or after a few washings.

Aiding in resin formation are such materials as chlorinated rubber or chlorinated paraffin, and some formulations have had antimony oxide to help evolve antimony chloride, which helps extinguish the flame. As it happens, antimony oxide is to be limited by federal regulation as a hazardous substance around households, so this helpful material may be lost. Other substitutes are available.

902.1 Solvent-Reduced Types. Alkyds, recently developed vinylated-amino alkyds, and vinyl toluene acrylate copolymers are among resins used for solvent-reduced intumescent coatings.

902.2 Water-Reduced Types. The resin used most frequently for water-reduced intumescent paints is polyvinyl acetate.

902.3 Effectiveness. Tests of solvent-reduced and water-reduced intumescent coatings have shown them to be effective. In one test, a Bunsen burner, which produces an intense high-temperature flame, set up a charred foam on coated pieces of thin wood. In tests of vinyl toluene acrylate-based coatings and those based on polyvinyl acetate, the intumescent coatings were so effective that for 20 minutes it was possible to place a hand behind the thin pieces of wood without discomfort.

On several occasions, small test structures with walls coated with intumescent coatings have been burned beside houses coated with conventional finishes. The treated houses burned far more slowly, and it was obvious that personnel could have been evacuated far more safely had they been full-scale houses containing people. See Fig. 902.3.

902.4 Performance as Paints. It must be remembered that in addition to helping prevent fire-spread, these materials must also function as paints. Hence, they have to be formulated to resist washing and some degree of wear. Unfortunately, some compromises must be made to improve their serviceability.

For example, the formulators working on the vinyl toluene

Fig. 902.3. Fire Retardancy Test.
Top: *Two identical structures, one protected with a fire-retardant coating and one painted with an ordinary coating, are set afire.* Middle: *About ten minutes later, the structure on the right, with ordinary paint, is aflame. On the other structure, the fire-retardant paint has begun to char and protect the wood surface.* Bottom: *The unprotected structure is completely consumed within 40 minutes, while the protected structure has sustained minimal fire damage.*

acrylate-based paint found that they could get a high intumescent insulating charred foam with a pigment PVC above 70. However, the serviceability dropped markedly when they went above a PVC of 50.

At a PVC of 50, they found (using the methods outlined in TTP 26; same as GPC 76) that the paint could withstand more than 5,000 serviceability strokes. When the PVC was increased to 55, serviceability strokes dropped below 2,000, the number required to pass the federal specification. As PVC increased, though, the flame-spread rating falls. In this instance, the formulator had to compromise. He had to put the PVC between 50 and 55, where both intumescence and characteristic serviceability met federal specifications.

All manufacturers contacted report trouble meeting the rigid leaching test in the federal specification, which, according to federal officials, is designed to permit the coatings to function in highly humid tropical environments, where it was found that the phosphate catalyst was often dissolved out of the coating, leaving it ineffective.

903 Specifying Fire Retardant Paints. Unless the surface to be protected will be subjected to extremes of humidity, Federal Specification TTP 26 (GPC 76) may be too rigid. Very few companies claim to meet its requirements for resisting leaching. Actually, many coatings serve quite well in all U.S. environments unless special conditions prevail.

903.1 Number of Topcoats. Adding a second nonintumescent coat to a coat of intumescent paint may not improve the flame-spread rating, but, on the contrary, may hurt it. However, this is sometimes necessary to improve the serviceability characteristics of the coating. Some firms provide compatible semigloss enamels for second coats.

903.2 Spread Rate. One of the disadvantages of intumescent paints is their low spread-rate, which when added to high cost means that the expenditure per square-foot-of-coverage is high indeed. Spread rates vary from 120 to 175 sq.ft. per gallon, as compared with 350 to 450 sq.ft. for conventional coatings. Cost of the paints runs from $9 to $14 a gallon.

904 Intumescent Mastics.

Intense heat is capable of buckling steel beams and plates making up the structural backbones of large buildings. For that reason, most major cities require encasement of steel in some form of fireproof mastic. Two factors detract from these fireproof materials. First, they are usually made of asbestos blown onto the surface, which has been drawing the attacks of environmentalists because of air pollution. Second, while asbestos will not burn, it fails to hold heat from steel as effectively as an insulating foam.

To overcome the second factor, manufacturers of intumescent coatings have perfected mastics for encasement of steel. These materials, of troweling consistency, intumesce upon exposure to flame and delay the reach of intense heat to the steel surface for as much as a half-hour.

chapter 10

Economics of Structural Coatings

1000 General. The claim is often made that once a structure has been erected the single most expensive element of maintenance throughout its life is painting. That has become virtually a truism, but it does not have to be true at all. The materials and methods used have been getting undeserved blame. The real blame should be assigned to the decision makers who save fractions of a cent per square foot by buying "economy" paints for initial application, or who hire incompetent applicators who skimp—through ignorance or greed—on surface preparation; or who, for the same reasons, may try to squeeze 500 sq.ft. out of a gallon of paint intended only to cover 400 sq.ft.

Builders who "save" in that short-sighted fashion today lay the groundwork for wasteful expenditures for decades.

The fact that it costs as much to apply second- or third-rate paint as it does top quality material is hard to drive home. It is also difficult to convince builders that the difference in material costs is ridiculously low compared with the extra value of added life expectancy. Finally, the lesson has to be repeated and repeated that the big costs—the money-consuming costs—are application and surface preparation, wherever they, as so often happens, are factors.

Table 1000 shows the cost of coating a wall with flat wall paint, the simplest of jobs. A premium flat latex paint, or an alkyd flat, costs about $0.03 per sq.ft. However, applying these somewhat inexpensive materials costs $0.15 cents per square

Economics of Structural Coatings

foot. For a 50,000 sq.ft. job, the total cost of material and application comes to $9,000 ($7,500 for application; $1,500 for paint).

An economy latex (the type known as "project" grade because it's used by many painters who win housing project bids) costs $0.008 per sq.ft. less than the premium grade. Yet, the same $0.15 per sq.ft. must be spent for application. The total cost reduction for 50,000 sq.ft. comes to a modest $400. Total cost would be $8,600.

The premium grade latex, or the alkyd, should provide three to five years service life, compared with a probable two years of attractive service life for the project grade. This takes the cost per year of the applied premium materials to $1,800 (five years) or $3,000 (three years), as compared with $4,300 for each of the two years of the project grade.

Table 1000. *Material Cost for Flat Wall Paint.*

	Painter's Price* (gal.)	Coverage (per gallon)	Cost (per sq.ft.)	Application Cost (per sq.ft.)
Premium flat latex	$10.00	350 sq.ft.	$0.028	$0.15
Economy (project) latex	6.00	300 sq.ft.	$0.02	$0.15
Alkyd, flat	9.00	275 sq.ft.	$0.33	$0.15

*Usually 20% off list price. These are relative price levels and change with the economy.

John Ballard, technical director of Kurfee's Coatings, Inc., of Louisville, Ky., ran experiments for the author, which indicate the life expectancy of high, medium, and low grade paints. A vinyl acrylic flat latex, representative of those used by many paint manufacturers, was tested in three quality levels, low, medium, and high.

The low-grade had a Pigment Volume Concentration of 66.7 percent and vehicle solids of only 13.02 percent. It withstood 270 brush strokes in a washability test with Ajax cleaner. The medium grade with a lower PVC of 59 percent and 17.04 percent vehicle solids withstood 500 strokes. The high grade, with lowest PVC (52.5 percent) and highest vehicle solids (27.75 percent) withstood 1,346 strokes.

As we went from the low-grade paint to high grade, and as we decreased our pigment content and increased our vehicle solids, we got substantial improvement in the washability and abrasion resistance of the material and in the indicators of life expectancy.

Mr. Ballard's tests, made at the same time, showed that an alkyd flat with a PVC of 60.7 percent and 26.9 percent vehicle solids withstood about 76 less strokes than the highest grade vinyl acrylic.

Since vinyl acrylic, or polyvinyl acrylate, which is a variant of polyvinyl acetate, uses a member of the acrylic family as a means of keeping it flexible, some confusion exists as to its relationship with a pure acrylic. For this reason, Mr. Ballard included in his test a pure acrylic latex with a PVC of 37.9 percent and 32.2 percent vehicle solids. The big difference between vinyl acrylic and pure acrylic was some 2,454 strokes additional for the pure acrylic. However, it must be noted that the pure acrylic was an off-the-shelf product of Mr. Ballard's company and was formulated to be a prime quality product. Remember, even pure acrylics can also be formulated to save money and cut quality at high PVC's and low vehicle solids. The fact that it is a pure acrylic does not eliminate the possibility of its being shoddily formulated.

But, if the specifier stipulates a low PVC and a high vehicle solids content, that helps assure top quality.

Referring to Table 1000 again and to the figures on savings per year by investing an extra $400 in the premium latex, it should be observed that any reasonably intelligent builder should know that it pays to spend a little more money—once—instead of paying and paying and paying; and that's what happens when the big costs of painting—application and surface preparation—are ignored, and poor paints are used instead of excellent ones.

1001 High Quality Coatings. Table 1001 shows the relatively low cost of high performance coatings. The table shows the film thicknesses recommended by leading paint manufacturers and indicates how much coverage is obtained per gallon. Using chlorinated rubber or silicone alkyds will cost $0.034 per sq. ft.,

Table 1001. Material Cost for High Quality Coatings.

	Painter's Price* (gal.)	Coverage (per gallon)	Cost per sq.ft.	Application Cost per sq.ft./coat	Paint as % of Cost
Chlorinated rubber (2-mils dry)	$14.00	255 sq.ft.	$0.054	$0.15	26%
Epoxy polyamide (3-mils dry)	$11.00	200	0.055	0.15	26%
Polyester epoxy	$15.65	130	0.12	0.15	44%
Vinyl chloride (3-mils dry)	$11.00	75	0.146	0.15	49%
Silicone alkyd (2-mils dry)	$14.00	270	0.05	0.15	25%

*20% off list price.

Source: Joseph Lattimer, architectural consultant, Glidden-Durkee Div., SCM Corp., and industry price lists. These are not absolute or representative of any one firm's list and change with the economy. The relative costs, only, are applicable.

or roughly $0.02 more than the general purpose alkyd or the premium latex described in Table 1000.

Using an epoxy polyamide at 3-mils film thickness will give even greater longevity and costs about $0.025 more per sq.ft. than the alkyd or premium latex.

Where the circumstances of exposure to abrasion and impact demands an epoxy polyamide, this additional $0.025 per sq.ft., compared with alkyd or premium latex, may mean the difference between ten years' life and only three or four. In the case of alkyds or a premium latex, this would mean repainting with at least one coat three or four times in ten years as against once in ten years for the epoxy. The total cost for ten years with alkyds would be the initial cost plus three or four one-coat applications at about $0.183 per sq.ft. for material and application costs. (We deliberately omitted the cost of surface preparation in this instance to avoid confusion.)

For ten years, then, we would pay out $0.205 per sq.ft. for epoxy material and application as against $0.18 to $0.19 for alkyds. If circumstances required the alkali resistance of chlorinated rubber, or the weather resistance of vinyl, or the ultra durability of polyester epoxy, then their higher one-time cost would probably bring even greater economies, under the circumstances, than using alkyds.

1002 Systems for Corrosion. Architects and engineers are often called on to specify coatings for metal in corrosive environments. Understanding the economics of the systems used for these situations can mean large returns for your clients.

1002.1 Anticorrosive Coatings, Material Costs. The elements of a paint system for corrosive environments are shown in Table 1002.1, together with cost per gallon and cost per sq.ft. The total material cost of each system per sq.ft. is shown in the lower right-hand figure for each system as the sum of the system's figures.

The vinyl chloride system, with a high ranking for withstanding rough weather conditions, costs $0.232 per sq.ft. for the three coats recommended. The epoxy polyamide-zinc dust system, highly regarded for all around use, particularly where impact and abrasion resistance are required along with corro-

Economics of Structural Coatings

Table 1002.1 *Material Costs for Anticorrosion Coating Systems.*

	Painter's Price (per gal.)	Coverage (sq.ft.)	Cost (per sq.ft.)
Vinyl chloride system			
Vinyl chloride primer (red lead) (sprayed only) (2 mil)	$9.10	200 sq.ft.	$0.046
Vinyl chloride intermediate coat (4 mil)	12.50	134	$0.093
Vinyl chloride topcoat (4 mil)	12.50	134	$0.093
Vinyl chloride system		134	$0.232/sq.ft.
Epoxy polyamide and organic zinc dust system			
Epoxy polyamide-zinc dust primer (3-mil dry, sprayed)	$24.50	331	$0.074
Epoxy polyamide topcoat (rolled) 3-mil	13.25	232	$0.057
Epoxy polyamide and organic zinc dust system			$0.131/sq.ft.
Urethane, catalyzed, system			
Epoxy-polyamide zinc dust primer (3 mil)	$19.15	235	$0.081
Polyurethane, catalyzed (2 mil)	28.90	400	$0.055
Urethane, catalyzed, system			$0.136/sq.ft.

Note: These figures represent averages for leading suppliers and are not to be construed as selling prices for any one brand.

sion resistance, is about the same as the cited urethane system, which is also highly regarded. Each of these systems cost about $0.13–13-1/2 per sq.ft.

1002.2 Application. Table 1002.2 shows the cost of applying these systems in the field, or in the shop, if this is feasible.

1002.3 Surface Preparation. Table 1002.3 shows the cost of preparing surfaces for these systems.

1002.4 Shop Work vs. Field Work. Both Tables 1002.2 and 1002.3 show the savings that can be had by specifying shop

Table 1002.2. *Application Costs (50,000 sq.ft.).*

Field Work	
1st coat	$0.12–.15/sq.ft.
2nd coat	$0.10–.12/sq.ft.
3rd coat	$0.08–.09/sq.ft.
Shopwork	
Deduct 25% from above figures.	

Source: See Table 1002.3.

Table 1002.3. *Metal Surface Preparation Costs.*

	White Metal Blast (50,000 sq.ft.)
Field work	$0.40–.55/sq.ft.
Shop work	$0.25 or less
	Near-White
Field work	$0.30–.45/sq.ft.
Shop work	$0.20 or less
	Commercial Blast
Field work	$0.25–.35/sq.ft.
Shop work	$0.15/sq.ft.
	Wire Brush
Field or shop	$0.10/sq.ft.

Source: Edward Gleason, vice president, MCP Facilities Corp., Glen Head, N.Y., personal communication.

surface preparation and application where size of metal parts and circumstances permit. Field work is given with a range of prices, because conditions determine costs. Preparing a structure rising 100 ft. above the ground will cost more than a one-level building. Also, in some situations, the abrasive can be reclaimed easily; in others it is destroyed, or is retrieved only with difficulty. These conditions require higher charges.

1003 Cost of Material vs. Cost of Preparation and Application.

To back up the contention that paint is cheap, painting is expensive, we present Table 1003, which shows that even with three coats of vinyl costing $0.23 per sq.ft., altogether, material constitutes only 21 percent of the entire job. Paint

Table 1003. Cost of Coatings Systems vs. Cost of Surface Preparation and Application for Metal Protection.

	Coatings Cost (sq.ft.)	Surf. Prep'n (White Metal Blast) (50,000 sq.ft.)	Cost of Application	Total	Coatings as % of Total
High-build vinyl (3-coats)	$0.232	$0.48	$0.36 (3 coats)	$1.07	21%
Epoxy polyamide (2-coats)	$0.131	$0.48	$0.26	$0.87	15%
Urethane (2-coats)	$0.136	$0.48	$0.26	$0.87	15½%
Alkyd (2-coats)	$0.04	$0.48	$0.26	$0.78	5%

Source: The first two paint costs estimated by Joseph Lattimer, architectural consultant, Glidden Coatings and Resins. Applications and preparation figures by Edward Gleason, Consultant, MCP Corp., specialists in these fields. Alkyds and urethanes were estimated by Mobay Chemical Co.

on this basis costs as little as 5 percent when an alkyd system is used and 15 percent when epoxy polyamide or urethane systems are selected.

1004 High Cost of Using Inexpensive Systems. Other than alkyds, each of the systems in Table 1003 is capable of at least ten years' life when applied properly on a suitably prepared surface. Actually the epoxy and urethane coatings can probably last from fifteen to twenty years with minimum touch up after ten to twelve years. Therefore, ten years is a conservative life expectancy. Table 1004 shows the ten-year cost of each of these high performance coatings compared with the alkyd system in Table 1003.

Since experience has shown that an alkyd system under corrosive conditions can be expected to need a new topcoat every two years, the cost of application ($7,500) each of the four times required in the ten years (Years 3, 5, 7, and 9) has been totaled at $31,000 (four times $7,500 for applying one coat plus the cost of material).

That puts the ten-year cost of the alkyd at $70,500 for 50,000 sq.ft., as against $48,000 for vinyl systems, and $43,000 for either epoxy or catalyzed urethane.

1005 Return on Investment Through Use of High-Quality Paint. The saving reflected in Table 1004 is presented in terms of return on investment in Table 1005. For the extra investment of $3,500 to buy either epoxy or urethane systems instead of an alkyd system, the consumer by the third year has earned $4,250, or more than 100 percent on an added investment of $3,500. (Table figures are rounded to simplify the examples.)

By the fifth year, he has more than doubled his extra investment, and by the seventh year he has a return of almost six times. Finally, by the tenth year, he has returned his investment almost eightfold.

Stated another way, the client would have saved 39 percent on the ten-year cost of $70,500 for the applied alkyd system if, instead, he chose to spend $43,000 at the outset for either the epoxy polyamide or urethane systems. Clients, however, are accustomed to thinking of return on investment as a means of deciding on expenditures. A return of 780 percent, or an average of 78 percent a year for ten years, is considered a

Economics of Structural Coatings

Table 1004. Cost Comparisons for Various Systems for Corrosion Prevention (50,000 sq.ft.).

System	Material Cost	Surface Preparation (White Metal Blast)	Cost of Initial Application	Cost of Application after Initial for 10 years	Total Cost for 10 years	Savings over Alkyd
Vinyl (3 coats)	$12,000	$24,000	$12,500	0	$48,500	31.6%
Epoxy polyamide (2 coats)	9,500	24,000	9,500	0	43,000	38.0%
Urethane (2 coats)	9,500	24,000	9,500	0	43,000	38.0%
Alkyd* (2 coats)	6,000	24,000	9,500	$31,000	70,500	—

*One coat every two years ($.06 per sq.ft. material cost plus $.095 application of one coat).

Table 1005. Return on Investment Using Extra Quality Paint for Corrosion Inhibition.

Year of Expenditure	Item	A Alkyd	B Epoxy or Urethane	Return on Investment
1st year	Cost of application, surface preparation, and materials	$39,500	$43,000	−3,500
2nd year	No work	0	0	−3,500
3rd year	Repaint A (one coat)	7,750	0	4,250
4th year	No work	0	0	4,250
5th year	Repaint B (one coat)	7,750	0	12,000
6th year	No work	0	0	12,000
7th year	Repaint C	7,750	0	19,750
8th year	No work	0	0	19,750
9th year	Repaint D	7,750	0	27,500
10th year	No work	0	0	27,500
		$70,500 minus	$43,000	Savings—$27,500

Note: $27,500 represents a saving of 39% on $70,000 expended for alkyds. $27,500 represents a return on an investment of $3,500; which is equal to 7.8 times cost or 780%, or an average of 78% per year for 10 years (Actually, to be fair, the time value of money should be taken into account as well as income tax factors.)

high return in any accountant's opinion. And remember, a ten-year life for either of these high performance systems is conservative.

Finally, consideration should be given to the depreciation for tax purposes granted for the more expensive material, which further adds to the benefits of using it.

1006 Bonus Dividends of High Performance Systems. Clients have other reasons than high return on investment to encourage them to use high performance systems. Some value is derived from the intangible benefits of having a plant or other facility that can function without interruption for painting in a period of ten years or more.

In planning an industrial structure, it may be wise to consult production management to get an estimate of the time lost when painting is done, and how often it is needed with the present system. Estimating the loss to the firm because of shutdowns due to painting can be carried out by reference to management and personnel figures on output and the overtime paid to workers to make up the lost time.

1007 Noncorrosive Atmospheres. Table 1007 shows that savings are less dramatic when high performance coatings are used in noncorrosive atmospheres. Translating these figures into terms of 50,000 sq.ft., we find that the ten-year cost for the alkyd system would have been $64,000; the chlorinated rubber-modified alkyd would have been $44,500; the latex, $43,500; and the urethane alkyd, $38,000.

Using the urethane alkyd (or uralkyd) as the best buy, if it has the proper performance characteristics for the particular situation, the ten-year savings for 50,000 feet, as against the $64,000 alkyd cost, would be $26,000, or 104 times the extra investment involved in specifying the uralkyd, which cost a mere $0.005 more per sq.ft. of material (or $250 for 50,000 sq.ft.).

A brief note is in order here. While these savings are large and significant, the actual savings must take after-tax profits into account. Inasmuch as these economies will be reflected in after-tax profits, they may be decreased by as much as 50 percent. Even at that, in all instances, the savings are of consequence, particularly since the examples used are based

Table 1007. *Cost Comparisons for Exterior Steel or Wood Surface Coatings in Noncorrosive Atmosphere.*

	Material Cost/ Sq.Ft.	Labor Cost/ Sq.Ft.*	Life Expectancy	Recoat Material Cost/ Sq.Ft.	Recoat Labor Cost/ Sq.Ft.†	Total Cost Per 10 Yrs./ Sq.Ft.
Alkyd	$0.009	$.395	2 yr.	$.0045	$.17	$1.276
Alkyd (Cl. Rubber Modified)	$0.030	$.395	4 yr.	$.0150	$.17	$.888
Latex	$0.020	$.395	4 yr.	$.0100	$.17	$.865
Uralkyd	$0.014	$.395	5 yr.	$.0070	$.17	$.763

System consists of 1/2-mil corrosion inhibiting primer and one coat of topcoat enamel for total topcoat build of 2 mils. Recoat consists of light sand and/or scrape to clean to adherent substrate, reprime bare spots and one coat of topcoat at 1-mil dry film thickness.

Total costs for 10-year period with all repairs consisting of brush cleaning and one complete recoat, 1-mil dry film thickness: (All figures are $/sq.ft.)

Years	Alkyd	Cl. Rubber Alkyd	Latex	Uralkyd
0	$.4040	$.425	$.415	$.409
1				
2	.1745			
3				
4	.1745	.185	.180	
5				.177
6	.1745			
7				
8	.1745	.185	.180	
9				
10	.1745	.093 (1/2 × .185)	.090 (1/2 × .180)	.177
Total ($/sq.ft.)	$1.2765	$.888	$.865	$.763

*Labor cost per sq.ft. = $.065—miscellaneous labor; $.19—hand cleaning (brush); $.14—two-coat brush application = $.395.

†Recoat labor cost per sq.ft. = $.19 × 1/2 = $.10—spot cleaning (brush); $.07—brush application = $.17.

Sources: Derived from C. H. Kline Report, Stanford Research Institute *Chemical Economics Handbook,* market surveys, Kenneth Tator Associates. Chart developed by Mobay Chemical Co.

Table 1008. *Cost Comparisons for Coatings for Wood Flooring.*

	Material Cost/ Sq.Ft.	Labor Cost/ Sq.Ft.*	Life Expectancy	Recoat Material Cost/ Sq.Ft.	Recoat Labor Cost/ Sq.Ft.†	Total Cost Per 5 Yrs./ Sq.Ft.
Alkyd	$.008	$.345	1/2 yr.	$.004	$.21	$2.493
Latex	$.016	$.345	1 yr.	$.008	$.21	$1.451
Moisture-Cure Urethane	$.020	$.345	2 yr.	$.010	$.21	$.915
Phenolic Spar	$.010	$.345	1 yr.	$.005	$.21	$1.430
Urethane oil	$.012	$.345	2 yr.	$.006	$.21	.897
Two Component Urethane	$.020	$.345	3 yr.	$.010	$.21	$.732

System consists of two clear coats for total build of 2-mil dry film thickness. Recoat consists of light sand and one coat at 1 mil dry film thickness. Substrate is wood or plastic flooring.

Total costs for five year period with all repairs being light and an additional full coat of original quality at one mil dry film thickness:

Six-month Periods	Alkyd	Latex	Moisture Cure Urethane	Phenolic	Urethane Oil	Two-Component Urethane
0	$.353	$.361	$.365	$.355	$.357	$.365
1	.214					
2	.214	.218		.215		
3	.214					
4	.214	.218	.220	.215	.216	
5	.214					
6	.214	.218		.215		.220
7	.214					
8	.214	.218	.220	.215	.216	
9	.214					
10	.214	.218	.110 (1/2 × .220)	.215	.108 (1/2 × .216)	.147 (2/3 × .220)
Total ($/sq.ft.)	$2.493	$1.451	$.915	$1.430	$.897	$.732

*Labor cost per sq.ft. = $.065—miscellaneous labor; $.15—sanding; $.13—two-coat application by roller or brush = $.345.

†Recoat labor cost per sq.ft. = $.15—light sand; $.06—one coat by hand or spray = $.21.

Sources: Derived from C. H. Kline Report, Stanford Research Institute *Chemical Economics Handbook*, market surveys. Chart developed by Mobay Chemical Co.

on 50,000 sq.ft., which is probably only a fraction of a realistic company situation.

1008 Wood Flooring. Table 1008 points out clearly the costs for various types of coatings for wood flooring. Note particularly that here again initial savings on material can be quite deceptive. The two highest cost coatings, both of them urethane at $0.02 per sq.ft., have the lowest total cost per five-year period per sq.ft., $0.915 and $0.732 per sq.ft., respectively, for the two-year and three-year urethanes. On the other hand alkyd, the cheapest material for initial application, is the most expensive coating over a five-year term at $2.493 per sq.ft.

In the matter of coating materials, initial economies can be false economies indeed. When the additional costs, not to speak of aggravations, are computed, specifiers and clients will readily recognize that this type of skimping is quite ill advised. For any given application, it is important to compute carefully the cost per sq.ft. for the total coating system over the time period involved; but it is also essential that other costs, tangible and intangible, be considered as well. Such costs include the cost of shutdown, interruption of production, and the intangible value associated with long-life, attractive coated surfaces for interiors and exteriors.

appendix A
Classification, GPC Specifications

GPC #	Type	Binder	Spec. Source	Relative Cost (H—High; M—Medium; L—Low; +, up to next range)	Cementitious, Exterior	Cementitious, Interior	Metal, Exterior	Metal, Interior	Wood, Exterior	Wood, Interior
1	Flat latex	Unspecified	TTP 29	L		x		x		x
2	Alkyd flat	Alkyd	TTP 30	L		x		x		x
3	Primer-sealer, topcoat	Oil	TTP 47	L		x		x		x
4	Acrylic	Acrylic	Rohm & Haas	L+		x		x		x
5	Acrylic flat	Acrylic	Rohm & Haas	L+		x				x
6	High gloss	Soya alkyd	TTE 505	M°		x		x		x
7	High gloss	Alkyd	TTE 506	M		x		x		x
8	Semigloss	Alkyd	TTE 508	M		x		x		x
9	Semigloss	Soya alkyd	TTE 509	M°		x		x		x
10	Gloss undercoat	Alkyd	TTE 543	M		x		x		x
11	High gloss & semigloss	Latex	TTP 1511	M		x		x		x
12	Gloss undercoat	Soya alkyd	TTE 545	M°		x		x		x

345

appendix A (Continued)
Classification, GPC Specifications

GPC #	Type	Binder	Spec. Source	Relative Cost (H—High; M—Medium; L—Low; +, up to next range)	Cementitious, Exterior	Cementitious, Interior	Metal, Exterior	Metal, Interior	Wood, Exterior	Wood, Interior
13	High gloss	Epoxy polyamide	TTC 535	H	x	x	x	x	x	x
14	All gloss	Polyester epoxy	PPG, Glidden, etc.	H	x	x	x	x	x	x
15	Gloss & semigloss	Polyester epoxy	TTC 545	H		x		x		x
16	White enamel	Epoxy polyamide	Ciba-Geigy	H	x	x	x	x	x	x
17	White enamel	Epoxy polyamide	Ciba-Geigy	H	x	x	x	x	x	x
18	Clear gloss	Polyurethane	TTC 540	H	x	x	x	x	x	x
19	Chemical-resistant gloss	Epoxy polyamide	Glidden	H	x	x	x	x	x	x
20	High gloss, clear & pigmented	Polyurethane	TTC 542	M	x	x	x	x	x	x
21	Glaze	Unspecified	TTC 550	H		x		x		x

Appendix A

#	Description	Type	Hughson									
22	High gloss	Polyurethane Acrylic		H	x	x		x			x	x
23	Flat, chalk-resistant			L+	x	x					x		
24	Eggshell	Oil or varnish	TTP 24	L+	x	x	x						
25	Flat, for wind-driven rain	Unspecified	TTP 35	L	x	x	x						
26	Flat	Polyvinyl acetate	TTP 55	L	x							
27	Flat, fast-drying	Vinyl chloride acrylate or PVAcetate	TTP 96	L	x			x		x			
28	Flat or gloss	Chlorinated rubber or styrene	TTP 95	M	x		x						
29	Semigloss	Styrene butadiene	TTP 97	M	x								
30	Texture	Unspecified	TTC 555	H↑	x		x						
31	Primer for chalking masonry	Alkyd and linseed oil	TTP 620	L+	x	x							
32	Primer & topcoat	Latex	TTP 650	L	x		x						
33	Filler for porous surfaces	Styrene butadiene	TTF 1098	H↑	x		x						
34	Concrete curing compound	Unspecified	TTC 0800	L	x		x						
35	Tints & deep tones	Styrene acrylate	TTP 1181	M	x								
36	Waterproofing	Vinyl toluene butadiene	TTP 1411	M	x		x						
37	Surface sealer for plaster & wallboard	Oil or varnish	TTS 179	L	x		x						

appendix A (Continued)
Classification, GPC Specifications

GPC #	Type	Binder	Spec. Source	Relative Cost (H—High; M—Medium; L—Low; +, up to next range)	Cementitious, Exterior	Cementitious, Interior	Metal, Exterior	Metal, Interior	Wood, Exterior	Wood, Interior
38	Moisture & chemical resistant	Acrylic	Glidden	L+						x
39	Water-repellent	Silicone	SS W 110	L+	x					
40	Flat, white lead & zinc oxide	Alkyd	TTP 81	L	x		x		x	
41	Flat, lead-free, chalk resistant	Linseed oil	TTP 105	L			x		x	
42	Aluminum, leafing	Spar varnish	TTP 320	L+			x		x	
43	Primer-surfacer	Alkyd, oil-modified	TTP 659	L+			x		x	
44	Enamel	Alkyd	TTE 489	M			x	x	x	
45	Semigloss	Silicone alkyd	TTE 490	M			x	x	x	
46	Enamel, low-lustre	Tung oil phenolic	TTE 522	M			x		x	

Appendix A

#	Name	Binder	Spec	Cat				
47	Semigloss	Alkyd	TTE 529	L+	x	x		
48	Aluminum, reflective	Phenolic	TTP 320	M	x	x		x
49	Primer (red lead)	Linseed oil	TTP 86, Type I	L+	x	x		
50	Primer (red lead)	Alkyd & linseed	TTP 86, Type II	L+	x	x		
51	Primer (red lead)	Alkyd	TTP 86, Type III	L+	x	x		
52	Primer-topcoat (red lead)	Tung oil phenolic	TTP 86, Type IV	M	x	x		
53	Primer-topcoat (red lead)	Vinyl chloride	Glidden	M	x	x		
54	Semigloss or eggshell for galvanized metal	Linseed, alkyd, phenolic	TTP 641, Types I, II, III	M	x	x		
55	Primer (basic silico lead)	Linseed, alkyd, phenolic	TTP 645, Types I, II, III, IV, V	M	x	x		
56	Primer-topcoat (basic silico lead chromate)	Tung oil phenolic linseed oil alkyd, or vinyl chloride	NL Industries	M		x		
57	Primer-topcoat (basic silico lead chromate)	Epoxy ester	Ciba Geigy	M+	x	x		

appendix A (Continued)
Classification, GPC Specifications

GPC #	Type	Binder	Spec. Source	Relative Cost (H—High; M—Medium; L—Low; +, up to next range)	Cementitious, Exterior	Cementitious, Interior	Metal, Exterior	Metal, Interior	Wood, Exterior	Wood, Interior
58	Primer-topcoat metal & galvanized (zinc dust)	Chlorinated rubber	TTP 1046	M+			x	x		
59	Primer-topcoat	Epoxy polyamide	Mathiesen & Hegeler Zinc Co.	H			x			
60	Primer inorganic zinc rich	Lithium, ethyl, or ammonium silicate	New Jersey Zinc Co.	H			x	x		
61	Primer-topcoat	Polyurethane	Mobay Chemical	H			x	x		
62	Semigloss, rust-inhibiting, for sheet metal. Also used as primer	Alkyd	TTE 485, Type II	M			x	x		

Appendix A

No.	Name	Composition	Specification	Cost					
63	Primer (lead-based)	Linseed oil, coumarone, or ester gum	TTP 25	L		x			
64	Semigloss trim	Alkyd	TTP 37	L+		x		x	x
65	Primer, lead-free	Alkyd	HUD-HM HMG 7482	L+		x		x	x
66	Primer,	Acrylic	Rohm & Haas	L+		x			
67	Flat, shake & shingle	Alkyd & linseed oil	TTP 52	L+		x			
68	Sealer, varnish	Oil or unspecified varnish	TTS 176	L		x			x
69	Stain	Oil	TTS 711	L+		x		x	x
70	Spar varnish, flat, clear	Alkyd	TTV 85	L		x		x	x
71	Spar varnish, clear	Alkyd	TTV 109	L		x			x
72	Heavy duty	Vinyl urethane	Baker Castor Oil	H		x			
73	Enamel, floor and deck	Unspecified	TTE 487	M	x		x	x	x
74	Concrete floor coating	Styrene butadiene	TTP 91 Goodyear Tire & Rubber Co.	M	x				
75	Coal tar epoxy finish	Cured epoxy & coal tar	Glidden	M	x		x	x	
76	Fire-retardant	Unspecified	TTP 26	H	x				x
77	Primer-topcoat	Soya alkyd	TTP 636	L+			x	x	x

*Depends on cost of soya oil.
†Due to low spread rate.

appendix B
Detailed Specifications Charts

GPC 1
Specification Source: TTP 29
Product type: Interior flat; primer-sealer and topcoat.
Outstanding characteristics: Color fast; quick drying; good washability and scrubbability, flexibility, hiding and good alkali and fungus resistance.
Resin type: Latex (usually either polyvinyl acetate copolymer, styrene butadiene, or acrylic).

PERFORMANCE
Dry-hard: Within 1 hr. (2.5 mils wet film).
Reflectance of mixture: Not less than 86%.
Dry opacity: 0.94 C.R. for white at 630 sq.ft./gal.
Gloss: 5-15 (specular)
Yellowing: Accelerated test, 0.07.
Flexibility: 1/4" rod.
Mildew resistance: No growth after test.
Water resistance: After 4 hrs. of distilled water wetting and 2 hrs. wait, no wrinkling, softening, or re-emulsification.
Scrubbability: 3,000 scrubs, no undercoat visible.
Washability: Reflectance at least 95% of pre-wash; 85° gloss not over 15. Use 100 cycles, recharging the sponge after each cycle.
Primer: TTS 179.

GPC 2
Specification Source: TTP 30
Product type: Interior flat, alkyd, odorless.
Outstanding characteristics: Smooth finish; good coverage; good washability; low absorption.

Resin type: Phthalic anhydride alkyd; long oil. At least 23% phthalic anhydride in alkyd.
Special formula notes: Total pigments—53% by weight; nonvolatile vehicle—26% of total vehicle; odorless solvents only; and odorless octoates for driers.

PERFORMANCE
Set-to-touch: Within 1/2 hr.
Dry-hard: Within 7 hrs.
Dry opacity: 0.94 contrast ratio for white at spread rate of 6.0 mil./sq.ft.
Gloss: Not over 8 (specular).
Water vapor transmission rate of dry film: Absorption into #2 Whatman Paper: Not over 1/16" migration in 3 hr. period.
Spread rate: 450 sq.ft./gal.
Brushing: No seeds, runs, sags, or streaks. Should brush easily.
Spraying: No dusting, mottling, or color separation; no seediness.
Color retention: (accelerated) After 72 hrs. in atmosphere saturated with potassium sulfate at 49° C., not over 0.10 yellowing.
Flexibility: 1/2" rod.
Washability: After 35 double rubs, retain 95% of original reflectance value and not over 125% of specular gloss.

GPC 3
Specification source: TTP 47
Product type: Interior flat, primer-sealer and topcoat.
Main uses: Self-priming and sealing paint.
Binder type: Treated oil, varnish, alkyd.
Special formula notes: Pigment—56% by weight of paint; nonvolatile vehicle—30% by weight of vehicle.

PERFORMANCE
Set-to-touch: One-half to 2 hrs.
Dry-hard: Within 7 hrs.
Dry opacity: Dry film contrast ratio of 0.95 at 630 sq.ft./gal. spread rate.
Gloss: Not over 6 specular at grazing angle.
Water vapor transmission rate of dry film: Absorption into #2 Whatman paper: Not more than 1/8" migration.
Brushing: No running, sagging, streaking, or dusting, mottling, or color separation. On rolling, no lapping, localized floating, or unevenness.
Spraying: Same as brushing, plus no orange-peel.
Color retention: Accelerated ultraviolet exposure, 72 hrs., not more than 0.12 yellowing. No change after 6 hr. exposure in sun.

Washability: Reflectance not less than 95% or more than 150% of original.

GPC 4
Specification Source: Rohm & Haas, Formula 341
Product type: Interior flat paint, acrylic latex.
Outstanding characteristics: High hiding; good scrub resistance; good burnish resistance; excellent roller application.
Main uses: All interior flat usage where good hiding and good scrub resistance is required.
Resin type: Acrylic latex, 44.5% solids, 364 lbs. per 100 gallons of paint.
Special pigments: Nonchalking rutile titanium dioxide, Ti Pure R 901 or equivalent 250 lbs. per 100 gallons of paint.
Special formula notes: Pigment volume concentration, 52.3%; nonvolatiles, 54.8%; clay, 200 lbs. thin flat plate anhydrous, calcined; silica 50 lbs.; water added in manufacture not to exceed 204.5 lbs.; other water found in materials used not counted.

PERFORMANCE
Gloss: 85° sheen 14.
Scrub resistance: 2,300 scrubs.
Note: Other characteristics meet or exceed TTP 29.

GPC 5
Specification Source: Rohm & Haas, Formula 344
Product type: Interior flat paint, acrylic latex.
Outstanding characteristics: Outstanding scrub resistance (durability); reasonable cost; heavy film build for one coat hiding.
Main uses: For high durability interior flats.
Resin type: Acrylic latex, 44.5% solids, 364 lbs. per 100 gallons of paint.
Special pigments: Nonchalking rutile titanium dioxide Ti Pure R-901 or equivalent, 225 lbs. per 100 gallons of paint.
Special formula notes: Pigment volume concentration 51%; clay content, 125 lbs. of flat, plate anhydrous, calcined; water not to exceed 149 lbs., other than that found in materials used in accordance with accepted manufacturing practice; nonvolatiles 54%.

PERFORMANCE
Scrubbability: 3,800 scrubs.
Note: Other characteristics equal or exceed TTP 29.

GPC 6
Specification Source: TTE 505
Product type: High gloss enamel, white and tints.
Outstanding characteristics: Excellent leveling and hiding; good tint retention.
Resin type: Soya alkyd with phthalic anhydride not less then 24% of nonvolatiles.
Special pigments: Pigment volume content not less than 19% of total nonvolatiles; zinc oxide at least 10–15% of pigment by weight.

PERFORMANCE
Tack-free: 5–6 hrs.
Dry-through: 8–12 hrs.
Reflectivity: Apparent (white)—at least 84%.
Dry opacity: Not over 7 ml/sq.ft. for dry film contrast ratio of 0.99.
Gloss: 60-degree—at least 85 after 48 hrs. and 70 after 7 days.
Accelerated yellowness: (Reflectivities over 80% only), not more than 0.12.
Flexibility: 1/8" rod.
Recoating: No lifting after 16 hrs.

GPC 7
Specification Source: TTE 506
Product type: Enamel, alkyd, tints and white, interior.
Outstanding characteristics: Tint retention and flexibility.
Main uses: Wood trim paint, general purpose.
Resin type: Alkyd.
Special pigments: Titanium dioxide, at least 2.5 lbs./gal.
Special formula notes: Nonvolatiles at least 50% by weight, with phthalic anhydride at least 23% of that.

PERFORMANCE
Set-to-touch: 1/2 to 2 hrs.
Dry-hard: Within 16 hrs.
Contrast ratio: For white, 0.95 at spread rate of 540 sq.ft./gal.
Gloss: 60-degree specular, at least 80.
Flexibility: 1/8" mandrel.

GPC 8
Specification Source: TTE 508
Product type: Semigloss enamel, interior.

Outstanding characteristics: Washability, flexibility, tint-retention.
Main uses: Protection and decoration of wood surfaces, where a semigloss is desired.
Resin type: Alkyd with at least 23% phthalic anhydride and at least 60% drying oil.
Special pigments: Rutile titanium dioxide, at least 2.5 lbs./gal.; pigment volume, 37–42% of total nonvolatiles.
Special formula notes: Nonvolatile vehicle, at least 45%.

PERFORMANCE

Set-to-touch: 1/3 to 2 hrs.
Dry-hard: 18 hrs.
Dry opacity: Not over 7 ml./sq.ft. for dry film contrast of 0.95 for white and 0.95–1.0 for tints.
Gloss: 60-degrees specular, 40–70 after 48 hrs.; after 168 hrs. at least 40.
Water vapor transmission rate of dry film: Absorption: not over 1 in. on Whatman paper.

GPC 9

Specification Source: TTE 509
Product type: Soya alkyd, semigloss; white and tints.
Outstanding characteristics: Odorless, good leveling, tough, good tint retention.
Resin type: Soya alkyd.

PERFORMANCE

Set-to-touch: 2–6 hrs.
Dry-hard: Within 12 hrs.
Dry opacity: Not more than 7 ml/sq.ft. for dry film contrast ratio of 0.92 for white; for tints 82% reflectance, C.R. of 0.92 to less than 60% with C.R. of 0.99.
Flexibility: 3-mil film air dried 2 hrs. and baked 24 hrs. at 105 C—no cracking over 1/8" rod.
Recoating: After 12 hrs. air dry, no lifting or irregularity.

GPC 10

Specification Source: TTE 543
Product type: Undercoat for gloss and semigloss.
Outstanding characteristics: Good enamel holdout, easy brushing, rapid dry, high opacity.
Resin type: Alkyd.
Special formula notes: Alkyd to have at least 23% by weight phthalic anhydride; and drying oil acids not less than 60%.

PERFORMANCE

Set-to-touch: 1/2 to 2 hrs.
Dry-hard: 7 hrs.
Appearance: Smooth; no shiners, flashes, laps, or brush marks.
Reflectance of mixture: White—not under 82.
Dry opacity: Not over 6 ml/sq.ft. for dry film contrast ratio of 0.94 for white; tints—0.95 C.R. at 80% reflectance to 0.99 C.R. at 62% and under.
Gloss: Specular 3–10.
Enamel holdout: Specular gloss shall not decrease more than 15%.
Flexibility: Over 1/2" rod.

GPC 11

Specification Source: TTP 1511
Product type: Gloss (Type II) and Semigloss (Type I) latex enamel for interior use.
Outstanding characteristics: Easy wash-up; good flow and leveling.
Main uses: Over primer or previously painted plaster, wallboard, wood.
Resin type: Latex.
Special formula notes: Formulations developed by Rohm & Haas in cooperation with authorities call for acrylic resin binders and pigment volume content of 25.6% for semigloss and 22% for gloss.

PERFORMANCE

Set-to-touch: 12 min.
Dry-hard: 6 hrs.
Dry opacity: at 450 sq.ft., 0.95 CR.
Gloss: (60-degree specular) Type I, 5–20; Type II, 20.
Yellowness: After prescribed test—only 0.05.
Adhesion: No loss of adhesion to primer after 10,000 brushing strokes.
Flexibility: 1/8" mandrel.
Alkali resistance: After 2 hrs., exposure to aqueous sodium hydroxide—no blistering, softening, wrinkling; no change in hue and only slight change in reflectivity and gloss.
Washability: After test washing, 95% of original reflectivity; and between 80% and 120% of original gloss.
Working properties: Applied over surface sealer meeting Fed. Specif. TTS 179 by spray, brush, or short-nap roller should dry uniformly with no lap marks, streaks, pinholing, or other irregularities.

GPC 12
Specification Source: TTE 545
Product type: Enamel, odorless, alkyd interior undercoat.
Outstanding characteristics: Good enamel holdout.
Main uses: Undercoat for enamels.
Resin type: Alkyd, soya (25-28% phthalic anhydride).
Special pigments: 59-61% wt. of paint.

PERFORMANCE
Set-to-touch: 1/2-2 hrs.
Dry-hard: Not over 12 hrs.
Contrast ratio of mixture: 0.94 for white.
Dry opacity: 6 ml/sq.ft.
Gloss: 3-10.

GPC 13
Specification Source: TTC 535
Product type: High gloss, catalyzed cure, high performance.
Outstanding characteristics: Extremely washable and resistant to stains and abrasion.
Resin type: Epoxy (epichlorohydrin bisphenol A type), cured with polyamide resin at 1:1 ratio.
Special pigments: Not less than 41.25 of total solids must be titanium dioxide.
Special formula notes: Epoxy must not be more than 58.75% of solids, by weight; solvent—not more than 59.5% of total; to consist of xylol 15.9-17.9%; diacetone alcohol 18.1-20.1%; ethylene glycol monobutyl ether 12.6-14.6%.

PERFORMANCE
Set-to-touch: 1/2 hr.
Dust-free: 1 hr.
Dry-hard: 6 hrs.
Water vapor transmission rate of dry film: (thickness at 1.5 mils, 0.28-0.36; 3 mils, 0.19-0.25).
Pot life of mixture: Not less than 8 hrs.
Brushing: Comfortable flow.
Spraying: No orange peel.
Abrasion resistance: No evidence of complete removal of 3-mil film after 1,000 strokes.
Fire resistance: Flame spread no more than 5, smoke density 0; fuel contribution 0.
Stain resistance: After 16 hr. test for various stains, no remaining stains on 3-mil film.

GPC 14

Specification Source: Glidden, PPG Industries, Ashland Chemical, et al.
Product type: Low gloss, semigloss, high gloss enamel.
Resin type: Polyester Epoxy.
Special formula notes: Using Aroflint 858 in Ashland Formula H-45 for low gloss; H-55 for semigloss; and Aroflint 1010 for high gloss. For interior high performance, Aroflint 505 should be specified.

PERFORMANCE
Set-to-touch: 1-6 hrs.
Dry-hard: At 70° F., 5 hrs.
Gloss: 60 degree—initial 97 for high gloss; semigloss—40-50; low gloss—30-40.
Spraying: Must follow instructions.
Abrasion resistance: 60 milligrams, maximum weight loss, 3-mil film, 1000 cycles using a Taber CS17 wheel under 500-gm. load.
Chalk: With 10 as perfect—rated 8 after 12 months Florida South exposure; and 6 after 18 months (similar products).
Fire resistance: Flame spread—10; fuel contributed—0; smoke density—0.
Weather resistance: Exposed 10 years on Minnesota homes; no sign of needing repainting; Weather-Ometer—100 hrs.; Florida South 45-degree reflectivity of 80 after 18 months (97 initial); acceptable chalking; 18 months yellowness from 3.8 unwashed, 1.6 washed.
Adhesion: To most surfaces.
Chemical resistance: Most acids and alkalies normally used in tests.
Hardness: (Sward) 28-36.
Water resistance: Permeance of 0.00271 perm inches per sq.ft./hr.*
Stain resistance: Outstanding—no effect for merthiolate, lipstick, tomato paste, iodine, etc.; slight difficulty with mustard; resisted silver nitrate.
Scrubbability: Outstanding.

GPC 15

Specification Source: TTC 545
Product type: Gloss and semigloss, high performance.
Outstanding characteristics: High-build film; outstanding washability for sanitary clean-up; excellent leveling; good impact resistance; outstanding chemical resistance.

*See P. 261.

Resin type: Polyester epoxy, 2-component (polyester acid number 90-140; epoxidized oils epoxide equivalent 175-325).
Special formula notes: Total solids—high gloss—71, semigloss—72; vehicle solids—48% by weight of coating for high gloss, 40% for semigloss.

PERFORMANCE
Set-to-touch: 1-6 hrs.
Dry-hard: Within 16 hrs.
Contrast ratio of mixture: 0.94.
Hiding power: White 0.94 C.R.
Reflectance of mixture: Not less than 86 for white.
Gloss: 60-degree specular—high gloss, 80; semigloss, 70-40.
Hardness: (Sward) at least 28.
Brushing and spraying: Shall dry to a smooth, uniform coating.
Abrasion resistance: Weight loss not more than 75 mg.
Fire resistance: Flame spread, 15.
Smoke development: Density, 25.
Weather resistance: Accelerated yellowness test, not more than 0.10.
Impact flexibility: At least 24 in.lbs.
Chemical resistance: Dry film—6 mils; no irregularities after 24 hrs. of various chemicals; acids: phosphoric, hydrochloric, nitric, sulfuric, acetic; and ammonium hydroxide; sodium hydroxide.
Hardness: (Sward) 28.
Perspiration resistance: 4 hrs. in stained synthetic perspiration; remove by wiping with cheesecloth or use 25 wash cycles; no more than 25% of gloss is permitted.
Water resistance: 18 hrs.—distilled water, 90% of gloss remaining, no loss of characteristics.
Stain resistance: Type I and II—allow specified food stains, catsup, tea, etc., to stay on surface 16 hrs.; wash and remove with less than 25% loss of gloss; same with stamp pad ink, merthiolate, and yellow mustard; see specification for wash details.
Washability: Initial reflectance at least 98%; gloss after washing at least 75%.
Primers: Plaster, wallboard, gypsum board, TTP 650; wood, ferrous metal, TTP 636; painted surface, no primer required.
Instructions for use: Allow 1/2 hr. between mixing and application.

GPC 16
Specification Source: Ciba-Geigy 7071.
Product type: Enamel, white; Araldite 7071 with Polyamide 815 X-70.

Main uses: High performance coating where moisture and chemical exposures are problems.
Resin type: Epoxy polyamide.
Special pigments: Rutile titanium dioxide 5.83 lbs./gal.
Color limitations: White.
Special formula notes: Pigment/binder ratio 35/65; epoxy/hardener ratio, 100/54; nonvolatiles, 60%.

PERFORMANCE
Cure schedule: 7 days at 50 relative humidity and 77° F.
Film thickness: 2 mils.
Gloss: 60-degree, 95.
Adhesion: Excellent.
Chemical resistance: Most acids and alkalies used for tests.
Hardness: Pencil, HB.
Water resistance: 28 days, with few blisters; boiling water, 6 hrs., unaffected; 100% humidity, 28 days at 100° F., blisters.

GPC 17
Specification Source: Ciba-Geigy 7072
Product type: Enamel, white; Araldite 7072 with 835 hardener.
Outstanding characteristics: High impact resistance; water resistance.
Main uses: High performance coating where traffic and bruises are heavy.
Resin type: Epoxy Polyamide; 2 package.
Special pigments: Rutile titanium dioxide 7 lbs./gal.
Color limitations: White.
Special formula notes: Pigment to binder ratio, 40/60; epoxy to hardener ratio, 100/50; nonvolatiles, 60.3%.

PERFORMANCE
Cure: 10 days at 77° F. and 50 relative humidity.
Film thickness: 4 mils.
Gloss: 60 degrees—98.
Weather resistance: Weather-Ometer, 250 hrs.
Adhesion: Excellent.
Flexibility: 1/8" mandrel.
Impact resistance: Direct—140 in./lbs.; reverse—160 in./lbs.
Chemical resistance: 10% hydrochloric acid, 8 hrs., unaffected.
Salt fog resistance: 200 hrs., unaffected.
Water resistance: 7 days, unaffected.
Solvent resistance: MBK, 3 days, unaffected.

GPC 18

Specification Source: TTC 540
Product type: Clear gloss.
Outstanding characteristics: Good flexibility and adhesion; and abrasion and impact resistance.
Resin type: Polyurethane, linseed-modified, air-dry.
Color limitations: Clear, transparent.

PERFORMANCE
Set-to-touch: 1 hr.
Tack-free: 3 hrs.
Dry-hard: 8 hrs.
Brushing: Free-working and dried with good leveling.
Spraying: Good leveling.
Adhesion: Knife test; no chipping or flaking.
Abrasion resistance: Coefficient of at least 50.
Impact resistance: Reverse, at least 160 lbs./in.
Hardness: (Sward) at least 20.
Flexibility: No cracking or chipping over 1/8" mandrel.
Spotting resistance: Using a 3-mil coat, air-dried 168 hrs., no effect should be observed when the following are put on it: mineral spirits—24 hrs.; xylol—4 hrs.; 5% ammonium hydroxide—4 hrs.; 1% acetic acid—8 hrs.; 1% sodium hydroxide—8 hrs.; Clorox—24 hrs.
Weather resistance: 300 hrs., accelerated weathering; gloss at 60-degree specular, at least 90.
Water resistance: 72 hrs. in distilled water; 15 mins. in boiling distilled water; no whitening, dulling, or other defects.

GPC 19

Specification Source: Glidden Coatings & Resins
Product type: Low gloss, chemical resistant finish.
Outstanding characteristics: Chemical, moisture, and abrasion resistant coating.
Main uses: Walls, floors, machinery, storage tanks; for food-processing, chemical, petroleum, paper mills, and industrial plants.
Resin type: Epoxy polyamide, 1:1 ratio.
Special pigments: Pigment, 48.2%.
Special formula notes: Solids, 58.3% by weight.

PERFORMANCE
Set-to-touch: 3 hrs.
Full cure: 7 days.

Gloss: 60-degree, 65 minimum.
Humidity: 1,200 hrs., no change.
Spread rate: 200 (brush), 170 (spray, airless), 140 (conventional spray) at 3 mils.
Brushing: Level.
Fire resistance: Flame spread, 3; fuel contribution, 0; smoke density, 14.
Weather resistance: Weather-Ometer, gloss dropped from 65 (initial) to 58 at 13 hrs.; to 30 at 280 hrs.; to 8 at 400 hrs.; and to 4 at 600 hrs.
Chemical resistance: Most acids—resisted splashes, spills and fumes; most alkalies—withstood immersion.
Hardness: 2H pencil.
Salt spray resistance: 312 hrs.—O.K.: 551 hrs.—slight creepage.
Salt water immersion: 3,000 hrs.—no change.
Water resistance: Fresh water immersion (80° F) 1,872 hrs.—no change; 2,304 hrs.—tiny blisters.
Scrubbability: Outstanding.
Washability: Outstanding.

GPC 20
Specification Source: TTC 542
Product type: Gloss, clear and pigmented, high performance.
Outstanding characteristics: Quick dry-through; hard, improved color retention; very abrasion-resistant in pigmented form.
Resin type: Polyurethane, moisture-curing (isocyanate portion of resin combines with moisture in air), one-package.
Special formula notes: Nonvolatiles: Type I (class 1 and 2)—34% minimum; Type II—50%.

PERFORMANCE
Tack-free: Type I (clear), Class 1—1 hr., Class 2—2-1/2 hrs.; Type II (pigmented)—2 hrs.
Dry-through: Type I (clear), Class 1—1.5 hrs., Class 2—4 hrs.; Type II (pigmented)—4-1/2 hrs.
Contrast ratio of mixture: 0.99 for grey, red, green (at 1-mil); 0.94 for yellow and white at 1.5 mil.
Color: (Gardner Hellige) Type I—3.
Flexibility: 1/8" mandrel.
Abrasion resistance: Type I, Class 1—20, Class 2—30; Type II—60.
Hardness: (Pencil) Type I, Class 1—HB-H, Class 2—2B-H; Type II—HB-H.

GPC 21

Specification Source: TTC 550

Product type: Glaze finish; white and colors; high performance.

Outstanding characteristics: Tile-like finish; resistant to moisture, heat, chemicals, and cleaning agents.

Main uses: For outstanding durability and sanitary cleanup in laboratories, dairies, hospitals, and food plants.

Resin type: Wide latitude permitted; likely types are epoxy polyamides, epoxy polyesters, silicone alkyds, urethanes.

PERFORMANCE

Curing time: 7 to 15 days; hardness rating 40–85.

Dry opacity: White—2-mil dry film, at least 0.92 contrast ratio.

Gloss: Specular 60-degree—not less than 25.

Abrasion resistance: Abrasion coefficient—at least 40.

Adhesion: Resist 50-lb. pull on dry masonry; 20 lbs., wet masonry; 30 lbs. on re-dried masonry.

Moisture resistance: No blistering or adhesion loss; loss of gloss not to exceed 3 NBS units.

Perspiration resistance: No staining.

Fungus: No growth.

GPC 22

Specification Source: Hughson Chemical Corp.

Product type: high gloss, color-fast; and clear.

Resin type: Moisture-cure urethane.

Color limitations: Most colors available.

PERFORMANCE

Set-to-touch: 1/2 hr., clear and most colors; white and blue, 1 hr.

Dust-free: 1 1/2–2 hrs.

Dry-hard: 5–14 hrs.

Dry-hard for recoating: 3–4 hrs. at 30–60% relative humidity and good air circulation; heat will shorten time, as will extra catalysts.

Gloss: 94.0 initial 60-degree; after 1,200 hrs., Weather-Ometer—88.1; 20 months at 45-degree South in Florida—only 5 points drop.

Spread rate: 480 sq.ft.—clear; 700 ft., depending on color; all at 1-mil rate theoretical; heavier build is usual.

Abrasion resistance: (Taber) Clear, 12; white, 56; black, 15; blue, 13; up to 48 for pastel yellow.

Weather resistance: See *Gloss.*

Elongation: Passed 60% elongation, direct and reverse.

Flexibility: 1/8" mandrel.

Impact resistance: At 1.5 mils, reverse blow on steel and aluminum panels, passed 160 in.lbs.; same on direct blows.

Chemical resistance: With recommended primers, suitable for splash and spillage and fumes of most acids and alkalies; satisfactory for immersion in most.

Salt spray resistance: No effects after 1,000 hrs. on steel or aluminum, with 5% salt concentration.

GPC 23

Specification Source: TTP 19

Product type: Acrylic emulsion, exterior masonry.

Outstanding characteristics: Flat, resistant to chalking and fading.

Resin type: Acrylic emulsion.

Special pigments: Pigments not over 38%; TiO_2 not under 2.5 lbs./gal. (30% anatase; 70% rutile); tinting pigments.

Special formula notes: Total solids, not under 56%.

PERFORMANCE

Set-to-touch: Not over 15 mins.

Dry-through: Not over 1 hr.

Daylight apparent reflectance of mixture: Not under 87% (0.005 in. wet film).

Dry opacity: (0.005 in. wet film) Not under 0.98 for white.

Chalk-resistance: Excellent.

Abrasion resistance: Wet, not under 200 oscillations.

Condition in container: Pounds per gal.: not under 11 lbs.; viscosity: 65-86 KU.

GPC 24

Specification Source: TTP 24

Product type: Oil paint, exterior eggshell finish; concrete and masonry; Type I—white; Type II—tint base, white: Type III—pastel; Type IV—deep tones.

Resin type: Drying oil or varnish, or mixture of both.

Special pigments: Zinc oxide, 25-35% of total pigment: TiO_2 for Type I, not less than 30%, Type II and III, not less than 15%; silicate extenders, not more than 40% for Type I; others not over 55%.

Special formula notes: Vehicle not under 40%; volatiles consist of turpentine or mineral spirits or mixture; not over 1% water; pigment volume, 40-48.

PERFORMANCE
Set-to-touch: Not over 4 hrs.
Dry-through: Not over 18 hrs.
Reflectance of mixture: White, not less than 75%.
Gloss: 15–30 (60°).
Consistency: 89–104 KU.
Flexibility: No cracking or flaking of 0.0015-in. film when bent over 1/4" mandrel.
Water resistance: After 18 hrs. immersion, good adhesion, hardness and toughness.

GPC 25
Specification Source: TTP 35 and TTP 21 for Tints.
Product type: Exterior, portland cement powder paint (interior also).
Special formula notes: Not less than 73% portland cement by weight of powder; TiO_2, rutile, 3–5% (TiO_2 with aluminum oxide or silica surface treatment); (Note: TTP 21, Type II has 80% portland cement for thicker coats and TTP 21 may be used for textured walls).

PERFORMANCE
Reflectance of mixture: Apparent daylight (white), not less than 80.
Spread rate: 40–60 sq.ft./gal.
Hardness: 7th to 21st day of cure, 30–60; after 21st day, no chipping with 12 in. lb. load.
Chalking: Not over #6 chalking after accelerated weathering (Fig. 1 of ASTM-D-659).

GPC 26
Specification Source: TTP 55
Product type: Polyvinyl acetate emulsion.
Main uses: Exterior masonry.
Resin type: I, PVA, with dibutyl phthalate; II, PVA, self-plasticised.
Special pigments: Not more than 10% anatase TiO_2; not less than 90% rutile TiO_2 plus tinting pigments when required; extender, mica plus others.
Special formula notes: Total solids not less than 50%; not less than 30% pigments; not less than 0.3 lbs./gal. mica.

PERFORMANCE
Set-to-touch: Not over 15 mins.
Dry-hard: Not over 1 hr.

Reflectance of mixture: Not less than 87%, daylight apparent @ 0.005 in. wet film.
Dry opacity: Not less than 0.98 for white @ 0.005 wet film; not less than 0.99 for tints.
Abrasion resistance: Wet: Type I, not less than 500 oscillations; Type II, not less than 1,000 oscillations.
Viscosity: 68–82 KU.

GPC 27
Specification Source: TTP 96
Product type: Exterior latex, whites and tints.
Outstanding characteristic: Fast dry.
Main uses: For exterior wood, masonry, metal and weathered paint.
Resin type: Vinyl chloride acrylate or vinyl acetate with acrylate, maleate, or fumarate copolymer.
Special pigments: Titanium dioxide, zinc oxide, and extenders.
Special formula notes: Nonmercurial fungicide; at least 57% solids by weight for Type I (white), 55.5% for Type II (light tints), 54.5% for medium tints; pigment content not over 38.5%, Type I; 37%, Type II; 35%, Type III; nonvolatile vehicle, Type I, at least 20%; II and III, 19%.

PERFORMANCE
Set-to-touch: Not more than 5 mins.
Dry-hard: Within 1 hr.
Dry for recoating: After 6 hrs.
Reflectance of mixture: Type I, at least 86.
Dry opacity: at 630 sq.ft./gal., contrast ratio at least 0.92.
Gloss: Specular 60-degrees, 8–15.
Water resistance: No blistering, wrinkling, or emulsification.
Abrasion resistance: Not under 750.
Flexibility: 1/8" mandrel.
Accelerated weathering: Show limited chalking to conform to ASTM D-659 standard #8.

GPC 28
Specification Source: TTP 95 Type I
Product type: Concrete and masonry paint—Class A gloss; Class B semigloss; Class C flat.
Main uses: For concrete and masonry exposed to moist conditions; and swimming pools.
Resin type: Type I chlorinated rubber; Type II styrene-acrylate.

Special pigments: Any compatible alkali and acid-resistant pigments.

PERFORMANCE
Set-to-touch: 1/4 to 3/4 hr.
Dry-hard: 24 hrs.
Gloss: Specular 60°, Class A, 70; B, 25-55; C, 10 maximum.
Spread rate: 250 sq.ft./gal.
Brushing: No pull, no quick-set; dry free of sags, runs, pinholes and brush marks.
Consistency: 67-82 KU.
Flexibility: Over 1/4" mandrel.
Chemical resistance: No evidence of detrimental action by trisodium phosphate or brushing action.
Water resistance: Partial immersion on cement block for 10 days, then total immersion for 48 hrs; previously thin to 400 sq.ft. spread rate; no discoloration, blistering, cracking, or flaking.
Streaking: None when spread on glass plate as indicated in specification.

GPC 29
Specification Sourc: TTP 97
Product type: Exterior masonry; styrene butadiene, solvent type, white only, semigloss.
Main uses: For concrete, stucco, and masonry.
Resin type: Styrene butadiene.
Pigment volume concentration: 57.1.
Special pigments: Rutile TiO_2 33-37% of pigments' weight; zinc oxide 9.5-10.5%; extenders, 52.5-57.5%, consisting of silica, diatomaceous, and talc.
Special formula notes: Pigment (% by wt. of paint)—43.5-48.5; volatiles and vehicle 51.5-56.5; nonvolatile vehicle 22.5-27.5; fineness of grind—3; vehicle to be 50% high styrene butadiene resin (83-89% bound styrene); solids content, percent 59.1.

PERFORMANCE
Set-to-touch: 1/4 to 3/4 hr.
Dry-hard: 2 hr. maxim.
Reflectance of mixture: 80.
Gloss: 85° specular (sheen), 5.
Absorption: 1/4 in. maxim.
Spread rate: 400 sq.ft./gal.
Brushing: Good flowing and leveling without pull or sag.
Skinning: None after 48 hrs. in a 3/4 full container.
Consistency: 65-78 KU at 77° F.

Weight per gallon: 10.8–11.2 lbs.
Weather resistance: After 300 hrs. accelerated weathering, no more than two points drop in reflectance and chalk rating no less than 8.
Flexibility: No cracking of dried film over 1/8" mandrel.
Recoatability: No lifting or film irregularities.
Surface preparation: New concrete, cinder block or other porous material should be filled or grouted, using TTF 1098; for painted surfaces, remove loose material. (See GPC 33)
Application methods: For 2-coat system, reduce first coat with 1 pt. thinner to a gallon of paint.

GPC 30
Specification Source: TTC 555 Type I and Type II
Product type: Masonry coating, heavy bodied for hiding irregularities and to protect against wind-driven rain.
Main uses: Textured to hide surface defects and to provide exterior protection against rain driven by strong wind.
Resin type: Unspecified.
Special formula notes: Total solids, no less than 60%, nonvolatile vehicle not under 35%.

PERFORMANCE
Dry-hard: 24 hrs.
Dry opacity: Not under 0.99 C.R.
Water vapor transmission rate of dry film.: Not under 0.4 perms of moisture vapor permeability.
Spread rate: 40–60 sq.ft./gal.
Weather resistance: After accelerated test no checking or loss of film integrity; only slight color change; must withstand equivalent dynamic pressure of water driven by 98 m.p.h. wind; see specification for procedure.
Adhesion: Withstand at least 250.0 grams applied pressure.
Flexibility: Over 1" mandrel.
Impact resistance: Resistance to rapid blow of 6 in.lbs.
Mildew resistance: For Type II, no mildew allowed when tested under Fed. Standard 6271.
Water resistance: No blistering or loss of adhesion; only slight color change; no change in hardness.
Chalk resistance: Not under 8 ASTM after test.

GPC 31
Specification Source: TTP 620
Product type: Primer for chalking exterior painted masonry.

Binder type: Alkyd and alkali-refined linseed oil.

Special formula notes: Vehicle 47-49% of which alkyd resin is to be at least 29% of total and alkali refined linseed oil not under 23.8%; Pigment—51-53% of total; TiO_2 12-14%; extenders, 86-88%.

PERFORMANCE

Set-to-touch: Within 4 hrs.

Dry-hard: Within 24 hrs.

Absorption: 1/4-3/4 in.

Brushing: Easy brushing; no laps or brush marks or unevenness of color or gloss.

Skinning: None within 48 hrs. in 3/4 filled container.

Consistency: 77-89 KU.

Weight: 11.25 lbs./gal. minimum.

Adhesion: Over synthetic chalky surface, no removal when masking tape is pulled from dry film.

Flexibility: Over 1/8″ mandrel.

Chemical resistance: Alkali—no appreciable dissolving after 2 hrs. in 2% NaOH at 20° C.

Water resistance: No more than slight dulling and softening after 7 days in distilled water.

Self-lifting: No bubbles, blisters, pinholes, or lifting.

GPC 32

Specification Source: TTP 650

Product type: Interior primer and topcoat for gypsum walls and ceilings.

Outstanding characteristics: Quick-drying; 4-hr. recoat time.

Main uses: For plaster and wallboard.

Resin type: Latex, unspecified.

PERFORMANCE

Set-to-touch: 1/2 hr.

Dry-hard: Recoating within 4 hrs.

Reflectance of mixture: Not under 75%.

Dry opacity: Not less than 0.85 C.R. at 630 sq.ft./gal.

Gloss: Specular gloss, 60°, 5.

Appearance: As primed, no suction spots, flashes, or fuzz; as part of system, enamel topcoat should be free from flat spots, shines and color change.

Mildew resistance: Must meet test described in specification; no fungus growth after fungus innoculation.

GPC 33
Specification Source: TTF 1098
Product type: Filler for porous surfaces.
Outstanding characteristics: Ability to close pores of cementitious material and withstand water pressure.
Main uses: Cinder block, concrete block, concrete, stucco, masonry.
Resin type: Styrene butadiene.
Special pigments: Total pigments—54.2-58.2% by weight, of which asbestos shorts will be 7.5-11.5%, and the remainder silica, calcium carbonate, talc, or some mixture thereof.
Special formula notes: Total vehicle 41.8-45.8% by weight of which nonvolatile vehicle will be 17.2%.

PERFORMANCE
Set-to-touch: 1/4-3/4 hr.
Dry-hard: 2 hrs.
Consistency: 125 K.U., 50 sq.ft./gal.
Condition in container: Wt./gal.: 11.5 lbs.

GPC 34
Specification Source: TTC 0800; also ASTM, C 309-53T*
Product type: Concrete-curing compound.
Oustanding characteristics: Ability to cause new concrete to hold moisture and to serve as a base for paint and adhesives.
Color limitations: Clear; white; light gray; and black.

PERFORMANCE
Set-to-touch: 4 hrs. in 73.4° ± 3° F, and at 50° ± 10% relative humidity.
Reflectance of mixture: White after 3 days in sun, with 60% magnesium oxide; light gray with 50% magnesium oxide.
Moisture retention: Not more than 0.055 gm per sq.cm. of surface to be lost in test by ASTM, C 156.
Spread rate: 200 sq.ft./gal.
Spraying: Shall be sprayable at 40° F.
*Courtesy W. R. Grace, Sonneborn-DeSoto; Koppers Co.; Davis-Barlow; and Goodyear Chemical.

GPC 35
Specification Source: TTP 1181
Product type: Exterior masonry, tints and deep tones, styrene acrylate, solvent type.
Main uses: Concrete, stucco, and masonry.

Pigment volume concentration: 52-53%.
Special pigments: Nonchalking rutile TiO_2; extenders are calcium carbonate, diatomaceous silica, talc, and mica; tinting pigments.*
Color limitations: Black, brown, red, blue, green, yellow, and combinations.
Special formula notes: Pigment and extender, 46-51% by wt. of paint of which 20% should be TiO_2; styrene acrylate, 50% by wt of nonvolatile vehicle; volatile and nonvolatile vehicle shall be 49-54% by wt. of paint.

PERFORMANCE
Set-to-touch: 1/4-3/4 hr.
Dry-hard: 2 hrs.
Gloss: 85° specular (sheen), 5.
Absorption: 1/4 in. maximum. (See P. 271).
Weather resistance: Same as TTP 97.
Flexibility: Over 1/8" mandrel.
Consistency: 70-82 K.U. at 77° F; wt./gal.; 11.1-11.8 lbs.
Surface preparation: Same as TTP 97.
Application: Same as TTP 97.
*Yellow, brown, and red iron oxides; chromium oxide; lampblack; phthalo blue and green, nickel titanate yellow, carbon black.

GPC 36
Specification Source: TTP 1411
Product type: Masonry waterproofing compound.
Outstanding characteristic: Ability to withstand hydrostatic pressure.
Main uses: Interior and exterior protection against water both above and below grade for concrete and cinder block.
Resin type: Vinyl toluene-butadiene.
Special pigments: 6% of total pigments must be titanium dioxide; portland cement, 40%; extender, 49% (sand, mica, magnesium silicate, and asbestos may be used).
Color limitations: Permanent colors.
Special formula notes: Pigment, 60.5-64.5% by wt.; nonvolatile vehicle, 21.5-25.5% of total vehicle, which shall be 35.5-39.5% by wt. of paint.

PERFORMANCE
Set-to-touch: 1/4-1 hr.
Dry-hard: 3 hrs.
Consistency: Krebs Stormer, shearing rate (250 r.p.m. grams, 350-600), or 100-120 K.U.

Spread rate: 75 sq.ft./gal.
Flexibility: 3/4" mandrel.
Hydrostatic pressure: 9 ft. of pressure.
Water resistance: 4 lb. pressure/sq.in.—24-hr. immersion; no blisters, softening, or discoloration.
Scrubbability: 6,000 strokes.
Surface preparation: Sand blast to remove loose, powdery, or flaking old paint; porous or coarse surfaces may require two coats; large openings should be filled before application.

GPC 37
Specification Source: TTS 179
Product type: Surface sealer for plaster and wallboard.
Outstanding characteristics: Low absorption; good hiding, flexibility, and adhesion.
Main uses: To seal porous interior cementitious materials.
Binder type: Processed oils or varnish with drier and any suitable thinner.

PERFORMANCE
Set-to-touch: 1/2-4 hrs.
Dry-hard: For recoating, within 24 hrs.
Dry opacity: 0.84-0.90 dry-film contrast ratio for minimum apparent reflectivities of 74 to 60, when spread at rate of 7.0 ml/sq.ft.
Water vapor transmission rate of dry film: Absorption into #2 Whatman paper; not more than 1/8" in 3 hrs.
Spread rate: 450 sq.ft./gal.
Brushing: No pull or quickset.
Adhesion: Shall ribbon under knife point, without adjoining flaking.
Flexibility: 1/8" mandrel.
Appearance: No suction spots on succeeding coats.

GPC 38
Specification Source: Glidden, Division of SCM.
Product type: Moisture and chemical resistant latex paint.
Outstanding characteristics: Reasonably priced; color retention; chemical and moisture resistant.
Main uses: For areas where moisture and chemical splashing are a problem.
Resin type: Acrylic latex, 78%.
Special pigments: Pigment volume, 22%.
Color limitations: White.

PERFORMANCE

Set-to-touch: 20–30 mins.
Dry-hard: Suitable for recoat, 2 hrs.; full cure, 24 hrs.
Spread rate: 2-mil dry coat: 270 sq.ft. with brush or roller, 180–210 sq.ft. with spray.
Brushing: Good.
Weather resistance: 350 hrs.; initial gloss, 55 hrs.; final, 37 hrs.; color retention excellent.
Adhesion: Excellent, no blistering.
Chemical resistance: Good for alkalies and acids, splash, spillage, and fumes; not for immersion.
Salt spray resistance: 250 hrs., with light rust.
Water resistance: Splash satisfactory; no report on immersion.

GPC 39

Specification Source: SS W110
Product type: Water repellent.
Outstanding characteristics: Penetrating; repels in depth.
Resin type: Silicone.
Special formula notes: At least 5% by wt. of silicone.

PERFORMANCE

Storage: Must be capable of being stored 6 months at range of 0–100° F with no settling, separation, or solidifying.

GPC 40

Specification Source: TTP 81
Product type: Exterior, oil-based, medium shades on a lead-zinc base.*
Outstanding characteristics: Durable coating with white lead and zinc oxide.
Resin type: Linseed oil modified alkyd (70% linseed); 10% thinner and drier.
Special pigments: No organic colors or sulfides.
Special formula notes: Grind not under 4; vehicle, not over 45% by weight; not less than 80% solids in vehicle; water, not over 1%; color match—Federal Standard 595.

PERFORMANCE

Dry-hard: Not over 18 hrs.
Gloss: Not under 70.
Condition in container: Lbs./gal., not less than 12-3/4.
*Cannot be used on homes after Jan. 1, 1973.

GPC 41

Specification Source: TTP 105

Product type: Exterior paint, chalk-resistant, lead-free.

Outstanding characteristics: Chalk-resistant; rich in bodied linseed oil and lead-free zinc oxide.

Oil type: Raw or refined linseed oil, not less than 50%; bodied linseed oil 20–26%.

Special pigments: Zinc oxide, not less than 34% of pigment; titanium dioxide, not less than 23%; talc, clay, silica—together—not more than 43%; aluminum stearate, not less than 0.5%; thinner and drier, no more than 24%.

Special formula notes: Pigment, 59–61%; vehicle, 39–41% by weight of paint; total pigment in total nonvolatile, by volume—32–34%; nonvolatile vehicle, not less than 76% of vehicle-thinner combination; fineness of grind, not under 3.

PERFORMANCE

Dry-hard: Within 18 hrs.

Reflectance: Daylight 45°, 0° directional reflectance, white only not less than 75%.

Gloss: 60°, not under 70.

Water absorption: 1 in. (See P. 271)

Brushing: Good on wood, metal—no running or sagging, and dry to a uniform smooth finish.

Skinning: None within 48 hrs. in a 3/4 filled container.

Chalking: Accelerated weathering till chalking occurs, or 200 hrs.; check with velvet cloth to see if chalking exceeds a paint selected as a control; if exceeds, failure.

Dilution stability: 8 vols. paint with 1 vol. of mineral spirits; no excessive settling; easy redispersing with a paddle to a smooth, homogeneous state.

GPC 42

Specification Source: TTP 320 and TTV 81, Type II

Product type: Aluminum paint, leafing, and spar varnish.

Outstanding characteristics: Smooth, lustrous, heat reflecting.

Resin type: Rosin-pentaerythritol ester.

Special pigments: Aluminum powder with only 0.5 retained on #100 sieve; and 1.0 on #325 sieve; Class A, 1-1/4 lbs. aluminum paste per gal.; Class B, 2 lbs.; Class C, 3 lbs.

PERFORMANCE

Set-to-touch: 1/2–4 hrs.

Dry-hard: 18 hrs.

Leafing properties: Class A, not less than 55%; Class B, not less than 50%; Class C, not less than 40%.
Surface appearance: Leafing, luster, and smoothness should match an agreed-upon sample.
Brushing: Free flowing and easy brushing.
Water resistance: No whitening or dulling of varnish film after 18 hrs. immersion.
Spraying: Readily sprayable.

GPC 43

Specification Source: TTP 659
Product type: Primer-coating and surfacer for metal and wood.
Main uses: As a primer for white enamels, interior and exterior, such as TTE 489.
Resin type: Oil-modified alkyd varnish with not less than 30% phthalic anhydride and not less than 45% drying oil acids.
Special pigments: Titanium dioxide—not less than 45%; barytes—not less than 30% of total extender; silicate extender.
Special formula notes: Vehicle solids 35-40%; total solids—60-65% by weight of primer; grind—not under 5.

PERFORMANCE
Set-to-touch: 10 mins. to 2 hrs.
Dry-through: 18 hrs.
Reflectance of mixture: Directional (white) not under 80.
Gloss: Specular, 6-30.
Weather resistance: No appreciable film deterioration after 2 yrs. at Washington, D.C., or 1 yr. at Miami, Fla.
Flexibility: No cracking over 1/8" mandrel.
Chemical resistance: After 4 hrs. in hydrocarbons, no blistering or wrinkling.
Sanding properties: After 200 strokes as prescribed in specification, coating should be smooth and uniform with no gumming of the paper.
Water resistance: 18 hrs. immersion, no blistering or wrinkling immediately upon removal of panel.

GPC 44

Specification Source: TTE 489, Class A
Product type: Interior and exterior alkyd gloss.
Outstanding characteristics: Color and gloss retention; water and solvent resistance; good dry; flexible.
Resin type: Alkyd.

Special pigments: Free of extenders.

PERFORMANCE
Set-to-touch: Within 2 hrs.
Dry-hard: Within 8 hrs.; full hardness, 48 hrs.
Reflectance of mixture: (White) Not under 87.
Gloss: Specular 60°—not under 87; aged 20° not under 70.
Weather resistance: Accelerated after 168 hrs. only 30% gloss reduction and no chalking 2 yrs vicinity of Washington, D.C.; or 1 yr. Miami, Fla.
Adhesion: Knife test.
Flexibility: Over 1/8" rod after heat and age cycle.
Hydrocarbon resistance: 4 hrs.
Water resistance: 18 hrs.

GPC 45

Specification Source: TTE 490
Product type: Semigloss, exterior, silicone alkyd.
Outstanding characteristics: Water and weather resistant.
Resin type: Silicone-alkyd copolymer (silicone content not less than 30% of vehicle solids; phthalic anhydride 14-17%; soy oil acid 41-55%).

PERFORMANCE
Set-to-touch: Within 2 hrs.
Dry-hard: Within 8 hrs.
Gloss: 60° specular, 40-60.
Weather resistance: Slight chalking permitted after 300 hrs. accelerated weathering; color change not to exceed 4 units, except yellow, which is not to exceed 6 units.
Adhesion: Satisfactory, after knife test, no flaking or cracking at edges.
Flexibility: Over 1/4" rod after 96 hr. bake at 221° F.
Water resistance: 18 hrs., distilled water.

GPC 46

Specification Source: TTE 522
Product type: Enamel, exterior, low lustre.
Outstanding characteristics: Tough, water and salt resistant; high resin content with low luster.
Main uses: Where durable, low luster coating is required.
Resin type: Tung oil (1 part) phenolformaldehyde (4 parts).
Special pigments: See TTE 522.

Special formula notes: Nonvolatiles, 60% by weight for colors; 50% for black; phenolformaldehyde—specific gravity 1.0-1.1 @ 25° C; softening point, 250-280° F (ball and ring); no benzene or chlorinated thinners.

PERFORMANCE

Dust-free: Not more than 30 mins.
Dry-through: Not more than 2 hrs.
Dry-hard: Within 16 hrs.
Gloss: 60° specular.
Dry hiding: Brown—800-1,000 sq.ft.; red—200 ft.; green—700-1,000; blue—700-1,400; gray—1,000; white—3,000.
Weather resistance: 300 hrs. of accelerated weathering.
Flexibility: No cracking of baked film on steel panel, elongation not less than 8%.
Salt spray resistance: No softening, blistering, or embrittlement after 300 hrs.
Water resistance: No blistering or adhesion loss after 21 days immersion in distilled water.

GPC 47

Specification Source: TTE 529 Class A
Product type: Interior semigloss (also exterior).
Outstanding characteristics: Water resistant; nonchalking; hydrocarbon resistant; flexible; weather resistant.
Resin type: Alkyd with not less than 30% phthalic anhydride and oil consisting of 45-50% of vehicle weight.

PERFORMANCE

Set-to-touch: 20 mins. to 2 hrs.
Dry-hard: 8 hrs.; full hardness, 72 hrs.
Weather resistance: Accelerated 168 hr. test: only slight chalking 18 months at 45° south in Washington or 9 months in Miami, Fla.
Adhesion: Knife test: no cracking or flaking; cut must show beveled edge.
Flexibility: 1/4" rod.

GPC 48

Specification Source: TTP 320
Product type: Aluminum paint, leafing and phenolic spar varnish.
Outstanding characteristics: Resistant to alkali, gasoline, hot water, pitting, and wrinkling.

Main uses: Durable reflective and protective coating for metal and wood.
Resin type: Phenolic spar varnish, 33 gal. oil length, 45% tung oil; 4% castor oil and alkali-refined linseed oil.
Special pigments: Aluminum paste (20 oz./gal. of varnish plus 1/5 gal. of turpentine).
Special formula notes: Varnish nonvolatiles, not less than 57% by wt.; not less than 25% volatile aromatics, boiling range 265°-376° F; 75% mineral spirits, pass copper corrosion test, no blackening; flash point, 38° C.

PERFORMANCE
Set-to-touch: 1-2 to 1/2 hrs. on steel.
Dry-hard: 8 hrs. over steel; no after-tack within 24 hrs.
Gloss: Lustrous and free of streaks and blisters.
Spread rate: Good flow and leveling.
Brushing: Good flow and leveling.
Skinning: Dipentene may be added to prevent skinning.
Draft-proof: After drying in a draft of air moving at 550 ft./in., no pitting, wrinkling, or crowsfeet, or other defects.
Adhesion: Must not be separated from primer (TTP 636) by means of a diagonally applied knife or razor blade. No self-lifting when coated after 5 or 18 hrs. air dry.
Chemical resistance: 7 hrs. in sodium hydroxide.
Water resistance: Hot water resistance of varnish, no whitening or dulling or other visible defects when 1.5-mil wet film, dried 48 hrs. and immersed 7 hrs. in boiling water; 5 minutes recovery; 18 hrs. recovery hardness, gloss, toughness, and adhesion equal to unexposed film.
Gasoline resistance: Of varnish—4 hrs. immersion in Iso-octane, after 24 hrs. recovery gloss, toughness, and adhesion equal to unexposed film.
Condition in container: No skinning after 48 hrs. in 3/4 filled container.

GPC 49
Specification Source: TTP 86 Type I
Product type: Metal paint, red lead-based.
Main uses: Protection of structural steel where recoating is to be done after 36 hrs.
Binder type: Type I—linseed oil; 35-50% raw; 15-30% heat polymerized, low acid type, maximum acid No. 3 for viscosity of Q through Z9 and conformance to U.S. Govt. Spec. TTL 201, Type I.

Special pigments: 77% by weight, total pigments, of which 99.6% red lead and 0.3-0.4 suspending agent.

PERFORMANCE
Set-to-touch: 6 hrs.
Dry-through: 36 hrs.
Spread rate: 500 ft./gal.
Brushing: Good as received.
Spraying: Dilute 8 parts to 1 part mineral spirit.
Flexibility: 1/4" mandrel.
Condition in container: No skinning within 48 hrs., in a 3/4 filled closed container.

GPC 50
Specification Source: TTP 86 Type II
Product type: Metal paint, red lead based alkyd-linseed oil.
Main uses: Protection of structural steel where recoating is after 16 hrs.
Binder type: Alkyd varnish and linseed oil.
Special pigments: Total pigment, 66% by weight, of which 65% red lead (97% lead oxide); 15% red iron oxide (85% Fe_2O_3); 14.7% magnesium silicate; and 4-6% silicate.
Special formula notes: Alkyd must have 15% phthalic anhydride, minimum; not less than 28% raw linseed oil; not less then 28% alkyd; not more than 44% thinner and drier.

PERFORMANCE
Set-to-touch: 4 hrs.
Dry-through: 16 hrs.
Spread rate: 500 ft./gal.
Brushing: Satisfactory without sag.
Spraying: No orange peel or creep.
Skinning: None within 48 hrs. in a 3/4 filled closed container.
Flexibility: 1/4" mandrel.

GPC 51
Specification Source: TTP 86 Type III
Product type: Red lead, alkyd varnish.
Outstanding characteristics: Quick dry, but requires meticulous surface cleaning before application.
Main uses: Protection of metal where drying within 6 hrs. is required; and for touch-up.
Resin type: Alkyd.

Special pigments: Total pigments, 67% by weight, of which 99.6% red lead (97% lead oxide) and 0.3-0.4% suspending agent.

Special formula notes: Vehicle not less than 40% alkyd resin or more than 60% thinner and drier; alkyd—30% phthalic anhydride.

PERFORMANCE

Set-to-touch: 1/4-1 hr.
Dry-through: 6 hrs.
Spread rate: 500 ft./gal.
Brushing: Good as received; no sags or runs.
Spraying: Dilute 8 parts to 1 part mineral spirits.
Skinning: None within 48 hrs. in a 3/4 filled closed container.
Flexibility: 1/4" mandrel.

GPC 52

Specification Source: TTP 86 Type IV

Product type: Red lead, tung oil phenolic paint, primer and topcoat.

Outstanding characteristics: Water resistance; alkali resistance; quick drying.

Main uses: Protection of metal where high humidity is expected, or for water immersion where acids or alkalies will be present.

Resin type: Tung oil (25-gal. length) phenolic—not less than 44% phenolic solids, not more than 56% thinner.

Special pigments: 65% by weight, minimum of which 85% red lead (grade 97); 6-7% talc; 7.5-8.5% diatomaceous silica; and 0.3-0.4% pigment-suspending agent; fineness of grind, 4.

Special formula notes: Phenolic resin to be 100% para tertiary amyl phenol formaldehyde tested for water resistance; alkali resistance to meet the indicated specification.

PERFORMANCE

Set-to-touch: 1/4 to 1 hr.
Dry-through: 6 hrs.
Spread rate: 500 ft./gal.
Brushing: Good as received; no sags or runs.
Spraying: Dilute 8 parts to 1 part mineral spirits.
Flexibility: 1/4" mandrel.
Water resistance: Boiling water for 7 hrs., no whitening, blistering or dulling; cold water, 14 days, no softening, minor whitening permitted.
Condition in container: Weight: 16.7 lbs./gal.

GPC 53
Specification Source: Glidden Coatings & Resins
Product type: Vinyl solution finish.
Outstanding characteristics: Chemical and moisture resistant, tough, flexible, excellent for industrial or marine environments.
Main uses: Chemical and petroleum industry structures where water and chemical exposures are a problem.
Resin type: Vinyl chloride solution.
Special pigments: Available with red lead as primer to go over phosphoric acid wash primer.
Color limitations: Topcoats—eggshell white or light gray, or aluminum.
Special formula notes: Primer—76.5% vinyl chloride by weight; pigment, 23.5%; white eggshell or light gray—87% vinyl chloride; 13% pigment.

PERFORMANCE
Set-to-touch: 20 mins.
Recoat: 1 hr.
Spread rate: Primer, 200 sq.ft./gal.; topcoat, 270 sq.ft./gal.
Abrasion resistance: Taber, 200 revolutions, 19.2 mgm.
Weather resistance: 1 yr., Florida; 2 yrs., Cleveland, Ohio; trace of chalk; humidity chamber—1,000 hrs.; 400 hrs. accelerated.
Impact resistance: Passes 80 lbs. direct and reverse.
Chemical resistance: Immersion or splash—most acids, solvents, alkalies.
Hardness: 2-B pencil.
Salt spray resistance: 700 hrs.
Water resistance: 3,000 hrs., fresh or salt.
Surface preparation: White metal or near-white.
Application methods: 6 mils for topcoat over 9-mil red lead primer or 4-mil epoxy ester primer.

GPC 54
Specification Source: TTP 641 Types I, II and III
Product type: Primer and topcoat for galvanized surfaces.
Outstanding characteristics: Types I and II are for ordinary atmospheric exposure; Type III is for severe moisture or partial or complete water immersion; (this is a 2-package product; follow manufacturer's instructions).
Main uses: Undercoat for galvanized metal; also topcoat.
Binder type: I, linseed oil; II, phthalic alkyd; III, phenolic spar varnish vehicle.

Special pigments: Type I, total pigments 78-81%, of which total zinc dust (97.5% metallic zinc) should be 79-81%; and lead-free zinc oxide 19-21%; tinting pigments may replace 10% or less zinc oxide from following: iron oxide, chrome green, chromium oxide green, chrome yellow, burnt umber, zinc chromate yellow.

Special formula notes: Type I, linseed oil—19-20% raw linseed oil; Type II, total pigment 62-65%, volatile and nonvolatile vehicle, 35-38%, of which no less than 43% phthalic alkyd, of which no less than 23% phthalic, no rosin or derivatives allowed; Type III, 64-67% pigment, phenolic spar varnish, using equal proportion tung and linseed oil; drier consists of lead, manganese, and cobalt naphthenates.

PERFORMANCE

Set-to-touch: 1/2 to 4 hrs, Types II and III.
Dry-hard: Within 18 hrs.
Gloss: Eggshell or semigloss.
Brushing: No runs, streaks or sags.
Spraying: Dilute 8 parts to 1 mineral spirits.
Adhesion: Satisfactory; knife test; should remove as a ribbon with no flaking or loosening.
Flexibility: No cracking over 1/8" mandrel.
Water resistance: No damage after 24 hr. immersion at 20-30° C; and 6 hrs. at 75° C for a second panel.

GPC 55

Specification Source: TTP 645 Type I, II, IV
Product type: Metal primer (basic silico lead chromate).
Outstanding characteristics: Corrosion resistance, water resistance.
Main uses: Protection of metal.
Binder type: Type I, raw linseed oil (4 parts), alkyd (1 part) phthalic anhydride; Type II, raw linseed oil (1 part), alkyd (1 part); Type IV, phenolic varnish, 47% nonvolatile vehicle by weight of total vehicle.
Special pigments: Type I—64.5% pigment by weight (at least 99.3% basic silico lead chromate); Type II—57% pigment by weight (at least 93.2% basic silico lead chromate plus 5.7-7% iron oxide); Type IV—57% pigment by weight (at least 98.5% basic silico lead chromate).

PERFORMANCE

Water resistance: Type IV—2 coats, 14 days in distilled water, no blistering, wrinkling, whitening or softening.

GPC 56

Specification Source: NL Industries
Product type: Metal primer and topcoat.
Outstanding characteristics: Anticorrosion coating, chalk resistant.
Main uses: protection of structural metal in difficult exposures.
Binder type: Linseed oil alkyd, medium oil; also available in tung oil phenolic or vinyl chloride.
Special pigments: Basic silico lead chromate, 5.7 lbs./gal.; red iron oxide, 0.37 lbs./gal.
Special formula notes: Pigment volume concentration, 38.6%.

PERFORMANCE
Set-to-touch: 1/2-1 hr.
Dry-hard: 6 hrs.
Weather resistance: 2 yrs., South Jersey, no chalking; 9 yrs., no corrosion.
Salt fog resistance: 49 days.

GPC 57

Specification Source: Ciba-Geigy
Product type: Metal primer and topcoat.
Outstanding characteristics: Weather resistance, salt water resistance, excellent adhesion.
Resin type: Epoxy ester.
Special pigments: Primer—basic silico lead chromate 4 lbs./gal.; 40% PVC; red iron oxide, 1 lb.; topcoat—basic silico lead chromate 0.14 lbs./gal.
Special formula notes: Epoxy ester consists of 1 part of Araldite 6084 and 0.8 equivalent of linseed fatty acid, 4.75 lb./gal. of primer, and for topcoat, 1 part of Araldite 7098 with 0.5 equivalent of soya fatty acid, 6-1/2 lbs./gal.; the topcoat has a PVC of 20 with rutile and phthalo green at a ratio of 14/1.

PERFORMANCE
Weather resistance: 18 mos., no chalking; no rusting, no blisters in tidal zone; 3-1/2 yrs., South, 2 coats primer, plus color coat, no corrosion.
Impact resistance: Falling ball (6.3 lbs.) from 36 in., no fracture.
Surface preparation: Sandblast.
Application methods: Brush, roller or spray—primer, 2-3 mils; 2 coats primer, 4-6 mils; primer and color coat, 4-6 mils.

GPC 58

Specification Source: TTP 1046
Product type: Metal primer, zinc dust, chlorinated rubber, for steel and galvanized surfaces, can be used as a topcoat also.
Outstanding characteristics: Quick-drying, high solids content.
Main uses: For protection of steel and galvanized surfaces.
Resin type: Chlorinated rubber from 50–91% of vehicle solids by weight; nonsaponifiable resin or plasticisers, 9–50%.
Special pigments: Zinc dust, not less than 98.5% of total pigment, the remainder anti-gassing and anti-settling agents; total pigments, not less than 90% by weight of nonvolatiles.
Special formula notes: Total solids not less than 80% by weight of the total primer; water, not more than 0.2% by volume; thinnable by one part of thinner (toluene and mineral spirits) to 8 parts of primer.

PERFORMANCE
Set-to-touch: Within 15 mins.
Dry-hard: Within 30 mins.
Brushing: Dry without running, streaking or sagging.
Adhesion: No flake or crack after knife test.
Salt spray resistance: No blistering, wrinkling, or loss of adhesion.
Condition in container: Lbs./gal., not less than 21.5.

GPC 59

Specification Source: Matthiesen & Hegeler Zinc Co. & David Litter Laboratories
Product type: Metal primer, topcoat, zinc-rich organic.
Outstanding characteristics: Water resistance, corrosion-inhibiting, quick-dry.
Main uses: Corrosion inhibition of steel.
Resin type: Epoxy polyamide.
Special pigments: Zinc dust, 92.5%.
Special formula notes: 2-component product; the epoxy component to have as weight per epoxide 450–530; the polyamide component to have an amine value of 210–220; ratio 71.3 lbs. epoxy resin to 38.5 lbs. of polyamide; zinc dust to be fine-particle size, Mattheiesen & Hegeler 430 or equivalent.

PERFORMANCE
Set-to-touch: 10 mins.
Dry-hard: 30 mins.
Pot life: 5 days.

Abrasion resistance: 78 mgms., Taber, 1000 cycles; can be reduced to 26 mgms. by cutting zinc dust to 70%, but water resistance is drastically reduced; dropping to less than 85% markedly increases rust, but at 85%, abrasion is reduced to 48 mgms.
Salt fog resistance: 500 hrs.
Water resistance: 3 mos., fresh and salt.
Surface preparation: For severe exposure to water or salt fog, or for immersion, white sand blast is required.
Application methods: 2 coats are required.

GPC 60
Specification Source: New Jersey Zinc Co.
Product type: Inorganic zinc-rich paint.
Outstanding characteristics: Self-curing, abrasion resistance; salt fog resistance.
Main uses: Structural steel, above 5/8 in. flange thickness or over 30 ft. long or on piping more than 6 in. in diameter.
Resin type: Choice of lithium silicate; ethyl silicate; quaternary ammoniated silicate; butyl titanate.
Special pigments: Zinc dust (96% in lithium silicate).
Special formula notes: Lithium silicate to have silica to lithia ratio of 3 to 20 when obtained from Lithium Corp. of America; or 4.8 to 8.5 from duPont; various producers have patented formulations for other inorganic zinc-rich products.

PERFORMANCE
Set-to-touch: 1 hr.
Dry-hard: Varies 24–72 hrs., depending on type.
Pot life: 30 hrs.
Abrasion resistance: Excellent.
Weather resistance: Excellent; dry heat, 700° F; wet heat, 120° F.
Flexibility: Fair.
Impact resistance: Usually excellent.
Hardness: Excellent.
Salt spray resistance: Passed ASTM B-117-62.
Water resistance: Outstanding in fresh water.
Surface preparation: near-white sandblast with steel profile averaging 1.5 mils.
Application methods: Spray only; minimum of 2.5 mils, preferably 4.5–5 mils.

GPC 61
Specification Source: Mobay Chemical Corp.
Product type: Corrosion resistant metal coating.

Outstanding characteristics: Corrosion and impact resistance, water immersion.
Main uses: Corrosion-proofing and impact resistance and for water immersion.
Resin type: Polyurethane (2-package system) catalyzed.
Special pigments: Red iron oxide and talc in primer; topcoat—rutile titanium dioxide.

PERFORMANCE
Chemical resistance: Oil, crude oil, 8,000 hrs. immersion; satisfactory for most chemicals; not good in acetates or MIBK.

GPC 62
Specification Source: TTE 485 II
Product type: Enamel, semigloss, rust-inhibiting; for sheet metal.
Outstanding characteristics: Weather resistance and rust inhibition under hard conditions.
Main uses: As 1-coat system over cleaned and pretreated metals; or 2-coat system with 1st coat as primer.
Resin type: Medium-length phthalic anhydride alkyd.
Special pigments: By weight of total pigments: chemically-pure medium chrome yellow, not less than 32%; not more than 4% calcium carbonate; red lead (97% lead oxide), 14%; zinc oxide, 8%; channel-type carbon black and siliceous extender, not more than 38%.
Special formula notes: 60–65% solids in entire coating; vehicle solids, not less than 21%; drying oil acids (based on vehicle solids), not less than 45%; phthalic anhydride (based on vehicle solids), not less than 30%; water, not over 1%.

PERFORMANCE
Set-to-touch: Not more than 3 hrs.
Dry-through: Not more than 6 hrs.
Full hardness: Not more than 72 hrs.
Contrast ratio of mixture: 0.85 for natural color.
Gloss: 60 degree specular.
Brushing: Shall not require more than 5 parts mineral spirits to 95 parts enamel to impart good brushability.
Abrasion resistance: Lbs./gal.: 10.5.
Weather resistance: After 168 hrs. of accelerated weathering only slight chalking permitted and only slight color change; gloss shall be 65% of original; also Washington exposure 1 yr. or Miami 2 yrs.
Adhesion: Knife test.
Flexibility: 1/8″ mandrel.

Hydrocarbon resistance: 4 hrs.
Salt spray resistance: 192 hrs. in 5% salt spray.
Water resistance: 18 hrs. in distilled water.

GPC 63
Specification Source: TTP 25
Product type: Exterior wood primer
Outstanding characteristics: Contains white lead to combine with chemicals in raw wood to prevent migration of natural dyes in some woods to paint.
Main uses: For priming and sealing raw wood (Note: lead paints are banned for use in household applications).
Resin type: Coumarone, or ester gum, not more than 4%; raw linseed oil, not more than 32%; bodied linseed oil (Z), not less than 32%.
Special pigments: Lead carbonate (white lead), 25%; lead sulfate, 25%; titanium dioxide, 12% or more.
Special formula notes: Extenders and tinting colors, not more than 38%; thinner and drier, not more than 32%; nonvolatiles, at least 68% by wt. of vehicle; pigment volume, not less than 32%.

PERFORMANCE
Set-to-touch: 1/2-4 hrs.
Dry-hard: Within 18 hrs., ready for recoating.
Absorption: Shall not migrate more than 3/8" into Whatman #2 paper.
Sealing properties: Seal new wood uniformly.
Adhesion: Not over 1/8" of film picked up by tape.
Flexibility: 3 mil film on tin plate air-dried 18 hours, baked 5 hours at 105° C, cooled to 23° C, bent over 1/8" rod, no cracking or flaking.
Gallon weight: At least 14.4 lbs.

GPC 64
Specification Source: TTP 37
Product type: Wood trim for doors and shutters, alkyd.
Outstanding characteristics: lead-free, high-hiding, gloss.
Main uses: Wood trim.
Resin type: Alkyd.
Special pigments: Pigments by wt. 8% of paint for bright red (using toluidine or quinacridone), to 37% for bright green.

Special formula notes: Vehicle solids range from 43% for bright red to 60% for medium blue.

PERFORMANCE
Set-to-touch: 1/2–3 hrs.
Dry-hard: Within 18 hrs.
Gloss: At 60 degrees, not under 80.
Spraying: Dilute no more than 5% by volume of thinner.
Weather resistance: After 168 hrs. accelerated weathering in twin arc apparatus, no chalking and 60 degree gloss should be within 30% of original.
Adhesion: Knife test; shall ribbon, no cracking or flaking.
Flexibility: Over 1/8" mandrel.

GPC 65
Specification Source: HUD-HM Specific. #2 HMG7482
Product type: Lead-free exterior priming paint.
Outstanding characteristics: Primes wood and fixes natural stains; lead-free; developed by Housing and Urban Development department.
Main uses: To prime unpainted or painted wood.
Resin type: Alkyd (phthalic anhydride to constitute between 5 and 6% of paint by weight).
Special pigments: Titanium dioxide, 2 lbs./gal.; magnesium silicate, 2.5 lbs.; organophilic magnesium montmorillonite, 0.03 lbs.
Special formula notes: Flat alkyd solution (60% of solids) 2.88 lbs./gal.; long oil alkyd (60% solids), 1.92 lbs.; pigment volume concentration, 40%; vehicle, 56.8% by wt. of total paint.

PERFORMANCE
Dry-hard: 18 hrs.
Reflectance of mixture: Daylight 45-degree, white only, 75.
Gloss: 85-degree specular, 50.
Flexibility: 1/8" mandrel.

GPC 66
Specification Source: Rohm and Haas.
Product type: Lead-free stain inhibiting latex primer.
Outstanding characteristics: Excellent intercoat adhesion; long-term resistance to staining from wood dyes; humidity resistance.
Main uses: To control topcoat staining over woods with natural dyes.

Resin type: Acrylic latex.

Special pigments: Zinc oxide to use as a fortifying pigment; non-film forming acrylic emulsion E-726 is treated as a pigment for computing PVC.

Special formula notes: Pigment volume concentration, 22%.

PERFORMANCE

Spread rate: 450 sq.ft./gal.

Weather resistance: Overnight in a simulated rain consisting of water spray mist; 100% humidity at 140° F for 48 hrs.; two yrs. exterior exposure on staining-type woods; 18 months South 45-degree on yellow pine; face-down exposure on cedar and redwood panels, 18 mos. to simulate under-eave conditions.

GPC 67

Specification Source: TTP 52, Type I, II, III, IV

Product type: Alkyd-oil paint for wood shakes, siding, and shingles.

Outstanding characteristics: Thin-bodied; high hiding; water resistant.

Main uses: For protecting and decorating wood, shake, siding, and shingles.

Resin type: Alkyd, not less than 12% phthalic anhydride by wt. of vehicle nonvolatiles; and alkali-refined linseed oil, 47–53% by wt., of which titanium dioxide 42.5% (20% anatase and 80% rutile); zinc oxide, 10%.

Special formula notes: Siliceous extender, 47% by wt. of pigment; vehicle solids—61% by wt. of vehicle; total solids—81% by wt. of paint; Type I—white: Type II, tints: Type III, medium tones: type IV, deep tones.

PERFORMANCE

Set-to-touch: 3 hrs.

Dry-hard: Within 18 hrs.

Contrast ratio of mixture: 0.95 at 7 ml./sq.ft.

Water resistance: 18 hrs. in distilled water, no change.

Absorption: Not over 1/4" in Whatman paper. (See P. 271)

Adhesion: When scraped with test knife, shall leave beveled edges.

Flexibility: 1/2" mandrel.

Gallon weight: 11.8 lbs.

GPC 68

Specification Source: TTS 176

Product type: Sealer, varnish-type, for wood, cork, or floors.

Outstanding characteristics: Quick-drying, water resistant.

Main uses: To prevent penetration of wood by water or topcoats of paint or varnish.
Resin type: Wide latitude in selection of varnishes and oils.
Color limitations: Not over 13 Gardner color standard.
Special formula notes: Reducible with mineral spirits or turpentine; nonvolatiles, not less than 40% by weight.

PERFORMANCE
Set-to-touch: 1-2 hrs (A wet film of 0.015 in. on a clear glass panel).
Dry-hard: Not more than 3 hrs.
Gloss: Soft, even sheen; no cloudiness or grain.
Brushing: Capable of being applied by bristle brush or lambswool mop.
Water resistance: 24 hrs. immersion at room temperature, 1/2 hr. in boiling water—no whitening; slight dulling after 2 hr. recovery period.
Stain resistance: Able to resist blue-black ink.
Odor: Not abnormally offensive.
Touch-up: Must be able to recoat worn areas without showing lap marks.

GPC 69
Specification Source: TTS 711
Product type: Stain, oil-type, wood, interior.
Outstanding characteristics: Wood-simulation, solvent resistance.
Main uses: Decoration and protection of interior wood surfaces.
Resin type: Suitable drying oil.
Special pigments: Earth colors to match the following: cherry, light oak, dark oak, walnut, mahogany, and earth colors, plus organic toners to obtain desired match.
Special formula notes: Total solids not less than 30%; no rosin or derivatives permitted.

PERFORMANCE
Dry-hard: not over 8 hrs.
Miscibility: 80 ml. of stain mixed with 20 ml. of mineral spirits, no separation or precipitation.
Color stability: 24 hrs. accelerated weathering, only slight change allowed.
Dry opacity: None.
Solvent resistance: No staining when filter paper and cotton placed on surface and saturated in turn with turpentine, distilled water, ethyl alcohol, and mineral spirit.

GPC 70

Specification Source: TTV 85
Product type: Alkyd, spar varnish, flat.
Outstanding characteristics: Clear, air-drying varnish.
Main uses: General purpose protective coating for wood surfaces over clear primers or stains; also a satisfactory vehicle for aluminum paste as a water-resistant primer.
Resin type: Alkyd.
Color limitations: Clear; aluminum may be used for sealer.
Special formula notes: Nonvolatiles to be 44% minimum, containing 30% minimum phthalic anhydride and 45-55% oil acids.

PERFORMANCE
Set-to-touch: 1-3 hrs.
Dry-hard: 18 hrs.
Recoat time: 7 hrs.
Reflectance: Dried film must be clear, smooth, and glossy.
Gasoline resistance: 4 hrs. immersion.
Water resistance: Boiling water—19 mins.
Flexibility: 1/8" mandrel.

GPC 71

Specification Source: TTV 109
Product type: Spar varnish, clear, alkyd type.
Outstanding characteristics: Flexible, light-duty spar varnish.
Main uses: Clear film protection for wood.
Resin type: Alkyd; 30% phthalic anhydride, 45-55% oil acids.

PERFORMANCE
Set-to-touch: 1-3 hrs.
Dry-hard: Within 18 hrs.
Gloss: Glossy, smooth and clear when dry.
Adhesion: With aluminum paste—enamel version over primer, shall not separate when tested with razor blade.
Flexibility: No cracking over 1/8" mandrel.
Chemical resistance: Hydrocarbon immersion, no effect after 4 hrs.
Water resistance: Boiling water immersion, 19 mins.; no effect.
Recoatability: No film irregularities when recoated after 7 hrs.
Weight per gallon: 7.3 lbs. minimum.

GPC 72

Specification Source: Baker Castor Oil Co. EC 167
Product type: Exterior finish for wood.

Outstanding characteristics: Water and weather resistance, flexibility, good adhesion.
Main uses: Protection of plywood and siding.
Resin type: Vinyl urethane.
Special pigments: Antimony oxide (not for household use).
Special formula notes: Vinyl copolymer—VAGH, Union Carbide; urethane prepolymer—Vorite 174; PVC, from 15–45 acceptable.

PERFORMANCE
Set-to-touch: 7 mins. at 77° F.
Weather resistance: 10 yrs., Weather-Ometer equivalent; 6 mos. at 45°, South, and continued; 8 hrs. ambient water immersion followed by 16 hrs. oven drying (25 cycles); 8 hrs. ambient water immersion, followed by 16 hrs. freeze and 24 hrs. oven drying (10 cycles).
Adhesion: Excellent.
Chalk resistance: Very good.

GPC 73
Specification Source: TTE 487
Product type: Enamel, floor and deck.
Outstanding characteristics: Abrasion resistance, water resistance, imprint resistance, flexibility.
Main uses: General purpose enamel for concrete, metal, and wood floors.
Resin type: Not specified.
Special formula notes: Total solids, at least 65% by weight; nonvolatile vehicle, at least 50% of vehicle by wt.

PERFORMANCE
Set-to-touch: 1/2–4 hrs.
Dry-hard: Within 16 hrs.
Dry opacity: Contrast ratio of 0.99 at wet film thickness of 3 mils.
Gloss: High.
Water resistance: 18 hrs. immersion; no whitening after 4-hour recovery period.
Abrasion resistance: An area of 4 sq.mm. to require at least 20 litres of Ottawa sand (20–30 mesh) before abrading through to the substrate.
Flexibility: 1/8" mandrel.
Imprint resistance: No mark after cheesecloth is held to surface under 1/2 lb. load per sq.in. for 1/2 hr.

GPC 74

Specification Source: TTP 91, Goodyear Tire & Rubber Co.
Product type: Concrete floor coating, interiors.
Main uses: Water resistance, durability, abrasion resistance, detergent resistance, and resistance to dilute acids, alkalies, oils and grease found on garage floors.
Resin type: Styrene butadiene.
Special formula notes: The specification calls for the resin to be made with "suitable plasticisers dissolved in and thinned with turpentine, volatile mineral spirits, volatile coal-tar solvent (no benzol) or a mixture thereof."

PERFORMANCE
Set-to-touch: 1/4–3/4 hr.
Dry-hard: 3 hrs. (48 hrs. dry is recommended for severe wear, but 24 hrs. are enough for ordinary use.)
Dry opacity: Dry-film contrast ratio of at least 0.98 when spread black and white carrara glass panels at 2-mil wet film thickness.
Gloss: Specular 75.
Cement-Water Test: Cement blocks are painted with two coats and partially immersed 10 days and totally immersed 2 days, with no discoloration, blisters, cracking or flaking permitted.
Spread rate: 400 sq.ft./gal.
Brushing: Easy brushing with good flow and leveling. Setting shall not be quick enough to leave brush marks.
Abrasion resistance: A wear-index of 60.
Application: Three coats are recommended for unpainted concrete. For surface preparation for smooth new concrete see Par. 307.3a. New concrete floors should age 2 months first.

GPC 75

Specification Source: Glidden Coatings & Resins
Product type: Coal tar epoxy finish.
Outstanding characteristics: Rugged, strongly adherent, resistant to chemicals, water, abrasion, and corrosion.
Main uses: Interior and exterior metal and masonry and for floors.
Resin type: Cured epoxy and coal tar.
Special formula notes: Vehicle at least 68% coal tar epoxy, pigment 32%, solids 81% by wt.; gallon weighs 11 lbs.; reduce with 5 oz. xylol for conventional spray; other uses—no reduction.

PERFORMANCE
Set-to-touch: 4-1/2 hrs.
Recoat: 16 hrs.

Dry-hard: 72 hrs.
Pot life of mixture: At 80° F—4 hrs.; at 100° F—1-1/2 hrs.
Spread rate: Brush or roller—110 sq.ft.; airless spray—90 sq.ft.; conventional spray—80 sq.ft.
Impact resistance: Direct—better than 44 in./lbs.; reverses—better than 4 in./lbs.
Salt spray: 5% salt spray at 15 degrees from vertical—400 hrs. with 1/32" creepage.
Abrasion resistance: Excellent.
Humidity: Accelerated in chamber—800 hrs., no change.
Fresh water immersion: 800 hrs., no change.
Salt water immersion: 800 hrs., no change.

GPC 76

Specification Source: TTP 26
Product type: Interior fire retardant paint.
Outstanding characteristics: A flame spread rating of 25, with red oak flooring as 100; and a smoke development of not more than 50; scrubbability.
Main uses: To protect interior surfaces by delaying progress of fire.
Resin type: Not specified. Usually alkyd; vinyl toluene acrylate; polyvinyl acetate; or vinylated-amino alkyd.
Special formula notes: A carbonic, usually pentaerythritol, is provided to provide char; a catalyst, usually a diammonium phosphate is required to trigger the carbonific; and gas producer to puff the foam, an amine, is needed; and finally a resin binder.

PERFORMANCE
Set-to-touch: 1/2-4 hrs.
Dry-hard: 18 hrs.
Dry opacity: No more than 9.5 ml./sq.ft. spread, contrast ratio of 0.91 for white; and 0.95 for tints.
Gloss: Specular gloss, 40.
Spread rate: For class rating (flame spread between 0 and 25—about 140-175 sq.ft.)
Brushing: No sagging permitted; brush tip should be able to deposit a film.
Spraying: No more than 1 pint of solvent per gallon for spray.
Abrasion resistance: Test consists of a brush described in the specification and a cake grit soap and weight to produce 1 lbs. pressure on the film; after 1,000 oscillations (2,000 strokes) no

breaks, wear, or detachment of film; pinpoint breaks will be allowed.

Fire resistance: A flame spread rating according to National Fire Protection Ass'n standards, of more than 25 or a smoke development of more than 50 shall constitute failure.

Water resistance: Coated panels like those tested for flame spread are to be immersed in distilled water at 120° F for 24 hrs.; then air-dried and placed in an oven at 120° F for 40 hrs.; and then tested for fire retardancy; a weight loss of more than 15 grams and a char volume of more than 4.50 cu.in. shall constitute failure.

Fire retardancy test: Pertinent test methods: for details of fire retardancy tests, see American Society for Testing Materials D 1360-58—*Test Methods for Fire Retardancy of Paints* (Cabinet Method); and E 84-60 T—*Test for Surface Burning Characteristics of Building Materials;* copies can be obtained from the Society at 1916 Race St., Philadelphia, Pa. 19003.

GPC 77

Specification Source: TTP 636
Product type: Primer, alkyd.
Outstanding characteristics: Water, gasoline and weather resistance.
Main uses: Primer for wood and ferrous metal.
Resin type: Alkyd.
Special pigments: Total 40–45%, of which iron oxide at least 50%; zinc yellow, at least 10%; zinc oxide, 10–15%.
Color limitations: Affected by presence of iron oxide and zinc yellow.
Special formula notes: Not more than 30% siliceous extender; nonvolatile vehicle, at least 37% with minimum 30% phthalic anhydride component in alkyd; oil acids at least 48%.

PERFORMANCE
Set-to-touch: 15 mins.
Dry-through: 18 hrs.
Full hardness: 72 hrs.
Gloss: (60 degrees) 5–30.
Spraying: 1 part mineral spirits to 5 parts, must spray satisfactorily.
Weather resistance: 18 mos. exposure Washington, D.C., 2 coats of primer, no loss in metal protection properties.
Flexibility: 1/4″ mandrel.
Chemical resistance: 2-mil film air-dried 96 hrs. and immersed in gasoline 18 hrs. at 25° C; no wrinkling during immersion; 4-hr. recovery, slight dulling only permitted.
Water resistance: Same test as gasoline.

appendix C
Ultra High Performance Coatings

Excellent high performance coatings were discussed in considerable detail in Chapters 1, 7, 8, 9 and 10. While these coatings will protect exposed surfaces under those harsh conditions most likely to be encountered, they are inadequate for some extremely severe circumstances. Often they will not endure sufficiently in situations where recoating would be a serious undertaking.

Circumstances in which highly concentrated chemicals, radiation, ozone, abrasives, continual impact, or the vagaries of underground or underwater burial are involved, demand specially resistant coatings which we shall call Ultra High Performance, or UHP, Coatings.

Previously, consideration was given only to liquid coatings. Here, in addition to advanced liquid coating, we will discuss powder coatings: those made of conventional resins and necessary additives and color, and applied to surfaces either pre-heated or heated after application; and, powdered frits made from metal or specially formulated ceramics and applied by special heat guns or in highly sophisticated plasma arcs, yielding porcelain-like coatings that protect in atmospheres that would soon disintegrate conventional finishes.

Liquid UHP coatings are often used in thick films. Very thick versions of these materials, in fact, are often described as *membranes* or *linings*, since they have sufficient cross-section and flexibility to be handled as sheets. Frequently, these protective membranes or linings are extruded off-site and are affixed to the surface by powerful adhesives. Heavy-gauged linings are really a hybrid born of coating and plastic-film technology.

Cost factors, of course, enter in. Equipment intended to last for two or three years hardly requires heavy film builds or costly protective materials that provide 8 to 40 years life expectancy.

Where chemicals must be fended off, the coating selected and the film build may also depend on temperature of the environment and duration of exposure. Surfaces reached by splashed chemicals will not require the same degree of protection as those to be immersed.

Users of UHP coatings without previous extensive experience as to these coatings' characteristics, would be wise to consult specialists in protection against problem environments. Certified corrosion engineers (usually members of National Association of Corrosion Engineers) are logical sources of objective help where corrosion is a problem. Technical advisers from any one of several manufacturers of UPH coatings would also be helpful, even if understandably less objective.

Knowing what's available and the circumstances in which various UHP materials should be used will certainly help to understand the problems of treating threatened surfaces and to know if proper steps are being followed by personnel engaged to assure this treatment. Moreover, knowing what's involved in selecting materials and procedures will help in discussing the job with those who will be called on to contribute to its fulfillment. And, finally, the chances of getting good results are enhanced when working from a base of some—even if modest—knowledge of the subject.

What should you know before talking to a corrosion engineer or UHP coating supplier? Here are some of the most important points:

1) What is the environment to which the surface will be exposed?
2) If corrosive chemicals are involved, what are they?
3) What are the highest temperatures at which these chemicals will be held?
4) Will ozone or radiation be present?
5) Will the surface be subject to abrasion or high impact?//
6) How long a life expectancy has the object which is to be coated?
7) What difficulties are likely in recoating?

Then when your supplier discusses his recommended products, here are some details to consider:

1) Which of the recommended coatings is likely to last the longest before recoating is necessary?
2) Total applied cost of the recommended coating system (primers, intermediate, and top coats, plus surface preparation and application) and of alternatives.

Ultra High Performance Coatings

3) Will the extra cost of the most durable coating be justified considering the cost of recoating including the high cost of application and surface preparation? (See Chapter 10, Economics of Structural Coatings.)
4) Will best results and/or greatest economy result from off-site factory application?

With this information, you can make an intelligent decision about a coating system without being entirely dependent upon your supplier, who may or may not be interested in selling what yields him the best return.

To help understand UHP coating systems and when they should be used, a description of the various UHP families follows. You will note that some families previously described as High Performance Coatings are included. These materials—acrylics, urethanes, and epoxies—under some circumstances may be considered Ultra High Performance Coatings, although under many conditions other families outperform them.

ULTRA HIGH PERFORMANCE VEHICLE FAMILIES

1101 Epoxies. These tough, alkali-resistant materials have been covered in considerable detail in Chapters 1, 7 and 8. While most epoxies properly qualify only as high performance coatings, they are included under UHP vehicle families because: (1) epoxy primers, applied over properly prepared surfaces, are often used in UHP systems; (2) polyester-cured epoxies, also described in previous chapters, really qualify in the ultra category; (3) epoxy powder coatings—always factory applied—certainly qualify for a UHP rating because they approach porcelain in hardness and resistance to the environment; and (4) coal tar epoxies, if appearance is not important, perform outstandingly where heavy-duty resistance to abrasion and harsh atmospheres is required, or where burial in soil or immersion in water is involved.

1101.1 Epoxy primers with zinc dust. These are also known as organic zinc primers. When epoxy primers are to be specified, it is important to distinguish between those cured by a polyamine resin, which is relatively hard and chemical-resistant, and one that is cured by a polyamide, which is more flexible and water-resistant.

1101.2 Epoxy powder coatings. These are especially important to specifiers working with steel reinforcement bars. Where salt air or salt for road de-icing are likely to be borne by melting snow or rain water through permeable concrete, experience has shown that top mats of reinforcement bars will rust, cause ugly spalling on the surface and eventually lead to concrete failure. Coating these reinforcement bars with epoxy powder, melted and then hardened, yields a porcelain-like protective cover that fends off salt or other chemicals.

1101.3 Water-borne epoxies. Cured by an acrylic resin, water-borne epoxies show promise as a replacement for conventional epoxies when surface conditions rule out use of a solvent-reducible coating. Moreover, these materials may some day replace epoxies that need solvents which fail to meet requirements of air pollution authorities.

1101.4 Coal tar epoxies with amine curing agents. Used primarily as concrete primers with polyamide-cured epoxy topcoats. Polyamide-cured versions are also used for protection of buried or immersed metals, particularly in sea water, sour or crude petroleum products, and in a wide variety of chemicals.

1102 Fluorocarbons. Members of this family give maximum resistance to most chemical environments. They are costly, but where long-term resistance to chlorine, bromine, sulfuric acid, and most other commonly used acids and solvents is required, they are selected. Because of resistance to gamma and other kinds of radiation, they are widely used in nuclear power plants.

Fluorocarbons have given many years of maintenance-free service in hard-to-reach areas of structures, such as on the sides and fronts of skyscrapers. This is also true of mixing tanks, valves, and piping where maintenance shutdowns can lead to business losses and inconvenience.

Fluorocarbons safely withstand temperatures from minus 80°F to 300°F in prolonged usage.

Versions of fluorocarbons now available require factory application with in-the-field touchup.

1103 Polyesters. Members of this versatile family impart desirable properties to coatings, mostly in combination with epoxies, urethanes, or silicones. Addition of polyesters aids toughness, and, in some instances, provides glass-like characteristics.

1104 Silicones. Addition of members of this family to epoxies improves chemical resistance and permits higher baking temperatures. Clear silicones provide long-term protection to decorative metals.

1105 Urethanes. Whenever impact and abrasion is likely in an envi-

Ultra High Performance Coatings

ronment with fairly harsh acid and/or alkali chemicals, urethanes are desirable. They do not measure up to fluorocarbons in chemical resistance but are significantly less expensive, although far from low-priced.

1105.1 Two-package urethanes. Their major benefit is long-life protection. One such product, aliphatic urethane, is preferred where nonyellowing is important and where graffiti is a problem. When properly compounded this urethane is practically impervious to dyes and solvents used in spray paint and marking crayons used by graffiti "artists"; available cleaning solvents can remove the "art" and leave the coating almost intact after hundreds of cleanings.

1105.2 Air-dried urethanes. Applied in the factory or the field, they have the hardness and durability of oven-dried industrial coatings and thus aid in saving energy.

1105.3 Elastomeric urethanes. These versions are compounded to have characteristics of tough rubber. As linings for tanks and other containers for acids, alkalis, and abrasive materials they have given outstanding service.

1106 Acrylics. This family has long had representatives in ordinary over-the-counter household coatings, in industrial maintenance products, and in excellent solvent-reduced, factory-applied materials.

Its inclusion in the UHP class results from a successful method by which an excessively permeable dam structure, deteriorating because of water penetration, was saved from almost certain destruction. To save the Dworshak Dam in Colorado, an acrylic resin was polymerized on the site under an improvised temporary wooden structure; the result was the equivalent of about a one-half inch mantle of clear Plexiglass laced with a matrix of in-place concrete. Intermediates for making Plexiglass were mixed and applied to the concrete surface, and heat was applied under cover of a temporary wooden structure. Concrete was made impermeable and deterioration was arrested.

Another UHP acrylic is a road-topping compound developed in Germany and to be introduced in the U.S. It will serve as an overlay on road and bridge decks threatened by erosion of steel reinforcement bars and, if the German experience is repeated, will result in the salvaging of bridges and concrete structures that otherwise would have to be replaced.

1107 Chlorosulfonated Polyethylene (Hypalon). When protection is needed against ozone and oxidizing chemicals, members of this tough, chemical-resistant family are selected. Various curing agents are used in order to achieve specific characteristics. Litharge (a lead-containing catalyst), for example, is used when maximum chemical resistance is required. If sulfuric acid and little water is present, magnesium oxide is selected.

1108 Polychloroprene (Neoprene). Protection against ozone attack and abrasion is provided by members of this family, which also withstand a variety of strong chemicals. Neoprene is vulcanized in somewhat the same way as natural rubber, but results are superior. By changing modifiers and production procedures, Neoprene can be varied to suit the user's needs. It may be obtained as a solvent-based or water-reducible coating, and as a lining. When Neoprene is to be exposed to sunlight, or atmospheric conditions in general, it is usually over-coated with Hypalon.

1109 Chlorinated Polyether. Coatings or linings made with this material are resistant to numerous harsh chemicals. Sizes of objects to be coated are usually limited by the oven sizes available to provide fusion heat, which is about 375 to 425°F (190-218°C).

Single coats of 20 or more mils are readily applied. As much as 30-35 percent solids may be present in dispersions in chlorinated hydrocarbons. Fusion should be by separate coats, and a handsomer surface will result if the final coat is quenched immediately after its application.

Chlorinated polyethers are also available as powders and linings, the latter extruded in cross-sections of 40 or more mils from molding pellets. Linings are applied to sandblasted metal over rubber adhesives that are heat-reactivated at 250°F (121°C), just before the liner is applied.

1110 Rubber. (Linings) Natural rubber enhanced by additives is used for linings at least one-eighth inch thick to protect against chemicals, abrasion and impact.

Rubber sheet is calendered and pre-cut prior to placement on sandblasted surfaces. Hard rubber is used over rigid surfaces; soft rubber is used where abrasion is a problem.

1111 Vinylidene Chloride. Monomers such as acrylonitrile or vinyl chloride are combined with vinylidene chloride to form polymers that are effective in preventing damage from water penetration and capable of withstanding attacks by inorganic acids. They are used in

Ultra High Performance Coatings

solutions or as linings and are applied four to ten mils thick. Material quality varies for some reason; corrosion engineers recommend trials under prevailing conditions before full-scale use.

1112 Metallizing Sealers. Surfaces exposed to extremely corrosive atmospheres have been protected for long periods by powdered metals, usually zinc or aluminum, applied by special spray guns fitted with heating devices to liquify metal for application. Surfaces successfully protected this way include bilges of salt or fresh water, and tanks used for strong acids or alkalis.

Using clear or aluminum-pigmented vinyl coating over the intimately bonded metal overcomes any porosity that may have developed and gives added protection.

Some spray guns feed metal wire instead of powder.

Exotic anti-corrosive metals such as tungsten carbide-cobalt, chromium carbide, nickel chrome, and magnesium zirconate are used when extreme corrosion conditions, radiation, and temperatures beyond 1000°F are encountered, especially on wear surfaces.

appendix D

Preventing Underground and Underwater Corrosion

Engineers frequently are called on to recommend methods for protecting buried or immersed metals which are difficult, or often impossible, to replace.

Methods for long-term corrosion protection of these threatened materials have been developed. Certified corrosion engineers should be consulted; nonetheless, an understanding of the physical and chemical aspects of this type of corrosion is valuable in selecting a specialist and in helping him do his job.

Like all corrosion processes, underground or underwater metal damage of this kind is akin to the working of an electric storage battery, in which a current flow is set up whenever a complete electrical circuit exists between two metals in the battery. Over time one of the metals erodes or is, in effect, eaten up. When the entire supply of that metal has been consumed, the battery is dead.

In underground or underwater corrosion, an electric current flows when a complete circuit is established between two dissimilar metals, usually connected by moisture. As in a battery, one of the metals is eaten up in the process, causing damage to the structure involved—unless something is done.

Various metal protective coatings can be used, but these are often inadequate. In these cases cathodic protection systems are needed.

Simply stated, these systems sacrifice a replaceable metal for the one to be protected. The replaceable one is usually less "noble" than the metal of the structure. Zinc, for instance, will corrode in a buried circuit with steel; it is less noble than steel and is therefore a sacrificial metal when connected to buried or immersed steel. (See P. 28, "Cathode Protection", and PP. 131-139, "Noble Metals".)

To provide cathodic protection, we may use an Impressed-Current System or a Sacrificial Anode System.

In the Impressed-Current System, an external source of direct current is imposed between the anode (or positive terminal) of the power source and the protected metal, which becomes the cathode (or negative).

Current flows whenever the intervening earth has enough moisture to make it an electric conductor. Without the imposed current, the dampness would have served to conduct the electric charge between the stray metals in the soil and the endangered metal object, leading to breakdown of the latter.

The positive terminal at the power source is sacrificed when the Impressed Current System is in operation. Even more is done, however; minerals in the damp earth are deposited in the process on the protected material, providing an additional protective barrier against corrosion, and eventually, the amount of current needed for the system is reduced, lengthening the life of the power source.

The simpler Sacrifical Anode System uses suitably sized pieces of metal capable of setting up a current, in the presence of a conductor, with the metal to be protected. But this method is more difficult to control than the Impressed Current System. The sacrificial metal is consumed, the current is reduced; therefore, frequent inspections are needed when this method is used. The metals more commonly used as sacrifices are magnesium, aluminum, and zinc or alloys of them.

The Galvanic Series of Metals determines what metals will be effective in protecting other metals. In the Galvanic series listed below, lead is higher than copper or brass. Lead near copper in the ground will be corroded, while copper will be intact. Silver, carbon, platinum and gold, as can be seen in the Series listing, are lowest so they are the most stable.

Galvanic Series of Metals

10. Magnesium
9. Aluminum
8. Zinc
7. Steel, Iron
6. Lead, Lead-tin Solder
5. Copper, Brass
4. Silver, Silver Solder
3. Graphite, Carbon
2. Platinum
1. Gold

The magnitude of the current that will flow between metals in the soil depends on soil resistivity, which may vary over a range of 1,000 to 1. Season and rainfall are two primary factors.

Testing for soil resistivity requires spacing of four electrodes at varying distances and impressing electric charges on them and then taking averages. Resistivities of about 1,000 ohms per

Preventing Underground and Underwater Corrosion 407

centimeter or less are considered severely corrosive. (Sea water's resistivity, for example, is 20 ohms per centimeter.)

Soils having more than 20,000 ohms per centimeter are considered slightly corrosive. Even this degree may lead to pits in steel.

Steel structural members passing through several layers of soil will corrode more than those in one uniform layer, because the various layers will set up complex differences in electric potential that are more destructive. To get a true understanding of the corrosion problem in buried metal it is necessary to take borings and check resistivity at various levels.

Moisture content plays a role. Well-drained soils are normally highly resistive because much of the soluble elements that could aid conductivity have been washed away. Damp soils with cinders, on the other hand, are more corrosive because carbon in the Galvanic Series has a lower number than steel; thus, steel becomes the sacrificial metal.

CORROSION IN WATER

Structures built partly in water will often have rusting pilings unless coatings or cathodic protection is used.

Particularly affected is the so-called Splash Zone, which is just above the high tide level; corrosion is also bad just below the mean low water level.

Differential Aeration is responsible for this corrosion. Differential aeration is a difference in the quantity of oxygen present in the water at corrosion points and elsewhere.

That portion of the pile between mean low water and high water, because it is partly or wholly in touch with air, naturally has more oxygen than those portions constantly immersed. So, corrosion currents flowing in the steel between areas of low oxygen (the immersed portion) and those above where oxygen is more abundant, result in a situation similar to a current flowing from a noble metal to a less noble one; in this case, the area with less oxygen is less noble and is corroded. Differential Aeration, therefore, describes the difference in available air.

Buried and immersed metals often utilize coatings and galvanic protection. The National Association of Corrosion Engineers reports that tests show extreme care is needed in these cases. Selected coatings must resist the alkali reactions that may take place at the protected metal's surface; the reactions can cause blisters and subsequent failure. Vinyl coatings and chlorinated rubber are reported to be effective in this situation. Even these can be damaged if excessive aklali accumulates. Very modest voltages and currents must be used in order to avoid blisters. NACE recommends that cathodic protection voltages be less than 1 volt and current density less than 10 milliamperes/sq. ft.

We repeat this caution: use only a certified, accredited corrosion engineer in setting up a cathodic protection system.

appendix E

Urethane Foam Insulation Elastomer Roof Coating
GUIDE SPECIFICATION D/U 4-01

Note: This is an example of an unusually clear, concise specification free of confusing terminology.

SECTION 07200/07541

PART 1–GENERAL

    1.01 SCOPE:
        A. This specification covers the preparation and application to roof surfaces of a monolithic, spray applied rigid urethane foam insulation and elastomer roof coating which shall be a composite roof system.
        B. This system shall provide a specified insulative value and shall provide a waterproof elastomeric weather barrier possessing excellent adhesion and physical bond strength to substrate. This system shall maintain hydraulic stability without age-hardening or slump.
        C. The applicator shall furnish all labor, materials and equipment and perform all operations required as specified.

    1.02 QUALIFICATIONS
        A. Manufacturer: For the purpose of defining the quality of the work and materials in this Section:
            1. Coating specifications are based on physical properties of _____ as manufactured by _____.
            2. Sprayed Urethane Foam specifications are based on general standards acceptable throughout the sprayed urethane foam industry.

B. Applicator: Application shall be by a properly qualified applicator with basic knowledge of these products and who has contacted the manufacturers for proper application procedures.

1.03 SUBMITTALS:
A. Manufacturer's Data: Submit tech data sheets, independent testing reports, application instructions and precautions, and manufacturer-contractor warranty.
B. Materials or formulation types other than that specified shall be submitted to the architect for approval not later than ten (10) days prior to bid date. Request shall be accompanied with notarized certification and test data delineating physical properties, coated urethane foam sample, and warranty.

1.04 WARRANTY
A. Refer to ___'s ten year insured roofing warranty or ___'s standard five-year roofing warranty, as required for project.

1.05 PRODUCT HANDLING:
A. Deliver products in manufacturer's original sealed containers, with seals and labels intact.
B. Store materials in an enclosed space protected from weather and out of the direct rays of the sun.
C. Do not ship or store materials unless protection against freezing (32°F.) is available.

PART 2–PRODUCTS

2.01 MATERIALS:
A. Primer: As required or recommended for jobsite conditions.
  1. Foam insulation over new substrate:
     a. Concrete or Masonry = _____.
     b. Lightweight Concrete = _____.
     c. Chalky or Punky Concrete = _____.
     d. Wood = _____.
     e. Metal = _____.
  2. Foam insulation over asphalt or coal tar BUR systems = _____.
  3. Recoating roof over existing foam and coatings = _____.

B. Sprayed Urethane Foam:
   1. Material for the roofing foam shall be a two-component liquid applied sprayable type. Application shall result in a high quality rigid urethane foam roofing as per following physical properties:

Physical Properties	Test Unit	Value	Method
Density (overall)	Pounds/cubic ft.	2 to 3 lbs. as determined by specific project requirements.	ASTM D-1622
"K" Factor (average)	Btu/hr/ft^{2}F./in.	.14 to .16	ASTM D-518
Open Cell Content	% by volume	10	ASTM D-2856
Water Vapor Permeability	Perm/In.	2.0 to 3.0	ASTM C-355
Water Absorption	Pounds/square ft.	.03 max	ASTM D-2127
Flammability*	Flame spread	75 maximum	ASTM E-84**
Compressive Strength Parallel to rise	Psi	30 minimum	ASTM D-1621
Tensile Strength Parallel to rise	Psi	40 minimum	ASTM D-1623
Shear Strength Parallel to rise	Psi	20 minimum	ASTM D-273

*Note: When tested for use on roof structures in combination with _____, will have passed all UL-790 Class A label requirements.

**Note: ASTM tests are used solely to measure and describe properties in response to heat and flame under controlled laboratory conditions, and are not intended to reflect hazards presented under actual fire conditions.

C. Elastomer Roof Coating:
1. Shall be a non-oxidizing high solids elastomer rubber with reinforcing laminar pigments, and non-migrating fire retardants.
2. Elastomer roof coating shall contain no solvents, migratory plasticizers, vegetable oils, marine oils, asphaltic or cementacious materials. Use of non-elastomeric resins are not permitted.
3. Material shall conform to the following minimum physical properties:

Physical Properties	Test Procedure	Value	Method
Ultra Violet (UV) Resistance	Atlas Twin Arc Weatherometer	No deleterious effects: no oxidation, surface checking, or fade —after 5,000 hours	ASTM D-822 ASTM G-23
Weather Resistance	Atlas Twin Arc Weatherometer	Retains elastomeric properties, equivalent to 15-20 years	ASTM D-822 ASTM G-23
**High Temperature Stability	Thermostatically controlled heat chamber	No age-hardening or slump up to 200°F.	ASTM D-794
Resistance to Wind Driven Rain	1) Pressurized water chamber 2) Moisture meter	Moisture recorded = 0%	Fed. Std. TTC-555-B
**Elongation Unaged	Instron Universal Testing Instrument	0°F = 120% 40°F = 200% 75°F = 280% 100°F = 360%	ASTM D-2370
**Elongation Retained After Aging	1) Atlas Twin Arc Weatherometer 2) Instron Universal Testing Instrument	95% Elongation retained after 2,000 hours	ASTM G-23 ASTM D-822 ASTM D-2370
**Hardness	Durometer Shore A	45	ASTM D-2240
Bond Strength	Instron Universal Testing Instrument	52 lbs./sq. in.	ASTM C-297
Resistance to Foot Traffic	Penetration Plate	4,176 lbs./sq. ft.	ICBO Roofing Standard

Urethane Foam Insulation/Elastomer Roof Coating

Hailstone Resistance	Steel Ball Velocity Procedure	No fracture of surface film	National Bureau of Standards
**Film Breathing Ability	Honeywell Water-Vapor Transmission Rate Tester	3.0 perms = 20 dry mils	ASTM E-398
Conformance to Environmental Pollution Standards	Air Pollution Control Dept. (APCD) State of California	No photochemically reactive solvents	Rule 442, 6, 3
Fire Resistance	1) Roof Structure Assembly Test	Class A Labels	UL-790
	2) Roof Structure Calorimeter	Class 1	Factory Mutual

**These tests conducted on free films as per ASTM requirements. All other tests conducted on Diathon/Urethane Foam composite.

2.01 MATERIALS:

D. Topcoat (optional): may be utilized at high visual impact areas where maximum soil resistance, cleanability, and beauty are of prime importance. Coverage: Topcoat shall be applied by roller or airless spray at a rate of 150 to 200 square feet per gallon. Color as selected by architect.

PART 3 – EXECUTION

3.01 CONDITION OF SURFACE:

The remainder of this specification covers surface preparation and application. Since the purpose of this appendix is merely to illustrate the simplicity of language and the specificity of detail desired, the balance of this specification has not been included.

appendix F

Comparative Corrosion Resistance Chart

The following chart has been compiled as a guide to the service of various liquid coatings in chemical fume and splash service. It should be used as a general guide only and not as a standard for specification or performance.

KEY FOR USE OF DATA

Number	— Maximum temperature (°F)
NR	— Not recommended
ID	— Insufficient data available
F&S	— Fume & Splash service only
L	— Product may be used as lining
>	— Greater than (i.e. > 50%—greater than 50%)
<	— Less than (i.e. < 50%—less than 50%)

The information compiled in this chart is based on Gates Engineering Company's research and tests made by others we believe to be reliable. Because of the wide variations of conditions that may be encountered in the field, these are not to be regarded as hard and fast recommendations.

Comparison Corrosion Resistance Chart

	EAA-35 Epoxy Amine-Adduct (F&S)	ECP-21 Epoxy (F&S)	Polyamide (F&S)	Neoprene-Bituminous Blend (F&S)	H-2 Hypalon (F&S)	N-700-A Neoprene (F&S)	Plastisol, Vinyl (L)	Phenolic, Baked (L)	UCP-21 Urethane Med. Chemical Resistant (F&S)	UM-31 Urethane High Chemical Resistant (F&S)	D-202-V-80 Plasticized Vinyl Coating (L)
Acetic Acid, 60%	NR	NR	NR	NR	NR	NR	NR	70	NR	NR	NR
Acetic Acid, 20%	ID	NR	NR	NR	NR	NR	110	90	NR	NR	NR
Acetic Acid, 5%	70	NR	NR	70	NR	NR	120	120	NR	NR	NR
Acetic, Glacial	NR	NR	NR	NR	NR	NR	NR	NR	NR	NR	NR
Acrylic Acid, Conc	NR	NR	NR	NR	NR	NR	NR	NR	NR	NR	NR
Adhesive, Aqueous	80	70	70	70	100	100	120	120	70	100	120
Adipic Acid, aq sol	70	70	NR	NR	100	100	120	120	70	90	100
Alcoholic Spirits	NR	NR	NR	ID	150	150	140	100	70	70	120
Aluminum Acetate, aq sol	70	70	70	120	200	175	140	100	NR	80	140
Aluminum Basic Acetate, aq	80	80	NR	120	NR	NR	140	100	70	100	140
Aluminum Chloride-anhy	NR	NR	NR	NR	150	150	140	100	NR	ID	70
Aluminum Chloride-solutions	NR	70	120	120	150	100	140	100	70	100	120
Aluminum Nitrate, aq sol	90	70	ID	ID	125	200	140	100	70	100	120
Aluminum Sulfate, aq sol	70	80	120	120	200	200	140	120	70	100	140
Alums	90	80	120	120	200	70	140	120	70	120	140
Ammonium 5%	70	80	ID	ID	90	NR	100	NR	NR	NR	100
Ammonium 15%	NR	NR	ID	ID	70	NR	90	NR	NR	NR	100
Ammonium 28%	NR	NR	ID	ID	NR	100	70	NR	ID	NR	70
Ammonium Acetate, aq sol	70	70	120	120	100	125	140	100	70	80	120
Ammonium Bromide, aq sol	80	70	120	120	150	NR	140	100	70	90	120

	EAA-35 Epoxy Amine-Adduct F&S	ECP-21 Epoxy Polyamide F&S	Neoprene-Bituminous Blend F&S	H-2 Hypalon F&S	N-700-A Neoprene F&S	Plastisol, Vinyl L	Phenolic, Baked L	UCP-21 Urethane Med. Chemical Resistant F&S	UM-31 Urethane High Chemical Resistant F&S	D-202-V-80 Plasticized Vinyl Coating L
Ammonium Chloride, aq sol	80	70	120	175	150	140	120	70	100	120
Ammonium Fluoride, aq sol	80	70	120	150	125	140	100	70	100	120
Ammonium Nitrate, aq sol	70	70	120	200	175	140	100	70	100	100
Ammonium Phosphate, aq sol	80	70	120	175	150	140	120	70	100	140
Ammonium Sulfate, aq sol	90	80	120	200	200	140	70	70	100	140
Amyl Acetate	NR	NR	NR	NR	NR	NR	70	NR	NR	NR
Aniline Hydrochloride, aq	70	NR	NR	ID	NR	ID	70	NR	ID	100
Animal Glue, aq	70	ID	70	100	100	120	120	ID	100	100
Animal Oils & Fats	NR	NR	NR	NR	NR	NR	100	NR	90	100
Anthracene Oil	80	NR	NR	NR	NR	NR	120	NR	70	NR
Asphalt	80	NR	NR	NR	NR	NR	120	NR	90	70
Beer	NR	NR	ID	NR	NR	NR	100	70	70	120
Benzene	70	NR	NR	NR	NR	NR	120	NR	70	NR
Benzoic Acid, aq sol	NR	70	NR	70	70	120	120	NR	70	70
Bismuth Subnitrate	70	70	120	150	100	140	120	70	80	120
Borax	100	120	120	200	200	140	150	100	120	140
Boric, dry or aq	100	70	100	150	120	140	100	NR	120	120
Borium Hydroxide 10%	90	100	120	175	175	140	100	NR	70	120
Bromine Water, Saturated	NR	NR	ID	ID	NR	NR	120	NR	NR	NR
Butane	90	80	NR	70	70	NR	NR	NR	100	ID

Comparison Corrosion Resistance Chart

	EAA-35 Epoxy Amine-Adduct	ECP-21 Epoxy Polyamide	Neoprene-Bituminous Blend	H-2 Hypalon	N-700-A Neoprene	Plastisol, Vinyl	Phenolic, Baked	UCP-21 Urethane Med. Chemical Resistant	UM-31 Urethane High Chemical Resistant	D-202-V-80 Plasticized Vinyl Coating
	F&S	F&S	F&S	F&S	F&S	L	L	F&S	F&S	L
Butanols	NR	NR	NR	NR	NR	NR	100	NR	NR	110
Butter	NR	NR	NR	NR	NR	NR	100	NR	70	NR
Butyl Acetates	NR	NR	NR	NR	NR	NR	70	NR	NR	NR
Butyric Acids, aq sol	NR	NR	NR	NR	NR	100	120	NR	NR	NR
Cadmium Chloride, aq sol	80	70	120	150	150	140	120	NR	90	120
Cadmium Nitrate, aq sol	80	70	ID	150	100	140	100	NR	90	120
Cadmium Sulfate, aq sol	90	80	120	200	175	140	120	70	100	140
Calcium Bisulfide	NR	NR	ID	100	NR	140	70	NR	ID	70
Calcium Carbonate, wet	120	120	120	200	200	140	150	80	120	140
Calcium Chlorate, aq sol	NR	NR	ID	150	100	140	ID	NR	ID	120
Calcium Chloride, aq sol	100	90	120	200	200	140	120	70	100	140
Calcium Hydroxide, any %	90	100	ID	175	175	140	120	NR	70	120
Calcium Hypochlorite, dil	NR	NR	ID	150	NR	140	100	70	70	100
Calcium Nitrate, sol	80	70	120	150	100	140	100	100	100	120
Calcium Sulfate, sol	100	90	120	200	175	140	150	100	120	140
Carbon Dioxide, gas	120	120	120	150	150	70	150	70	100	140
Carbonated Beverages	NR	NR	70	NR	NR	140	120	70	120	140
Carbonated Water	120	120	120	200	175	140	120	70	120	120
Cheese	NR	NR	NR	NR	NR	NR	100	NR	100	NR

Chemical	EAA-35 Epoxy Amine-Adduct F&S	ECP-21 Epoxy Polyamide F&S	Neoprene-Bituminous Blend F&S	H-2 Hypalon F&S	N-700-A Neoprene F&S	Plastisol, Vinyl L	Phenolic, Baked L	UCP-21 Urethane Med. Chemical Resistant F&S	UM-31 Urethane High Chemical Resistant F&S	D-202-V-80 Plasticized Vinyl Coating L
Chlorine Water, dilute	NR	NR	ID	70	NR	70	NR	NR	NR	NR
Chlorotrifluoroethylene	ID	ID	ID	ID	ID	ID	120	NR	ID	NR
Chromic 1%, (aq)	NR	NR	ID	100	NR	140	70	ID	ID	100
Chromic 10%, (aq)	NR	NR	ID	100	NR	140	70	NR	NR	70
Chromic 30%, (aq)	NR	NR	NR	90	NR	140	NR	NR	NR	NR
Chromic 40%, (aq)	NR	NR	NR	70	NR	130	NR	NR	NR	NR
Chromium Chloride, sol	80	70	120	150	150	140	120	70	80	120
Chromium Nitrate, sol	70	70	120	150	100	140	100	70	90	100
Chromium Sulfate, sol	80	80	120	200	175	140	120	70	100	120
Citric, aq sol	70	70	70	100	100	120	120	NR	90	120
Coal Tar	NR	NR	NR	NR	NR	NR	120	NR	70	NR
Coal Tar Oils	NR	NR	NR	NR	NR	NR	120	NR	70	NR
Cobalt Chloride, sol	80	70	120	150	150	140	100	70	100	120
Cobalt Nitrate, sol	70	70	120	150	100	140	120	70	100	100
Cobalt Sulfate	90	80	120	200	175	140	100	70	100	140
Copper/Ammonium Solutions	ID	ID	ID	ID	NR	70	120	ID	ID	70
Copper Chloride, sol	80	70	120	125	125	140	100	70	80	110
Copper Nitrate, sol	70	NR	ID	150	100	140	120	70	80	100
Copper Sulfate, sol	90	80	120	200	175	140	100	70	90	120
Corn Syrup	NR	NR	NR	NR	NR	100	100	NR	70	120

Comparison Corrosion Resistance Chart

	EAA-35 Epoxy Amine-Adduct	ECP-21 Epoxy Polyamide	Neoprene-Bituminous Blend	H-2 Hypalon	N-700-A Neoprene	Plastisol, Vinyl	Phenolic, Baked	UCP-21 Urethane Med. Chemical Resistant	UM-31 Urethane High Chemical Resistant	D-202-V-80 Plasticized Vinyl Coating
	F&S	F&S	F&S	F&S	F&S	L	L	F&S	F&S	L
Crude Oil	90	NR	NR	NR	NR	NR	120	NR	70	100
Cyclohexane	70	NR	NR	NR	NR	NR	120	NR	100	70
Dichlor Acetic, aq sol	NR	NR	NR	NR	NR	70	ID	NR	NR	NR
#12 Dichlorodifluoromethane	ID	ID	ID	70	70	ID	120	NR	ID	NR
#114 Dichlorotetrafluoroethane	ID	ID	ID	ID	70	ID	120	NR	ID	NR
Diethanolamine, aq	NR	NR	NR	NR	NR	70	ID	NR	NR	NR
Diethyl Maleate	NR	NR	ID	NR	NR	NR	70	NR	NR	100
Diethylene Glycol, aq<50%	70	70	ID	175	175	100	100	NR	70	NR
1,4 Dioxane	NR	NR	NR	NR	NR	NR	70	NR	ID	NR
Dowtherm	70	NR	NR	NR	NR	NR	120	NR	ID	NR
Ethanol, aq<50%	80	70	70	100	100	100	100	NR	70	100
Ethyl Acetate	NR	NR	NR	NR	NR	NR	70	NR	NR	NR
Ethyl Acrylate	NR	NR	NR	NR	NR	NR	70	NR	NR	NR
Ethyl Ether	NR	NR	NR	NR	NR	NR	70	NR	70	NR
Ethyl Methacrylate	NR	NR	NR	NR	NR	NR	70	NR	NR	NR
Ethylene Glycol, aq<50%	70	70	70	175	175	100	100	NR	70	100
Ethylene Oxide	NR	NR	NR	NR	NR	NR	70	NR	70	NR

	EAA-35 Epoxy Amine-Adduct F&S	ECP-21 Epoxy Polyamide F&S	Neoprene-Bituminous Blend F&S	H-2 Hypalon F&S	N-700-A Neoprene F&S	Plastisol, Vinyl L	Phenolic, Baked L	UCP-21 Urethane Med. Chemical Resistant F&S	UM-31 Urethane High Chemical Resistant F&S	D-202-V-80 Plasticized Vinyl Coating L
Ferric Chloride-anhy.	NR	NR	ID	NR	NR	ID	NR	NR	NR	100
Ferric Chloride-solutions	70	ID	120	100	70	140	100	70	90	120
Ferric Nitrate, sol.	70	NR	120	150	100	140	100	70	90	120
Ferrous Chloride-anhy.	70	70	120	150	150	140	120	70	100	140
Ferrous Chloride, dry	80	70	120	150	125	140	120	70	100	120
Ferrous Chloride-sol dil	80	70	120	150	125	140	120	70	100	120
Ferrous Sulfate, sol	90	80	120	200	175	140	120	70	100	120
Flavoring Extracts	NR	NR	NR	NR	NR	NR	100	NR	70	70
Flour, dry	100	100	NR	100	100	120	150	NR	70	140
Fluoboric, aq dil	70	70	100	NR	NR	140	100	ID	120	120
Formaldehyde 37%	NR	NR	ID	NR	NR	100	70	NR	ID	70
Formic, aq sol	NR	NR	NR	NR	NR	70	100	NR	NR	100
Fruit Juices	NR	NR	NR	NR	NR	NR	100	NR	70	120
Fruit Pulp	NR	NR	NR	NR	NR	NR	100	NR	70	120
Fumaric, aq sol	70	70	70	70	70	120	120	70	90	100
Gasoline	80	NR	NR	NR	NR	NR	120	NR	100	70
Gin	NR	NR	ID	NR	NR	NR	100	NR	70	NR
Glucose, aq	NR	NR	NR	NR	NR	120	100	NR	100	120
Glutaric, aq sol	70	70	70	100	100	120	120	70	90	120

Comparison Corrosion Resistance Chart

	EAA-35 Epoxy Amine-Adduct (F&S)	ECP-21 Epoxy Polyamide (F&S)	Neoprene-Bituminous Blend (F&S)	H-2 Hypalon (F&S)	N-700-A Neoprene (F&S)	Plastisol, Vinyl (L)	Phenolic, Baked (L)	UCP-21 Urethane Med. Chemical Resistant (F&S)	UM-31 Urethane High Chemical Resistant (F&S)	D-202-V-80 Plasticized Vinyl Coating (L)
Glycerine, aq<50%	70	NR	ID	150	150	100	100	NR	70	120
Glycol Diacetate	NR	NR	NR	NR	NR	NR	70	NR	NR	70
Halogens, Chlorine, Moist	NR	NR	NR	NR	NR	70	NR	NR	NR	NR
Heptane	90	80	NR	70	70	NR	120	NR	100	70
Hexane	90	80	NR	70	70	NR	120	NR	100	70
Hexanols	NR	NR	ID	NR	NR	NR	100	NR	NR	70
Hexylene Glycol, aq<50%	ID	ID	ID	70	70	100	100	NR	ID	70
Hydriodic Acid 5%	NR	NR	70	NR	NR	90	NR	NR	ID	NR
Hydriodic Acid 20%	NR	NR	NR	NR	NR	70	NR	NR	ID	NR
Hydriodic Acid 40%	NR	NR	NR	NR	NR	ID	NR	NR	ID	NR
Hydrobromic Acid 5%	NR	NR	70	70	70	90	90	NR	70	70
Hydrobromic Acid 15%	NR	NR	ID	NR	NR	70	70	NR	ID	NR
Hydrobromic Acid 30%	NR	NR	ID	NR	NR	NR	NR	NR	ID	NR
Hydrochloric Acid 5%	70	70	70	70	70	90	90	ID	70	70
Hydrochloric Acid 20%	70	NR	ID	NR	ID	70	70	ID	ID	NR
Hydrochloric Acid 36%	70	NR	ID	NR	ID	NR	NR	ID	ID	NR
Hydrocyanic Acid 10%	ID	ID	70	100	100	100	100	ID	70	100
Hydrofluoric Acid 5%	NR	NR	ID	NR	NR	90	NR	ID	ID	70
Hydrofluoric Acid 15%	NR	NR	ID	NR	NR	70	NR	NR	NR	NR

	EAA-35 Epoxy Amine-Adduct	ECP-21 Epoxy Polyamide	Neoprene-Bituminous Blend	H-2 Hypalon	N-700-A Neoprene	Plastisol, Vinyl	Phenolic, Baked	UCP-21 Urethane Med. Chemical Resistant	UM-31 Urethane High Chemical Resistant	D-202-V-80 Plasticized Vinyl Coating
	F&S	F&S	F&S	F&S	F&S	L	L	F&S	F&S	L
Hydrofluoric Acid 30%	NR	NR	NR	NR	NR	NR	NR	NR	NR	NR
Hydrofluoric Acid 50%	NR	NR	NR	NR	NR	NR	NR	NR	NR	NR
Hydrofluosilicic Acid, aq	ID	ID	100	120	120	140	120	70	120	70
Hydrogen Peroxide, 10–30%	ID	ID	ID	NR	NR	100	NR	ID	ID	70
Hydroquinone, aq	ID	NR	NR	70	70	100	100	NR	ID	100
Hydrosulfuric (H2S), aq	70	70	70	100	100	120	120	ID	70	100
Hydroxides, Heavy Metal Moist	100	100	120	175	175	140	140	70	120	140
Hypochlorous, dil	NR	NR	ID	NR	NR	70	NR	NR	NR	NR
Isobutyl Acrylate	NR	NR	NR	NR	NR	NR	70	NR	NR	NR
Isopropanol, aq<50%	80	70	70	100	100	100	100	NR	70	100
Isopropyl Acetate	NR	NR	NR	NR	NR	120	70	NR	NR	NR
Isopropyl Ether	NR	NR	NR	NR	NR	140	70	NR	70	NR
Kerosene	80	NR	NR	NR	NR	NR	120	NR	100	100
Lactic, aq sol	70	70	70	70	70	120	120	70	90	120
Lactose, aq or dry	NR	NR	NR	NR	NR	120	100	NR	70	100
Lead Acetate, sol	70	70	120	100	70	140	120	ID	80	110
Lead Basic Acetate, sol	80	70	120	125	100	140	120	70	100	120

Comparison Corrosion Resistance Chart

	EAA-35 Epoxy Amine-Adduct (F&S)	ECP-21 Epoxy Polyamide (F&S)	Neoprene-Bituminous Blend (F&S)	H-2 Hypalon (F&S)	N-700-A Neoprene (F&S)	Plastisol, Vinyl (L)	Phenolic, Baked (L)	UCP-21 Urethane Med. Chemical Resistant (F&S)	UM-31 Urethane High Chemical Resistant (F&S)	D-202-V-80 Plasticized Vinyl Coating (L)
Lead Chloride, sol	90	80	120	200	175	140	150	70	100	140
Lead Nitrate, sol	70	70	120	150	125	140	100	70	100	120
Lead Oxide, dry	90	90	120	200	175	140	150	100	120	140
Levulinic, aq sol	NR	NR	ID	NR	NR	120	120	ID	ID	100
Lime Water	90	100	120	175	150	140	120	NR	100	120
Lithium Hydroxide, aq 10%	90	100	120	175	150	140	100	NR	70	100
Lubricating Greases	70	70	NR	70	70	70	130	70	100	100
Lubricating Oils, Mined Type	70	70	NR	70	70	70	130	70	100	100
Magnesium Chloride, sol	80	70	120	175	150	140	120	70	100	120
Magnesium Fluosilicate, sol	70	ID	120	175	150	140	70	70	100	100
Magnesium Hydroxide	100	100	120	200	200	140	140	70	120	140
Magnesium Nitrate, sol	70	70	120	125	100	140	100	70	100	110
Magnesium Sulfate, sol	80	80	120	200	200	140	120	70	100	140
Maleic, anhydride	90	70	70	70	70	120	ID	70	90	100
Malic, aq sol	NR	NR	NR	NR	NR	NR	ID	NR	90	120
Malt Extract	70	70	120	150	150	140	100	70	70	100
Manganese Chloride, sol	NR	NR	120	150	125	140	120	70	100	120
Manganese Nitrate, sol	80	70	120	150	150	140	100	70	100	100
Manganese Sulfate, sol	90	80	120	200	175	140	120	70	100	140

Paints & Coatings Handbook

	EAA-35 Epoxy Amine-Adduct (F&S)	ECP-21 Epoxy Polyamide (F&S)	Neoprene-Bituminous Blend (F&S)	H-2 Hypalon (F&S)	N-700-A Neoprene (F&S)	Plastisol, Vinyl (L)	Phenolic, Baked (L)	UCP-21 Urethane Med. Chemical Resistant (F&S)	UM-31 Urethane High Chemical Resistant (F&S)	D-202-V-80 Plasticized Vinyl Coating (L)
Mercuous Nitrate, sol PH>6.5	70	70	120	150	100	140	100	ID	70	120
Mercuric Chloride, sol	80	70	120	150	150	140	100	ID	ID	100
Mercury	120	120	120	200	200	140	150	120	120	140
Mercury-Metal	100	100	120	150	150	140	150	100	120	110
Methanol, aq<50%	80	70	70	125	125	100	100	ID	ID	70
Methyl Styrene	ID	NR	NR	NR	NR	NR	120	NR	70	120
Milk	NR	NR	NR	NR	NR	NR	100	NR	70	70
Mineral Spirits (Naptha)	90	70	NR	NR	NR	120	120	NR	100	120
Molasses	NR	NR	ID	ID	ID	ID	100	NR	70	NR
#13B1-Monobromotrifluoromethane	ID	ID	ID	70	70	ID	120	NR	ID	NR
#22-Monochlorodifluoromethane	ID	ID	ID	70	70	ID	120	NR	ID	NR
#13-Monochlorotrifluoromethane	ID	ID	ID	70	70	ID	120	NR	ID	NR
Muriatic Acid-See Hydrochloric Acid										
Napthanene	70	NR	NR	NR	NR	NR	120	NR	70	NR
Neopentyl Glycol, aq<4.0	ID	NR	ID	70	70	100	100	NR	NR	70
Nickel Chloride, sol	80	70	120	150	150	140	120	70	100	120
Nickel Nitrate, sol	70	70	120	150	100	140	100	70	100	100
Nickel Sulfate, sol	90	80	120	200	175	140	120	70	100	140
Nitric Acid 5%	NR	NR	ID	100	NR	120	NR	NR	70	70

Comparison Corrosion Resistance Chart

	EAA-35 Epoxy Amine-Adduct (F&S)	ECP-21 Epoxy Polyamide (F&S)	Neoprene-Bituminous Blend (F&S)	H-2 Hypalon (F&S)	N-700-A Neoprene (F&S)	Plastisol, Vinyl (L)	Phenolic, Baked (L)	UCP-21 Urethane Med. Chemical Resistant (F&S)	UM-31 Urethane High Chemical Resistant (F&S)	D-202-V-80 Plasticized Vinyl Coating (L)
Nitric Acid 25%	NR	NR	NR	NR	NR	100	NR	NR	NR	NR
Nitric Acid 60%	NR	NR	NR	NR	NR	NR	NR	NR	NR	NR
Nitrous Acid	NR	NR	NR	NR	NR	100	NR	ID	ID	NR
Octane	90	80	NR	70	70	NR	120	NR	100	70
Oleic Acid, dry	NR	NR	NR	NR	NR	70	120	NR	70	120
Oxallic, aq sol	70	NR	70	125	125	120	120	70	90	120
Oxygen	120	120	120	200	150	140	150	120	120	140
Pentane	90	80	NR	70	70	NR	120	NR	100	70
Pentanols	NR	NR	NR	NR	70	NR	100	NR	NR	70
Penterythritol, aq<50%	80	70	70	175	150	100	100	NR	70	120
Petrolatum	70	70	NR	ID	NR	70	130	70	100	120
Petroleum Oils	90	NR	NR	ID	70	70	120	NR	100	100
Phosphoric Acid, 20%	NR	NR	100	NR	70	140	110	NR	70	70
Phosphoric Acid, 50%	NR	NR	ID	NR	140	100	100	NR	70	70
Phosphoric Acid, 85%	NR	NR	ID	NR	NR	140	70	NR	NR	NR
Phthalic Anhydride, dry	NR	NR	ID	NR	NR	100	120	ID	ID	120
Pinene	ID	NR	NR	NR	NR	NR	120	NR	70	NR

	EAA-35 Epoxy Amine-Adduct (F&S)	ECP-21 Epoxy Polyamide (F&S)	Neoprene-Bituminous Blend (F&S)	H-2 Hypalon (F&S)	N-700-A Neoprene (F&S)	Plastisol, Vinyl (L)	Phenolic, Baked (L)	UCP-21 Urethane Med. Chemical Resistant (F&S)	UM-31 Urethane High Chemical Resistant (F&S)	D-202-V-80 Plasticized Vinyl Coating (L)
Potassium—Same as corr. Sodium Salt										
Potassium Carbonate, aq	120	120	120	175	175	140	120	NR	100	140
Potassium Hydroxide 5%–15%	70	100	120	175	175	140	120	NR	70	120
Propanol, aq<50%	70	NR	70	100	100	100	100	NR	70	70
Propylene Dichlorine	NR	ID	NR	NR	NR	NR	110	NR	70	NR
Propylene Glycol, aq<50%	70	ID	ID	100	125	100	100	NR	70	70
Propylene Oxide	NR	NR	NR	NR	NR	NR	70	NR	70	NR
Salicylic, dry	NR	NR	NR	NR	NR	120	120	NR	NR	70
Salt Spray	80	70	120	200	200	140	120	120	120	120
Shortening (e.g. Lard)	NR	NR	NR	NR	NR	NR	100	NR	70	70
Silicone Oils	ID	ID	NR	ID	ID	ID	120	NR	ID	120
Soaps, aq dil	70	ID	NR	70	125	120	100	70	100	100
Sodium Acetate, sol	90	80	120	200	200	140	100	70	100	140
Sodium Aluminate, sol	90	90	120	200	200	140	100	70	100	140
Sodium Aluminum Sulfate	120	120	120	200	200	140	120	70	100	140
Sodium Bicarbonate, sol	120	120	120	175	150	140	120	100	120	140
Sodium Bisulfate, sol	70	70	120	100	NR	140	100	70	100	120
Sodium Bisulfite, sol	70	NR	120	200	NR	140	70	NR	80	120
Sodium Borate, sol	120	110	120	200	200	140	120	100	120	140

Comparison Corrosion Resistance Chart

	EAA-35 Epoxy Amine-Adduct	ECP-21 Epoxy Polyamide	Neoprene-Bituminous Blend	H-2 Hypalon	N-700-A Neoprene	Plastisol, Vinyl	Phenolic, Baked	UCP-21 Urethane Med. Chemical Resistant	UM-31 Urethane High Chemical Resistant	D-202-V-80 Plasticized Vinyl Coating
	F&S	F&S	F&S	F&S	F&S	L	L	F&S	F&S	L
Sodium Bromide, sol	100	90	120	175	150	140	120	90	120	140
Sodium Carbonate, aq	120	120	120	175	175	140	120	NR	100	140
Sodium Chlorate, sol	70	70	120	150	125	140	120	NR	80	140
Sodium Chloride, sol	120	120	120	200	200	140	120	100	120	140
Sodium Chromate, sol	70	70	120	200	150	140	120	70	90	140
Sodium Citrate, sol	90	80	120	150	150	140	120	70	100	140
Sodium Cyanide, sol	70	70	120	150	150	140	120	NR	80	140
Sodium Dichromate, sol	80	70	120	200	150	140	120	NR	90	140
Sodium Hydroxide 5%	110	120	120	200	150	140	120	NR	70	120
Sodium Hydroxide 15%	90	100	120	200	175	140	100	NR	70	120
Sodium Hydroxide 40%	70	70	120	200	200	140	ID	NR	ID	70
Sodium Hypochlorite, dil	NR	NR	ID	125	NR	140	100	70	ID	140
Sodium Nickel Sulfate, sol	120	110	120	200	200	140	120	100	120	120
Sodium Nitrate, sol	100	100	120	200	175	140	120	80	120	100
Sodium Nitrite, sol	100	70	120	175	150	140	120	80	120	140
Sodium Oleate, sol	80	70	ID	100	100	140	100	70	100	100
Sodium Permanganate, dil	NR	NR	ID	NR	NR	140	NR	NR	NR	140
Sodium Phosphate, sol	90	90	120	200	200	140	120	100	120	140
Sodium Silicate, sol	90	100	120	200	200	140	120	100	120	140
Sodium Stearate	70	70	120	125	125	140	120	NR	90	140

	EAA-35 Epoxy Amine-Adduct	ECP-21 Epoxy Polyamide	Neoprene-Bituminous Blend	H-2 Hypalon	N-700-A Neoprene	Plastisol, Vinyl	Phenolic, Baked	UCP-21 Urethane Med. Chemical Resistant	UM-31 Urethane High Chemical Resistant	D-202-V-80 Plasticized Vinyl Coating
	F&S	F&S	F&S	F&S	F&S	L	L	F&S	F&S	L
Sodium Sulfate	120	120	120	200	200	140	120	100	120	140
Sodium Sulfite	70	70	120	150	125	140	100	NR	80	140
Sodium Tartrate	100	90	120	175	150	140	120	100	120	140
Sodium Thiosulfate	100	90	120	150	150	140	120	100	120	120
Sorbitol, aq<50%	70	70	70	150	175	100	100	NR	70	70
Stannic (Tin) Chloride, aq	70	70	120	100	100	140	120	70	100	70
Stannous (Tin) Chloride, aq	70	70	120	125	125	140	120	70	100	NR
Stearic Acid, dry	NR	NR	NR	NR	NR	70	120	NR	ID	120
Styrene	ID	NR	NR	NR	NR	NR	100	NR	70	140
Succinic Acid, aq sol	70	70	70	70	70	120	70	70	90	100
Sugar, dry	NR	NR	NR	NR	NR	140	100	NR	100	NR
Sulfamic Acid, aq to 20%	NR	NR	ID	NR	NR	120	80	NR	70	120
Sulfonated Oils	NR	NR	ID	NR	NR	NR	70	NR	ID	120
Sulfur Dioxide-dry gas	ID	ID	ID	NR	NR	120	70	ID	ID	100
Sulfur Dioxide-wet gas	NR	NR	NR	NR	NR	120	70	ID	ID	70
Sulfuric Acid 10%	NR	NR	70	160	160	120	NR	ID	70	70
Sulfuric Acid 25%	NR	NR	ID	140	120	110	NR	NR	ID	NR
Sulfuric Acid 40%	NR	NR	ID	120	70	100	NR	NR	NR	NR
Sulfuric Acid 60%	NR	NR	NR	100	NR	70	NR	NR	NR	NR
Sulfuric Acid 98%	NR	NR	NR	70	NR	NR	NR	NR	NR	NR

Comparison Corrosion Resistance Chart

	EAA-35 Epoxy Amine-Adduct (F&S)	ECP-21 Epoxy Polyamide (F&S)	Neoprene-Bituminous Blend (F&S)	H-2 Hypalon (F&S)	N-700-A Neoprene (F&S)	Plastisol, Vinyl (L)	Phenolic, Baked (L)	UCP-21 Urethane Med. Chemical Resistant (F&S)	UM-31 Urethane High Chemical Resistant (F&S)	D-202-V-80 Plasticized Vinyl Coating (L)
Sulfurous + Water, dilute	70	NR	70	NR	NR	120	70	NR	ID	100
Tannic Acid	90	70	70	175	175	120	140	70	90	120
Tartaric, aq sol	90	70	70	125	125	140	120	70	90	120
Tetrachlorethane	NR	NR	NR	NR	NR	NR	110	NR	70	NR
Titanium Chloride, anhy	NR	NR	ID	ID	NR	ID	ID	NR	ID	ID
Toluene	80	ID	ID	ID	NR	NR	120	NR	70	NR
#11-Trichloromofluoromethane	ID	ID	ID	ID	ID	ID	120	NR	ID	NR
#113-Trichlorotrifluoroethane	ID	ID	ID	ID	ID	ID	120	NR	ID	NR
Triethanolamine, aq	ID	ID	NR	ID	70	120	100	NR	NR	NR
Trisodium Phosphate	90	100	120	200	175	140	120	NR	100	140
Turpentine	70	NR	NR	NR	NR	NR	120	NR	70	NR
Uranium Nitrate	80	70	120	125	70	140	ID	70	100	120
Urea, aq or dry	ID	ID	70	70	70	120	100	NR	NR	150
Vegetables Oils & Fats	NR	NR	NR	NR	NR	NR	100	NR	90	120
Vinegar	NR	NR	NR	NR	NR	NR	100	NR	ID	120
Vinyl Isobutyl Ether	NR	NR	NR	NR	NR	NR	70	NR	ID	NR

	EAA-35 Epoxy Amine-Adduct F&S	ECP-21 Epoxy Polyamide F&S	Neoprene-Bituminous Blend F&S	H-2 Hypalon F&S	N-700-A Neoprene F&S	Plastisol, Vinyl L	Phenolic, Baked L	UCP-21 Urethane Med. Chemical Resistant F&S	UM-31 Urethane High Chemical Resistant F&S	D-202-V-80 Plasticized Vinyl Coating L
Water-Deionized	70	70	70	70	70	140	120	70	80	120
Water-Distilled	70	70	70	70	70	140	120	70	80	100
Water-Mine (Coal, average)	70	70	70	70	70	140	120	70	80	120
Water-Salt	70	70	70	150	175	140	120	70	80	120
Water-Sea	70	70	ID	150	175	140	120	70	80	120
Water-Sewage (average)	70	70	70	70	70	140	120	70	80	100
Water-Tap (average)	70	70	ID	70	70	140	120	70	80	120
Whiskey	NR	NR	ID	NR	NR	NR	100	NR	70	NR
Wine	NR	NR	ID	NR	NR	NR	100	NR	70	100
Wort	NR	NR	ID	NR	NR	70	100	NR	70	120
Xylene	70	NR	NR	NR	NR	NR	120	NR	70	NR
Zeolite	120	120	120	150	150	140	150	100	120	140
Zinc Acetate	80	70	120	150	125	140	70	70	100	120
Zinc Bromide	80	70	120	125	125	140	100	70	100	120
Zinc Chloride	80	70	120	150	150	140	100	70	100	120
Zinc Nitrate	70	70	120	150	125	140	100	70	100	120
Zinc Sulfate	90	80	120	200	175	140	120	80	100	120

Index

A

Abrasion test for, 95-96
Abrasion resistance, 233, 400, 401, 402
Acid resistance, 293, 401, 402, 403
Acrylic, 49, 401; calks, 150; concrete impregnating, 401; performance, 229, 278-282; sealants, 145-147; semi gloss, 282, 285
Acrylonitrile, 402
Additives, 114
Adhesion, resistance to, 292; test for, 89
Airless spray, 175, 178
Alcohol resistance, 293
Alkali-resistance, 401, 402, 403
Alkyds, 25; phenolic, 31; specifying, 284
Aluminum, coatings for, 243
Application of coatings, 160-184; faulty, 165

B

Binders, solvent-reduced, 21-42; latex, 42
Blasting, sand, 112, 113, 134-137
Blistering, 186, 191
Bloom, 202
Brass, coatings for, 243
Bricks, surface preparation, 119
Brush, application by, 161, 163-167; marks, 205
Buried metals, 400, 405-407
Burnish resistance, 292
Butadiene, vinyl toluene, 32; styrene, 33
Butyl rubber calks, 148

C

Calcium carbonate, 72
Calcium molybdate, 60
Calks, 143-145
Cathodic protection, 28, 29, 405, 408
Cementitious surfaces, preparation, 118; defects, 198-201; specifying for, 227-236, 258-266, 294-297; problem areas, 229
Chalking, 101, 192, 203, 235
Chemical resistance, 92
Chlorinated polyether, 402
Chlorinated rubber, 31, 407; permeability of, 188; economics, 333
Clay, 71

Cleaning surfaces, 110-121; blast, 133
Color retention, 224
Concrete, block, 123; floors, 123, 306; peeling from, 199; surface preparation, 119
Condensation, prevention, 191
Contrast ratio (C.R.), 96
Copper, coatings for, 243
Corrosion, 28, 129-142, 403; economics, 334-341; resistance chart, 413-430; underground, 400, 405-407; underwater, 400, 407-408
Cor-ten, coatings for, 251
Cracking, 90, 203; repairing, 124
Critical pigment volume concentration, 7, 45, 186
Crosslinking, 191
Crystals on surface, 198
Curing, 16

D

Dairies, coatings for, 291
Defects, surface, 185-211
Degreasing, 115
Design, construction, as a fault, 198
Deterioration of coatings, 185-211
Differential aeration, 407
Discoloration, 196
Driers, 79
Drying time, test for, 83

E

Economics of surface coatings, 216, 287, 330-334
Efflorescence, 200; removal, 120
Elastomers, 401, 414-430; testing of, 412
Enamel holdout, 11, 237
Enamels, 11, 18, 49, 85, 166
End uses, 218
Epoxy, 27-28, 399-400; amine-cured 399, 414-430; coal tar, 400; effect on absorption, 191; effect of PVC, 244; performance of, 288; economics, 333, 337; polyamide-cured, 399, 414-430; polyester-cured, 399, 400; powder coating, 399; water-borne, 400
Ethylene vinyl acetate, 49

Evaluation of systems, 216-218, 220-224
Exposure data, 218
Extender pigments, 68-72
Exterior surfaces, selection of coatings, 212-273

F

Federal specifications; guide to specifications, appendix, 345-351; comparative charts, 258-273, 312-321; detailed charts, 352-396
Fillers, 236; wood, 297, 304
Film, properties, 217, 235, 278-292; thickness, 104
Fire retardant coatings, 322-329; mastics, 329
Flaking, 193; from galvanized, 201
Flat finishes, 18, 277
Flat pad applicators, 161, 172-175
Flexibility, 86
Floor finishes, 306-309
Fluorocarbons, 400
Food plants, coatings for, 291
Formulating, 2-10
Frits, 397
Fungicide, 194

G

Galvanic series of metals, 406
Galvanized steel, 243, 251-252; peeling, 201
Gasoline resistance, 293
Glaze removal, 121
Gloss, 85; uneven, 199; loss of, 205; specifying, 277
Graffiti, 233, 401
Green lumber, 198

H

Hospitals, 291
House paints, formulation, 5
Humidity, 210
Hypalon, 402, 413-430; sealants, 151

I

Immersed metals, protection of, 400, 407-408

431

Impact resistance, 88, 233, 292, 400
Impressed current, 406
Inorganic vehicles, 110
Interiors, specifying for, 274-321
Intumescence, 322
Isocyanates, 34

L

Lacquer, 17
Laitance, removal, 120
Latex paints, 42-50; permeability of, 189
Lead pigments, 58-59
Lifting of coats, 206
Lightfastness, test for, 89
Linings, 397, 401, 402
Linseed oil, 22; permeability of, 189

M

Mandrel, test, 86
Marine environment, 240, 243
Membranes, 397
Metal, coating defects, 201-202; durability of coatings, 131; floors, 309; formulating for, 10; mechanical preparation, 132; specifying for, 236-252; surface modification, 118
Metallizing protection, 397, 403
Mildew, 194; treatment, 127
Mill scale, 132, 238
Moisture; resistance, 296; vapor permeability, 94
Moisture-cure, 16, 35
Molybdate, calcium, 60

N

Neoprene, 402, 413-430; sealant, 149
Noble metals, 131-139, 405
Non-ferrous metal, preparation, 141
Nonflammability, 31
Nuclear plants, coatings for, 400

O

Oil paints, 22
Orange peel, 204
Ozone, protection against, 398, 402

P

Paint removers, 117
Passivation, 139-141
Peeling, 186-201
Performance, effect of formulating on, 3

Permeability, 94, 187
Phenolic, 413-430; alkyd, 31
Phosphate treatment, 118
Pigments, 50-72; effect on permeability, 189
Pigment volume concentration, 5-10, 24, 43, 147, 186
Pine, problems with, 254
Plasma arcs, 397
Plaster, 123
Plasticisers, 19
Polyamide, 27
Polyester, 399-400; epoxy, 333, 400
Polysulfide sealant, 153
Polyvinyl acetate, 48, 284
Pores, sealing, 126
Pot life, 16
Preservatives, 80
Primer sealers, 7
Primers, 18, 299

R

Radiation, protection against, 400, 403
Repainting, 203
Resins, 31, 42-50
Return on investment from quality coatings, 338
Rollers, 161, 167-172
Rubber, 402
Rust on walls, 186, 196

S

Sacrificial anode system, 406
Salt spray resistance, 91
Sandblasting, 124
Sanding, 125, 304
Scrubbability, 99, 285
Sealants, 147-159
Sealers, masonry, 297; wood pores, 126; joint, 128
Shop coatings, 250; economics, 335
Silicone, 32, 400; water repellency chart, 232; sealants, 158
Soluble dyes in wood, 196
Solvent resistance, 92
Solvents, 75
Sorption, water, 185
Spray gun, 162, 175-184
Spread rate, calculation, 98
Specifications, see Guide to Specifications, 345-351; comparative charts, 258-273, 312-321; detailed charts, 352-396
Soil resistivity, 406; guide for writing, 409-413
Stain; finishes, 17, 304; resistance, 99; from wood dyes, 253

Steel, preparation, 140
Stone, preparation, 119
Structural faults, 207
Stucco, 123
Styrene; acrylate, 31, 189; butadiene, 32, 189
Systems, evaluation of, 216-218, 245-249
Sward hardness, 88
Swelling, 190

T

Testing, 83-107; elastomers, 412
Texture paint, 31, 230
Titanium dioxide, 51
Tile-like coatings, 26, 297
Transparent finishes, 301

U

Ultra high performance coatings, 397-430
Uralkyds, 34
Urethane, 34-37, 400, 401, 413-430; elastomeric, 401

V

Varnish, 16
Vinyl, acetate, 48; chloride, 40, 413-430; toluene butadiene, 32
Vinylidene chloride, 402, 413-430

W

Wallboard, gypsum, 18
Washability, 213, 284
Wash primers, 48, 118
Water, defects due to, 185-203
Water vapor, 31; trapped, 188; transport, 188; immersion, 93; blasting, 112; resistance, 91, 92, 232
Water-thinned paint, 42-50
Weathering, tests for, 100-104
Whatman paper, 271 (note)
Wind-driven rain, 32, 230
Wood, chemicals in, 125; classification, open pores, 127; fillers, 304; floors, 309, 344; specifying for, 253-257, 270-273

Z

Zinc dust, 237-399; silicate primers, 29
Zinc primers, organic, 399
Zinc-rich paint, 29

REFERENCE	Date Due		
OCT 27 1987	OCT 27 1987		